SUPERVISION OF PETRI NETS

THE KLUWER INTERNATIONAL SERIES ON DISCRETE EVENT DYNAMIC SYSTEMS

Series Editor

Yu-Chi Ho
Harvard University

ANALYSIS OF MANUFACTURING ENTERPRISES: ***An Approach to Leveraging Value Delivery Processes for Competitive Advantage***
N. Viswanadham
ISBN: 0-7923-8671-X

INTRODUCTION TO DISCRETE EVENT SYSTEMS
Christos G. Cassandras, Stéphane Lafortune
ISBN: 0-7923-8609-4

OBJECT-ORIENTED COMPUTER SIMULATION OF DISCRETE-EVENT SYSTEMS
Jerzy Tyszer
ISBN: 0-7923-8506-3

TIMED PETRI NETS: ***Theory and Application***
Jiacun Wang
ISBN: 0-7923-8270-6

SUPERVISORY CONTROL OF DISCRETE EVENT SYSTEMS USING PETRI NETS
John O. Moody and Panos J. Antsaklis
ISBN: 0-7923-8199-8

GRADIENT ESTIMATION VIA PERTURBATION ANALYSIS
P. Glasserman
ISBN: 0-7923-9095-4

PERTURBATION ANALYSIS OF DISCRETE EVENT DYNAMIC SYSTEMS
Yu-Chi Ho and Xi-Ren Cao
ISBN: 0-7923-9174-8

PETRI NET SYNTHESIS FOR DISCRETE EVENT CONTROL OF MANUFACTURING SYSTEMS
MengChu Zhou and Frank DiCesare
ISBN: 0-7923-9289-2

MODELING AND CONTROL OF LOGICAL DISCRETE EVENT SYSTEMS
Ratnesh Kumar and Vijay K. Garg
ISBN: 0-7923-9538-7

UNIFORM RANDOM NUMBERS: THEORY AND PRACTICE
Shu Tezuka
ISBN: 0-7923-9572-7

OPTIMIZATION OF STOCHASTIC MODELS: THE INTERFACE BETWEEN SIMULATION AND OPTIMIZATION
Georg Ch. Pflug
ISBN: 0-7923-9780-0

CONDITIONAL MONTE CARLO: GRADIENT ESTIMATION AND OPTIMIZATION APPLICATIONS
Michael FU and Jian-Qiang HU

SUPERVISION OF PETRI NETS

by

Geert Stremersch
University of Ghent, Belgium

KLUWER ACADEMIC PUBLISHERS
Boston / Dordrecht / London

Distributors for North, Central and South America:
Kluwer Academic Publishers
101 Philip Drive
Assinippi Park
Norwell, Massachusetts 02061 USA
Telephone (781) 871-6600
Fax (781) 681-9045
E-Mail <kluwer@wkap.com>

Distributors for all other countries:
Kluwer Academic Publishers Group
Distribution Centre
Post Office Box 322
3300 AH Dordrecht, THE NETHERLANDS
Telephone 31 78 6392 392
Fax 31 78 6392 254
E-Mail <services@wkap.nl>

Electronic Services <http://www.wkap.nl>

Library of Congress Cataloging-in-Publication Data

Stremersch, G. (Geert), 1973-
Supervision of Petri nets / Geert Stremersch.
p. cm. – (The Kluwer international series on discrete event dynamic systems)
Includes bibliographical references and index.
ISBN –0-7923-7486-X
1. Petri nets. 2. Control theory. I. Title. II. Series

QA267.S765 2001
511.3--dc21 2001046195

Printed on acid-free paper.

Printed in the United States of America

To Heidi

Contents

Preface

The goal of this book is to present a unified, insightful and mathematically sound supervisory control theory for Petri nets. Petri nets are used to model discrete event systems, dynamic systems whose evolution is completely determined by the occurrence of discrete events. Control laws which guarantee that the system meets a set of specifications, are studied and constructed.

The requirement that the system behaviour always satisfies a number of conditions is expressed by a *legal set*, a subset of the state space. These conditions can express safety, avoidance of deadlock, the language is a subset of a given language, ... Because the controller can only disable events, and thus merely limit the set of all possible future paths of the system without being able to 'force' events, one speaks of *supervisory* control. The supervisory control goal is that the state of the Petri net always belongs to the legal set. Moreover, this needs to be done in an optimal way, namely by keeping the set of possible future evolutions of the Petri net as large as possible.

In practice, the presence of uncontrollable and unobservable events complicates control design. These events cannot be influenced or observed, respectively. Consequently, the controller should anticipate the worst-case uncontrollable behaviour of the system while its state may not be completely known. When all events are observable, and some are uncontrollable, the controller should keep the state within the *supremal controllable subset* of the given legal set.

Control design for such systems has been studied extensively over the last ten to fifteen years. Research is mainly motivated by automated manufacturing systems. Several theoretical problems remain open. The purpose of this book is to tackle the most fundamental of these:

- In the literature, different Petri net models and control laws have been used. It is not always clear which particular model or control paradigm is used. This often leads to confusion about the validity of results in other settings. In this book attention is given to different Petri net models, the differences between them and their common properties. A unifying definition of control laws and a general definition of optimality, or *maximal permissiveness*, is given.

- It is well known that for legal down-sets, in a lot of cases not the whole Petri net model needs to be considered when designing control laws. However, for more general legal sets, this problem has not been studied yet. Here, general reduction results are given and invariance properties are characterised.

- Modular design of supremal controllable subsets and control laws has hardly been studied. This book provides general decomposition results.

- The number of classes of Petri nets and legal sets for which the supremal controllable subset can be constructed is still limited. Moreover, different techniques have been used to obtain them. In this book a unifying approach is presented, which leads to more general algorithms and a better understanding of the problem.

- Control design of Petri nets with unobservable transitions has received little attention. Here, a technique is proposed which is based on Petri net supervisors.

The book is mainly meant as a research monograph. Researchers and graduate students will find in it a starting point and a source of insight provided by introducing an order theoretical viewpoint to the subject. Furthermore, several results are ready to be used in applications. Application oriented engineers will find algorithms suited to be put into practice.

The book is self-contained. Only some familiarity with basic notions from order theory and linear optimisation is assumed. A chapter by chapter overview of the book follows.

Chapter 1. Introduces Petri net models. Both the algebraic and the graphical representations are discussed. Attention is given to all possible 'speeds' of Petri net models. Each of them expresses a different degree of concurrency. A general Petri net definition which incorporates these concurrency assumptions, is proposed. Basic Petri net properties are formulated.

Chapter 2. Formulates the supervisory control problem for Petri nets with a legal set as the control goal. All transitions are assumed controllable. A unifying definition of control laws is proposed, together with a general definition of *permissiveness* and *maximal permissiveness.* The construction of control laws is discussed for different concurrency assumptions and control paradigms.

Chapter 3. Introduces supervisory control of Petri nets with uncontrollable transitions. This control problem is reduced to that for a fully controllable Petri net and another legal set, the so-called *supremal controllable subset* of the given legal set.

Chapter 4. Studies general legal sets and defines for each place in a Petri net one out of four possible invariance properties. It formulates general reduction theorems which show that it suffices to consider a sub-Petri net of the full model when constructing the supremal controllable subset of a given legal set or when designing control laws. Finally, invariance properties of legal sets given by a finite disjunction of linear inequalities are studied.

Chapter 5. Studies Petri nets which do not contain any cycles. The obtained reachability results are of great importance for the decomposition results of Chapter 6 and for the construction of (a subset of) the supremal controllable subset in Chapter 7. Further, for a general class of systems, the problem of deciding whether a given state belongs to the supremal controllable subset is translated into an integer linear programming problem.

Chapter 6. A practical control goal is that a finite conjunction and disjunction of conditions is satisfied. To limit the complexity of the control design it is interesting to be able to split the problem into several subproblems. General conditions are formulated under which the union of the supremal control subsets of two sets is equal to the supremal controllable subset of the union of these two sets. An analogous result for control laws is formulated.

Chapter 7. Is completely dedicated to the approximation, the construction and the structure of the supremal controllable subset for a given legal set. For a general class of Petri nets and a general class of legal sets, a subset of the supremal controllable subset is constructed, which can be used to design (sub)optimal control laws. For a number of classes of systems this same technique is applied to construct the supremal controllable subset itself. The general structure of the supremal controllable subset is discussed.

Chapter 8. Proposes a control law design method which is based on the structure of Petri net supervisors. These are controllers that consist of a number of additional places which are linked with the transitions of the original model. Some transitions can be uncontrollable or unobservable. To construct control laws it suffices to calculate the minimal —with respect to the componentwise partial order— vertices of a polyhedron, which are in addition minimal in this polyhedron.

GEERT STREMERSCH

This work presents research results of the Belgian Programme on Interuniversity Poles of Attraction, initiated by the Belgian State, Prime Minister's Office for Science, Technology and Culture. The scientific responsibility rests with its author.

Chapter 1

THE PETRI NET MODEL

In this chapter discrete event systems are defined and the Petri net model for these dynamic systems is introduced. Petri nets have both an algebraic and a graphical representation. We mainly focus on the algebraic representation as it is the one used most often throughout the book. Further, Petri net properties which are used in later chapters are formulated.

In the literature one finds Petri net models with different 'speeds', each expressing a certain degree of concurrency. The speed depends on the set to which transition vectors belong. The existence of different models make Petri net theory and supervisory control theory of Petri nets less accessible. Therefore, in this chapter three different speeds are discussed and compared, and a general Petri net definition is proposed.

Section 1 introduces discrete event systems. In Section 2 the used notation is specified. Section 3 summarises several definitions and results from order theory. In Section 4 Petri nets are defined algebraically and in Section 5 it is explained how they can model discrete event systems. Reachable sets of Petri nets are defined in Section 6. Section 7 describes the graphical representation of Petri nets. Reachability via subsets of transitions is studied in Section 8. The last two sections are devoted to other model speeds (Section 9) and a general Petri net definition (Section 10).

1. Discrete event systems

To define discrete event systems we quote Ramadge and Wonham [31]:

> *"A Discrete Event System is a dynamic system that evolves in accordance with the abrupt occurrence, at possibly unknown irregular intervals, of physical events."*

As discrete event systems evolve by the occurrence of events, they are at the other end of differential and difference equations within the set of dynamic systems. The evolution of differential and difference equations is completely determined by the evolution of time. Because not all dynamical systems can be modelled in a straightforward manner by difference or differential equations, discrete event systems form an important class of systems within system and control theory. For a general overview of discrete event systems (definition, difference with other dynamic models, place within system theory, importance and examples) we refer to Chapter 1 in [6]. We illustrate the definition of a discrete event system with an example which describes the dynamics of a railway system.

Example 1.1. Consider the section of a railway network in Figure 1.1. Trains are assumed to ride always from left to right. The two lines on the left of the figure come together to one line at the right. The junction is realised by a switch. For a train to be able to go from the upper left line to the right one, the switch should be in Position 0. For a train on the lower left line the switch should be in Position 1. Both lines to the left of the switch are divided into two zones, resulting in a total of four zones, 1 to 4. The line to the right of the switch forms a fifth zone. Trains can enter this section of the railway network via Zones 1 or 3 and leave it via Zone 5 only. Each time a train crosses the switch, the switch is automatically transferred to the opposite position. Thus if a train goes from Zone 2 to Zone 5, then the switch is in Position 0. As soon as this train is in Zone 5, the switch is set in Position 1. It is also possible to alter the position of the switch manually.

The dynamic system that describes the evolution of the numbers of trains in and around this railway section, can —at a certain modelling level— be seen as a discrete event system. It has six states:

m_1 : the number of trains in Zone 1.

m_2 : the number of trains in Zone 2.

m_3 : the number of trains in Zone 3.

m_4 : the number of trains in Zone 4.

m_5 : the number of trains in Zone 5.

m_6 : the position of the switch (1 if in Position 0, 0 if in Position 1).

The state of the system in Figure 1.1 is $m_1 = m_5 = m_6 = 1$ and $m_2 = m_3 = m_4 = 0$. The events considered in the model are:

e_1 : a train enters Zone 1.

e_2 : a train leaves Zone 1 and enters Zone 2.

e_3 : a train leaves Zone 2 and enters Zone 5 via the switch.

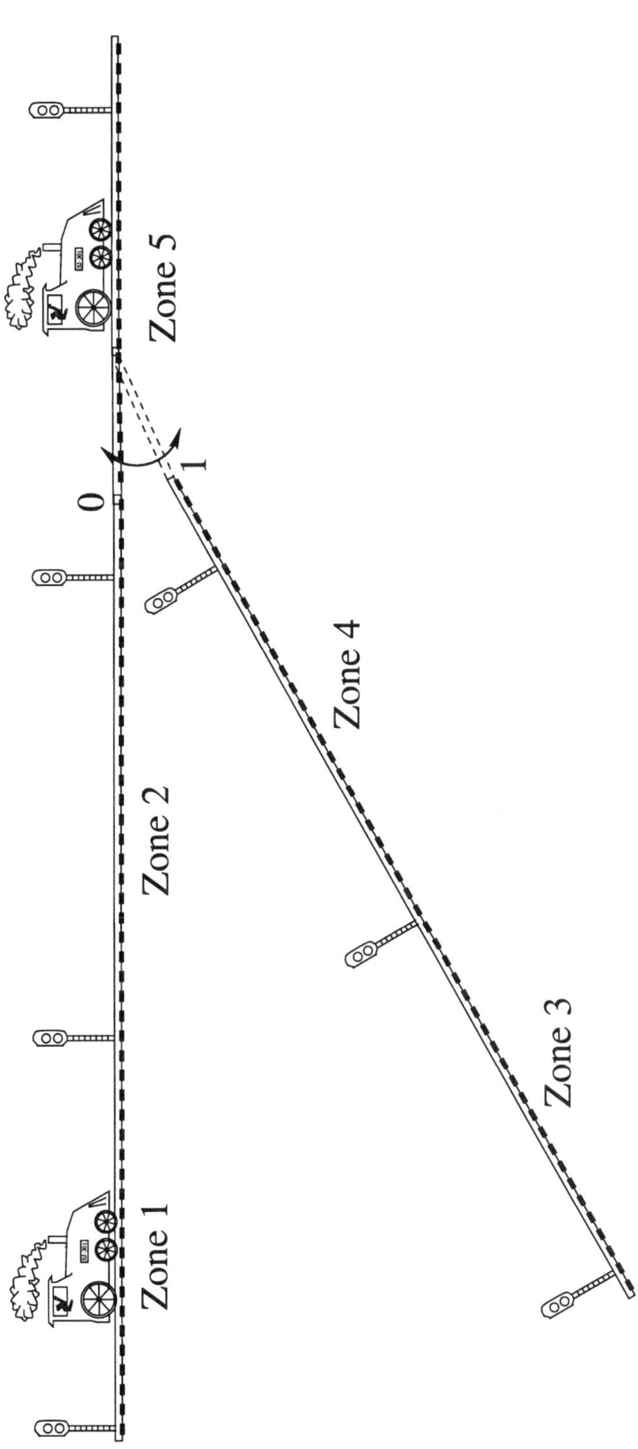

Figure 1.1. A section of a railway network. There are five zones and one switch. There is one train in Zone 1 and one in Zone 5. The switch is in Position 0, allowing a train to pass from the upper left line to the one on the right.

e_4 : a train enters Zone 3.

e_5 : a train leaves Zone 3 and enters Zone 4.

e_6 : a train leaves Zone 4 and enters Zone 5 via the switch.

e_7 : the switch is brought manually from Position 0 to 1.

e_8 : the switch is brought manually from Position 1 to 0.

e_9 : a train leaves Zone 5.

Not each event can take place at any time. The state of the system determines which events can occur. For instance,

e_2 : only if there is at least one train in Zone 1,

e_3 : only if there is at least one train in Zone 2 and the switch is in Position 0,

e_7 : only if the switch is in Position 0,

e_9 : only if there is at least one train in Zone 5.

In this model events e_1 and e_4 can always occur. This would be different if the rail lines leading to Zones 1 and 3 were modelled as well. The physical upper limit on the number of trains that can be present in one zone is not modelled here. The state of this system is determined by the value of m_1 to m_6, and it only changes when one of the events e_1 to e_9 takes place. △

There are several possibilities to model discrete event systems, e.g. automata and formal languages, process algebras, state charts and Petri nets. Petri nets, which are used here, have an attractive distributed state representation. In comparison to automata, this leads to more compact models, but often makes analysis and supervisory control design more difficult. In this book, we present results to cope with the latter problem and attempt to provide more insight in it.

In the model used in this book, time is not modelled explicitly. Only the order in which events can occur is determined by the model. It does not contain any information about the time interval between successive events. As a consequence, only the *logical behaviour* of the system can be studied. However, it often suffices to analyse this part of the system. A system without time information, called an *untimed system*, can also be the result of the abstraction of a system with time information, e.g. a hybrid system.

2. Notation

Matrices are set bold, their elements are not. For example, the $m \times n$ matrix $\mathbf{X}$ has elements $X_{i,j}$ ($i = 1, \ldots, m; j = 1, \ldots, n$). For these elements, also the notation $(\mathbf{X})_{i,j}$ is be used. All vectors are assumed to be column vectors. For notational brevity, a column vector $\mathbf{X}$ is also written as $\mathbf{X} = (X_1, \ldots, X_m)$. The following symbols and notations are used throughout the book:

$\mathbb{Z}$	$\{\ldots, -2, -1, 0, 1, 2, \ldots\}$, set of all integers
$\mathbb{N}$	$\{0, 1, 2, \ldots\}$, set of all nonnegative integers
$\mathbb{Q}$	set of all rational numbers
$\mathbb{Q}_+$	set of all nonnegative rational numbers
$\mathbb{R}$	set of all real numbers
V^m	set of all m-dimensional vectors with elements in V
$V^{m \times n}$	set of all $m \times n$ matrices with elements in V
$\mathbf{X}^T$	transpose of a matrix $\mathbf{X}$
$\mathbf{0}$	zero matrix with dimension determined by the context
$\mathbf{X} \leq \mathbf{Y}$	the inequality holds componentwise for all elements $(\mathbf{X}, \mathbf{Y} \in V^{m \times n})$
$A \subseteq B$	A is a subset of B
$A \subset B$	$A \subseteq B$ and $A \neq B$
$X = A \dot{\cup} B$	$X = A \cup B$ and $A \cap B = \emptyset$
$A \times B$	cartesian product of the sets A and B
2^X	set of all subsets of the set X, i.e. the power set of X
$\sharp X$	cardinality of the set X
$\vee, \bigvee$	logical OR
$\wedge, \bigwedge$	logical AND
$\neg$	logical NOT
$\lfloor x \rfloor$	the largest integer that is smaller than or equal to $x \in \mathbb{R}$
$\lceil x \rceil$	the smallest integer that is larger than or equal to $x \in \mathbb{R}$
$a \mid b$	a is a divisor of b $(a, b \in \mathbb{Z})$
$a \nmid b$	a is not a divisor of b $(a, b \in \mathbb{Z})$
$a!$	faculty of $a \in \mathbb{N}$
$\min(x_1, x_2)$	the minimum of the real numbers x_1 and x_2
$\max(x_1, x_2)$	the maximum of the real numbers x_1 and x_2
$(x_1, \ldots, x_n)$	finite sequence
$(x_1, x_2, \ldots)$	infinite sequence
$f\|_{D'}$	restriction of the function f with domain D to $D' \subseteq D$

When a new concept is introduced, its name is set *italic*. Examples end with '$\triangle$' and proofs with '$\square$'.

3. Order theoretical preliminaries

In this section we summarise a number of order theoretical definitions. These basic concepts allow to study supervisory control of Petri nets formally. In particular, up- and down-sets in Section 3.3 are important for what follows.

3.1. Partially ordered sets

A *partially ordered set* $(X, \leq)$ is a non-empty set X with a binary relation $\leq$ on X which is reflexive, antisymmetric and transitive. The relation $\leq$ is called a *partial order relation* on X. In this book, we mainly equip sets of vectors with the componentwise partial order. A vector is less than or equal to another vector if this relation holds for each pair of corresponding elements of these vectors.

Consider a partially ordered set $(X, \leq)$. Let $y, z \in X$ and $S \subseteq X$. We say that:

a. z is an *upper bound* for S if $s \leq z$ for all $s \in S$. The set S is *bounded above* if there exists an upper bound for S. Analogously, z is a *lower bound* for S if $s \geq z$ for all $s \in S$. The set S is *bounded below* if there exists a lower bound for S.

b. z is the *maximum* of S if $z \in S$ and $s \leq z$ for all $s \in S$. Analogously, z is the *minimum* of S if $z \in S$ and $s \geq s$ for all $s \in S$. The terminology *the* maximum and *the* minimum is used because these elements are unique when they exist.

c. If S is bounded above and the set of upper bounds of S has a minimum, then this element is called the *supremum* of S (in X) and is denoted $\sup(S)$. Analogously, if S is bounded below and the set of lower bounds of S has a maximum, then this element is called the *infimum* of S (in X) and is denoted $\inf(S)$.

d. z is *maximal* in S if $z \leq s$ implies $z = s$ for all $s \in S$. Analogously, z is *minimal* in S if $z \geq s$ implies $z = s$ for all $s \in S$. The set of maximal elements in $(S, \leq)$ is denoted $\max(S, \leq)$. Analogously, we write $\min(S, \leq)$ for the set of minimal elements in $(S, \leq)$.

e. Two elements x and y in X are *comparable* if $x \leq y$ or $y \leq x$. We call $(X, \leq)$ a *chain* if each pair of elements in X is comparable. We call $(X, \leq)$ an *antichain* if there do not exist two different, comparable elements in X.

3.2. Lattices

The partially ordered set $(X, \leq)$ is a *lattice* if each subset of X that contains precisely two elements, has a supremum and an infimum in X. We say that $(X, \leq)$ is a *complete lattice* if each subset of X has a supremum and an infimum in X.

A sequence $(x_1, x_2, \ldots)$ in a partially ordered set $(X, \leq)$ is *increasing* if $i \leq j$ implies $x_i \leq x_j$ for all i and j.

Proposition 1.1 ([33]). *Each increasing sequence in a complete lattice converges to its limit.*

As the exact definition of convergence in this proposition is of no importance to our work, we refer to [33] for it.

Proposition 1.2 ([33]). *Consider a set X. Then $(2^X, \subseteq)$, the set of subsets of X with the set inclusion as the partial order relation, is a complete lattice.*

3.3. Up- and down-sets

In this subsection $(X, \leq)$ is a partially ordered set.

Definition 1.1. *A set $X' \subseteq X$ is a* down-set *in $(X, \leq)$ if*

$$\forall x' \in X', \forall x \in X : x \leq x' \Rightarrow x \in X'.$$

Analogously, $X' \subseteq X$ is an up-set *in $(X, \leq)$ if*

$$\forall x' \in X', \forall x \in X : x' \leq x \Rightarrow x \in X'.$$

Define $\mathcal{O}^{\downarrow}(X, \leq)$ and $\mathcal{O}^{\uparrow}(X, \leq)$ as the set of down- and up-sets in $(X, \leq)$, respectively.

In words, a down-set is closed under going down with respect to the partial order. An up-set is closed under going up with respect to the partial order. Other ways to refer to a down-set are *down-closed set*, *lower set* and *left-closed set*, and *up-closed set*, *upper set* and *right-closed set* for an up-set.

Proposition 1.3 ([9]). *Both the intersection and the union of any set of down-sets in $(X, \leq)$ are down-sets in $(X, \leq)$.*

Equivalently, the set of down-sets is closed under both union and intersection.

Proposition 1.4 ([9]). *Both the intersection and the union of any set of up-sets in $(X, \leq)$ are up-sets in $(X, \leq)$.*

Equivalently, the set of up-sets is closed under both union and intersection.

Proposition 1.5 ([9]). *Let $X' \subseteq X$. Then $X' \in \mathcal{O}^{\downarrow}(X, \leq)$ if and only if $X \setminus X' \in \mathcal{O}^{\uparrow}(X, \leq)$.*

In words, if a set is a down-set then its complement is an up-set, and vice versa.

4. Petri net definition

Now we are ready to define Petri nets. First, they are introduced algebraically. The equivalent graphical representation is discussed later in this chapter.

Definition 1.2. *A* Petri net *is a triple* $\mathcal{N} = (\mathbf{F}^+, \mathbf{F}^-, \mathbf{m}_0)$ *with* $\mathbf{F}^+, \mathbf{F}^- \in \mathbb{N}^{n\times m}$ *and* $\mathbf{m}_0 \in \mathbb{N}^n$.

We refer to $\mathbf{F}^+$ and $\mathbf{F}^-$ as the *incidence matrices* of $\mathcal{N}$. Their elements are natural numbers. The vector $\mathbf{m}_0$ is the initial state of $\mathcal{N}$.

Define $\Delta := \mathbb{N}^m \setminus \{\mathbf{0}\}$ as the set of all m-dimensional column vectors with natural numbers as elements, minus the zero vector. Elements of Δ are referred to as *transition vectors.* The state of a Petri net changes stepwise as follows. If $\mathbf{m}_k$ is the state after step $k \in \mathbb{N}$, then all possible next states $\mathbf{m}_{k+1}$ are given by

$$\mathbf{m}_{k+1} = \mathbf{m}_k + (\mathbf{F}^+ - \mathbf{F}^-)\boldsymbol{\delta}, \tag{1.1}$$

where

$$\boldsymbol{\delta} \in \Delta \tag{1.2}$$

and

$$\mathbf{m}_k \geq \mathbf{F}^-\boldsymbol{\delta}. \tag{1.3}$$

We refer to the condition in (1.3) as the *enabling condition under* $\mathbf{m}_k$. The equation in (1.1) is the *state equation* of the Petri net. Given the present state, the Petri net model does not determine the next state. It solely determines the set of *all possible* next states. Because of the enabling condition, $\boldsymbol{\delta} \geq \mathbf{0}$ and $\mathbf{F}^+ \geq \mathbf{0}$, all possible next states belong to $\mathbb{N}^n$. Thus, $\mathbb{N}^n$ is the state space of a Petri net.

The difference of the incidence matrices of $\mathcal{N}$ is written as

$$\mathbf{F} := \mathbf{F}^+ - \mathbf{F}^-. \tag{1.4}$$

We refer to $\mathbf{F}$ as the *incidence matrix* of $\mathcal{N}$ (although we refer to them with the same name, there is in general no danger for confusion between $\mathbf{F}^+$, $\mathbf{F}^-$ and $\mathbf{F}$). With this matrix we obtain for the state equation in (1.1):

$$\mathbf{m}_{k+1} = \mathbf{m}_k + \mathbf{F}\boldsymbol{\delta}.$$

Define for all $\mathbf{m} \in \mathbb{N}^n$,

$$\Delta_{\mathbf{m}} := \{\boldsymbol{\delta} \in \Delta \mid \mathbf{m} \geq \mathbf{F}^-\boldsymbol{\delta}\} \tag{1.5}$$

as the set of transition vectors which satisfy the enabling condition under $\mathbf{m}$. Each $\boldsymbol{\delta} \in \Delta_{\mathbf{m}}$ is called *enabled* under $\mathbf{m}$. The set $\Delta_{\mathbf{m}}$ is not necessarily finite. Note also that $\Delta_{\mathbf{m}}$ can be empty. For example $\Delta_{\mathbf{0}} = \emptyset$, i.e. no transition vectors are enabled when the system's state is equal to $\mathbf{0}$.

Example 1.2. Consider a Petri net with $n = 3$, $m = 2$, incidence matrices

$$\mathbf{F}^+ = \begin{bmatrix} 0 & 0 \\ 1 & 0 \\ 0 & 1 \end{bmatrix} \quad \text{and} \quad \mathbf{F}^- = \begin{bmatrix} 1 & 0 \\ 0 & 1 \\ 0 & 0 \end{bmatrix},$$

and initial state $\mathbf{m}_0 = (2, 1, 1)$. In this case one has

$$\Delta_{\mathbf{m}_0} = \left\{ \begin{bmatrix} 1 \\ 0 \end{bmatrix}, \begin{bmatrix} 0 \\ 1 \end{bmatrix}, \begin{bmatrix} 1 \\ 1 \end{bmatrix}, \begin{bmatrix} 2 \\ 0 \end{bmatrix}, \begin{bmatrix} 2 \\ 1 \end{bmatrix} \right\}.$$

$\triangle$

The set Δ together with the componentwise partial order relation forms the partially ordered set $(\Delta, \leq)$.

Proposition 1.6. *For each* $\mathbf{m} \in \mathbb{N}^n$ *it holds that* $\Delta_{\mathbf{m}} \in \mathcal{O}^{\downarrow}(\Delta, \leq)$.

In words, for each $\mathbf{m} \in \mathbb{N}^n$, the set of enabled transition vectors $\Delta_{\mathbf{m}}$ is a down-set in $(\Delta, \leq)$. If $\boldsymbol{\delta} \in \Delta_{\mathbf{m}}$, then each transition vector $\boldsymbol{\delta}'$ which is less than or equal to $\boldsymbol{\delta}$, i.e. $\boldsymbol{\delta}' \leq \boldsymbol{\delta}$, also belongs to $\Delta_{\mathbf{m}}$.

We use the notation

$$\mathbf{m} \xrightarrow{\boldsymbol{\delta}} \mathbf{m}' \tag{1.6}$$

to indicate that both $\boldsymbol{\delta} \in \Delta_{\mathbf{m}}$ and $\mathbf{m}' = \mathbf{m} + \mathbf{F}\boldsymbol{\delta}$ hold. We say that $\mathbf{m}'$ is *reachable* from $\mathbf{m}$ via $\boldsymbol{\delta}$, or that $\mathbf{m}'$ *is reached* from $\mathbf{m}$ by the *execution* of $\boldsymbol{\delta}$. We emphasise that the notation in (1.6) implies that $\boldsymbol{\delta} \in \Delta_{\mathbf{m}}$.

The above definitions can be extended immediately to finite sequences of transition vectors. Consider a finite sequence of transition vectors

$$(\boldsymbol{\delta}_1, \boldsymbol{\delta}_2, \ldots, \boldsymbol{\delta}_l) \quad \text{with } \boldsymbol{\delta}_i \in \Delta \text{ for } i = 1, \ldots, l.$$

We say that $(\boldsymbol{\delta}_1, \boldsymbol{\delta}_2, \ldots, \boldsymbol{\delta}_l)$ is *enabled* under $\mathbf{m}$ or that $\mathbf{m}'$ is *reachable* from $\mathbf{m}$ via $(\boldsymbol{\delta}_1, \boldsymbol{\delta}_2, \ldots, \boldsymbol{\delta}_l)$ if

$$\mathbf{m} \xrightarrow{\boldsymbol{\delta}_1} \cdot \xrightarrow{\boldsymbol{\delta}_2} \cdots \xrightarrow{\boldsymbol{\delta}_l} \mathbf{m}'.$$

Each transition vector $\boldsymbol{\delta}_i$ is enabled under the state which is reached by successively executing the transition vectors that precede $\boldsymbol{\delta}_i$ in the sequence. This is also denoted

$$\mathbf{m} \xrightarrow{(\boldsymbol{\delta}_1, \boldsymbol{\delta}_2, \ldots, \boldsymbol{\delta}_l)} \mathbf{m}', \tag{1.7}$$

or as

$$(\boldsymbol{\delta}_1, \boldsymbol{\delta}_2, \ldots, \boldsymbol{\delta}_l) \in \bar{\Delta}_{\mathbf{m}}, \tag{1.8}$$

where $\bar{\Delta}_{\mathbf{m}}$ is the set of finite sequences of transitions vectors which are enabled under $\mathbf{m}$.

By using the following property, sequences of transition vectors can be decomposed.

Proposition 1.7. *Let $\mathbf{m}_1, \mathbf{m}_2 \in \mathbb{N}^n$ and $\boldsymbol{\delta}, \boldsymbol{\delta}_1, \boldsymbol{\delta}_2 \in \Delta$. If $\mathbf{m}_1 \xrightarrow{\boldsymbol{\delta}} \mathbf{m}_2$ and $\boldsymbol{\delta} = \boldsymbol{\delta}_1 + \boldsymbol{\delta}_2$, then $\mathbf{m}_1 \xrightarrow{(\boldsymbol{\delta}_1, \boldsymbol{\delta}_2)} \mathbf{m}_2$ or, equivalently, $\mathbf{m}_1 \xrightarrow{\boldsymbol{\delta}_1} \cdot \xrightarrow{\boldsymbol{\delta}_2} \mathbf{m}_2$.*

Proof. From $\boldsymbol{\delta}_1 \leq \boldsymbol{\delta}$ and Proposition 1.6 it follows that $\boldsymbol{\delta}_1 \in \Delta_{\mathbf{m}_1}$. Consider the state $\hat{\mathbf{m}} = \mathbf{m}_1 + \mathbf{F}\boldsymbol{\delta}_1$, which is reached from $\mathbf{m}_1$ via $\boldsymbol{\delta}_1$. It remains to prove that $\hat{\mathbf{m}} \geq \mathbf{F}^-\boldsymbol{\delta}_2$. From $\hat{\mathbf{m}} = \mathbf{m}_1 + \mathbf{F}\boldsymbol{\delta}_1$ and $\mathbf{m}_1 \geq \mathbf{F}^-\boldsymbol{\delta}$ it follows that

$$\hat{\mathbf{m}} - \mathbf{F}\boldsymbol{\delta}_1 \geq \mathbf{F}^-\boldsymbol{\delta} = \mathbf{F}^-(\boldsymbol{\delta}_1 + \boldsymbol{\delta}_2),$$

or that $\hat{\mathbf{m}} - \mathbf{F}^+\boldsymbol{\delta}_1 \geq \mathbf{F}^-\boldsymbol{\delta}_2$, which implies that $\hat{\mathbf{m}} \geq \mathbf{F}^-\boldsymbol{\delta}_2$. □

The following counter example shows, with the notation of Proposition 1.7, that the implication

$$\mathbf{m}_1 \xrightarrow{(\boldsymbol{\delta}_1, \boldsymbol{\delta}_2)} \mathbf{m}_2 \quad \Rightarrow \quad \mathbf{m}_1 \xrightarrow{\boldsymbol{\delta}} \mathbf{m}_2$$

does not hold in general.

Example 1.3. Take

$$\mathbf{F}^+ = \begin{bmatrix} 0 & 0 \\ 1 & 0 \end{bmatrix} \quad \text{and} \quad \mathbf{F}^- = \begin{bmatrix} 1 & 0 \\ 0 & 1 \end{bmatrix}.$$

For $\mathbf{m}_1 = (1,0)$ and $\boldsymbol{\delta}_1 = (1,0)$ it holds that $\boldsymbol{\delta}_1 \in \Delta_{\mathbf{m}_1}$. Let $\hat{\mathbf{m}} = \mathbf{m}_1 + \mathbf{F}\boldsymbol{\delta}_1 = (0,1)$ and $\boldsymbol{\delta}_2 = (0,1)$. One verifies that $\boldsymbol{\delta}_2 \in \Delta_{\hat{\mathbf{m}}}$. However, $\boldsymbol{\delta} = \boldsymbol{\delta}_1 + \boldsymbol{\delta}_2 = (1,1) \notin \Delta_{\mathbf{m}_1}$. △

5. Petri nets as discrete event system models

Petri nets can be used to model discrete event systems. Each component of the state vector $\mathbf{m}$ can represent the state of an elementary component of the discrete event system, e.g. resources, machines, valves, locations, ... Each component δ_i of the transition vector $\boldsymbol{\delta}$ corresponds to an event e_i $(i = 1, \ldots, m)$.

If $\mathbf{m} \xrightarrow{\boldsymbol{\delta}} \mathbf{m}'$, then the event e_i takes place δ_i times for $i = 1, \ldots, m$. In this way, one reaches the state $\mathbf{m}' = \mathbf{m} + \mathbf{F}\boldsymbol{\delta}$, starting from $\mathbf{m}$.

The enabling condition $\mathbf{m} \geq \mathbf{F}^{-}\boldsymbol{\delta}$ expresses that the collection of events which is represented by $\boldsymbol{\delta}$ can occur simultaneously when $\mathbf{m}$ is the state of the discrete event system. This collection consists of δ_1 times the event e_1, δ_2 times the event e_2, ... and δ_m times the event e_m. By looking at a Petri net as a discrete event system model, we get some intuition for the result of Proposition 1.6. If a collection of δ_i events e_i $(i = 1, \ldots, m)$ can take place simultaneously, then each collection of $\delta_i' \leq \delta_i$ events e_i $(i = 1, \ldots, m)$ can take place simultaneously as well. Thus, if a collection of events can occur, then each subcollection of events can occur as well. The Petri net model only determines, via the enabling condition (1.3) and assuming $\Delta_{\mathbf{m}}$ finite, the maximal collections of events that can occur.

Because transition vectors $\boldsymbol{\delta}$ can have any value in $\Delta = \mathbb{N}^m \setminus \{\mathbf{0}\}$ (see (1.2)), we say that the *transition bag assumption* is used. In each step different events can occur simultaneously and each event more than once. Petri net models for which transition vectors are restricted to a subset of Δ in (1.2) are discussed in Sections 9 and 10 of this chapter.

Example 1.4. Consider again the railway section of Example 1.1 and Figure 1.1. Because of the form of the enabling condition in (1.3), we introduce a seventh state component:

m_7 : The position of the switch (0 if in Position 0, 1 if in Position 1).

It is clear that $m_6 + m_7 = 1$ always holds. The sum of the two switch position states equals one at any time. The railway section can be modelled by a Petri net with incidence matrices:

$$\mathbf{F}^{+} = \begin{bmatrix} 1 & 0 & 0 & 0 & 0 & 0 & 0 & 0 & 0 \\ 0 & 1 & 0 & 0 & 0 & 0 & 0 & 0 & 0 \\ 0 & 0 & 0 & 1 & 0 & 0 & 0 & 0 & 0 \\ 0 & 0 & 0 & 0 & 1 & 0 & 0 & 0 & 0 \\ 0 & 0 & 1 & 0 & 0 & 1 & 0 & 0 & 0 \\ 0 & 0 & 0 & 0 & 0 & 1 & 0 & 1 & 0 \\ 0 & 0 & 1 & 0 & 0 & 0 & 1 & 0 & 0 \end{bmatrix},$$

$$\mathbf{F}^{-} = \begin{bmatrix} 0 & 1 & 0 & 0 & 0 & 0 & 0 & 0 & 0 \\ 0 & 0 & 1 & 0 & 0 & 0 & 0 & 0 & 0 \\ 0 & 0 & 0 & 0 & 1 & 0 & 0 & 0 & 0 \\ 0 & 0 & 0 & 0 & 0 & 1 & 0 & 0 & 0 \\ 0 & 0 & 0 & 0 & 0 & 0 & 0 & 0 & 1 \\ 0 & 0 & 1 & 0 & 0 & 0 & 1 & 0 & 0 \\ 0 & 0 & 0 & 0 & 0 & 1 & 0 & 1 & 0 \end{bmatrix}.$$

The initial state in Figure 1.1 is

$$\mathbf{m}_0 = (1,0,0,0,1,1,0).$$

The ith element of the state vector $\mathbf{m}$ is equal to m_i for $i = 1,\ldots,7$. The transition vectors $\boldsymbol{\delta}$ have nine rows. For $i = 1,\ldots,9$, the element δ_i corresponds to the event e_i. Componentwise, the enabling condition $\mathbf{m} \geq \mathbf{F}^-\boldsymbol{\delta}$ is equivalent to:

$$\begin{aligned} m_1 &\geq \delta_2, \\ m_2 &\geq \delta_3, \\ m_3 &\geq \delta_5, \\ m_4 &\geq \delta_6, \\ m_5 &\geq \delta_9, \\ m_6 &\geq \delta_3 + \delta_7, \\ m_7 &\geq \delta_6 + \delta_8. \end{aligned}$$

Some of these inequalities have been formulated in words in Example 1.1 (p. 2). $\triangle$

It is not necessarily so that each pair of different transition vector components corresponds to two different events. However for simplicity, throughout this book we assume that this is the case. At the end of Chapter 3 we discuss the (limited) consequences of this assumption for the results of this book.

In the definition of Δ, the set of all transition vectors and thus the set of all possible collections of events, the zero vector $\mathbf{0}$ has been excluded. From the above it is clear that the occurrence of the zero vector $\mathbf{0}$ would not have any physical meaning as it corresponds to an empty collection of events. Nevertheless, for our convenience we allow the notation $\mathbf{m} \xrightarrow{\mathbf{0}} \mathbf{m}$ to avoid that in the course of the book the nonzeroness of transition vectors needs to be checked continuously.

6. Reachable sets

In this section we define the sets of states which can be reached by successively executing transition vectors.

Definition 1.3. *Define for all $\mathbf{m} \in \mathbb{N}^n$ and $k \in \mathbb{N} \setminus \{0\}$ the set of states*

$$\mathcal{R}_k(\mathbf{m}) := \{\mathbf{m}' \in \mathbb{N}^n \mid \\ \exists \mathbf{m}'' \in \mathcal{R}_{k-1}(\mathbf{m}), \exists \boldsymbol{\delta} \in \Delta_{\mathbf{m}''} : \mathbf{m}' = \mathbf{m}'' + \mathbf{F}\boldsymbol{\delta}\} \cup \mathcal{R}_{k-1}(\mathbf{m}),$$

where $\mathcal{R}_0(\mathbf{m}) := \{\mathbf{m}\}$.

The sets $\mathcal{R}_k(\mathbf{m})$ are called *reachable sets* of $\mathbf{m}$. For each $k \in \mathbb{N}$ and $\mathbf{m} \in \mathbb{N}^n$, $\mathcal{R}_k(\mathbf{m})$ is the set of states which are reachable from $\mathbf{m}$ in at most k steps.

Proposition 1.8. *Let* $\mathbf{m} \in \mathbb{N}^n$ *and* $k \in \mathbb{N}$. *It holds that* $\mathcal{R}_k(\mathbf{m}) \subseteq \mathcal{R}_{k+1}(\mathbf{m})$.

Consequently, for each $\mathbf{m} \in \mathbb{N}^n$ the infinite sequence

$$(\mathcal{R}_0(\mathbf{m}), \mathcal{R}_1(\mathbf{m}), \mathcal{R}_2(\mathbf{m}), \dots)$$

is increasing in $(2^{\mathbb{N}^n}, \subseteq)$. From Propositions 1.1 and 1.2 we then have that the set of states

$$\mathcal{R}_\infty(\mathbf{m}) := \lim_{k\to\infty} \mathcal{R}_k(\mathbf{m}) \tag{1.9}$$

is well defined. It is the set of states which can be reached from $\mathbf{m}$ in any finite number of steps. Define for subsets X of $\mathbb{N}^n$ the sets of states

$$\mathcal{R}_k(X) := \bigcup_{\mathbf{m}\in X} \mathcal{R}_k(\mathbf{m}) \qquad (k \in \mathbb{N}), \tag{1.10}$$

and analogously for $k \to \infty$. The reachable set of a set of states is the union of the reachable sets of the states in this set. One verifies that

$$\mathcal{R}_k(\mathcal{R}_l(X)) = \mathcal{R}_{k+l}(X)$$

for all $k, l \in \mathbb{N}$.

Proposition 1.9 formulates a necessary condition for reachability in Petri nets. This condition is a direct consequence of the state equation in (1.1). In general it is not sufficient.

Proposition 1.9. *Let* $\mathbf{m}_0, \mathbf{m}_\mathrm{f} \in \mathbb{N}^n$. *If* $\mathbf{m}_\mathrm{f} \in \mathcal{R}_\infty(\mathbf{m}_0)$, *then there exists a vector* $\mathbf{y} \in \mathbb{N}^m$ *such that* $\mathbf{m}_\mathrm{f} = \mathbf{m}_0 + \mathbf{F}\mathbf{y}$.

Proof. For each $\mathbf{m}_\mathrm{f} \in \mathcal{R}_\infty(\mathbf{m}_0)$ there exist transition vectors $\boldsymbol{\delta}_1, \dots, \boldsymbol{\delta}_k$ such that

$$\mathbf{m}_0 \xrightarrow{\boldsymbol{\delta}_1} \cdots \xrightarrow{\boldsymbol{\delta}_k} \mathbf{m}_\mathrm{f}.$$

From the state equation (1.1) one obtains that

$$\mathbf{m}_\mathrm{f} = \mathbf{m}_0 + \mathbf{F}\sum_{i=1}^{k} \boldsymbol{\delta}_i.$$

□

As the above condition is in most cases not sufficient, the vector $\mathbf{y}$ in Proposition 1.9 is not necessarily enabled under $\mathbf{m}_0$.

Next, we gather a number of Petri net properties.

Definition 1.4. *Let $\mathcal{N} = (\mathbf{F}^+, \mathbf{F}^-, \mathbf{m}_0)$ be a Petri net. Then $\mathcal{N}$ is*

(a) in deadlock *if $\Delta_{\mathbf{m}_0} = \emptyset$,*

(b) k-safe *for $k \in \mathbb{N}$ if*

$$\mathbf{m} \leq \begin{bmatrix} k \\ \vdots \\ k \end{bmatrix} \quad \textit{for all } \mathbf{m} \in \mathcal{R}_\infty(\mathbf{m}_0),$$

(c) bounded *if $\mathcal{N}$ is k-safe for at least one $k \in \mathbb{N}$.*

When a Petri net is in deadlock, each further state evolution is impossible. A Petri net is k-safe if for all reachable states each state component is not greater than k.

7. Graphical representation

Apart from its algebraic definition, a Petri net also has an equivalent graphical representation; a bipartite graph with weighted and directed arcs. A *bipartite graph* is a graph of which the set of nodes can be partitioned into two nonempty sets, such that there are no arcs in the graph between nodes which belong to the same set. In a Petri net the two types of nodes are called *places* and *transitions*. Graphically they are represented by a circle and a bar, respectively. The Petri net graph consists of n places $p_1, \ldots, p_n$ and m transitions $t_1, \ldots, t_m$.

The $n \times m$ matrices $\mathbf{F}^+$ and $\mathbf{F}^-$ determine the interconnection between places and transitions. The matrix $\mathbf{F}^+$ is the incidence matrix from transitions to places and the matrix $\mathbf{F}^-$ is the incidence matrix from places to transitions. If $F^+_{i,j}$, the (i, j) element of $\mathbf{F}^+$, is positive, then there is an arc with weight $F^+_{i,j}$ from t_j to p_i, i.e. with orientation from t_j to p_i. If $F^-_{i,j}$, the (i, j) element of $\mathbf{F}^-$, is positive, then there is an arc with weight $F^-_{i,j}$ from p_i to t_j. In the graph, weights are shown next to the corresponding arc. Weights equal to one are not indicated. Consequently, we have represented the structure of a Petri net by a graph.

Before representing the state of a Petri net graphically, the above graph is illustrated by means of the railway example.

Example 1.5. The Petri net that models the railway section of Figure 1.1 is represented graphically in Figure 1.2. The incidence matrices $\mathbf{F}^+$ and

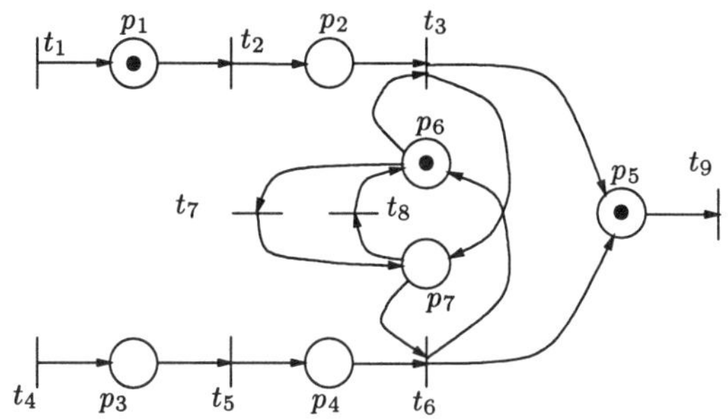

Figure 1.2. Graphical representation of the Petri net model of the railway section in Figure 1.1. The token in the place p_1 corresponds to the train in Zone 1, the token in the place p_5 corresponds to the train in Zone 5 and the token in the place p_6 corresponds to the switch in Position 1.

$\mathbf{F}^-$ were constructed in Example 1.4 (p. 11). All arc weights are equal to one and are therefore not indicated. Ignore for a moment the black dots within places. △

Next, we represent the Petri net state in the graph and graphically interpret the enabling condition in (1.3) and the state equation in (1.1).

In the Petri net graph, the state $\mathbf{m}$ of the Petri net is represented by m_i dots, so-called *tokens*, within each place p_i. When a transition vector $\boldsymbol{\delta}$ is executed, then, simultaneously, $(\mathbf{F}^-\boldsymbol{\delta})_i$ tokens are removed from and $(\mathbf{F}^+\boldsymbol{\delta})_i$ tokens are added to each place p_i $(i = 1, \dots, n)$. The enabling condition $\mathbf{m} \geq \mathbf{F}^-\boldsymbol{\delta}$ expresses that in each place there should be as many tokens as there are removed.

Example 1.6. The initial state of the railway example of Figure 1.1 is indicated in the graph in Figure 1.2. It is straightforward to interpret graphically the enabling conditions which were formulated in words in Example 1.1. △

In the course of the text, the algebraic and the graphical representations are used together and interchangeably. Therefore, we further use the following equivalent definition for a Petri net.

Definition 1.5. *A* Petri net *is a five-tuple*

$$\mathcal{N} = (P, T, f^+, f^-, \mathbf{m}_0).$$

Herein, P and T are finite, nonempty and mutually disjoint sets with respective cardinality n and m. We refer to P and T as the set of places *and* the set of transitions, *respectively. Further,*

$$f^+ : T \times P \to \mathbb{N} \quad \textit{and} \quad f^- : P \times T \to \mathbb{N}$$

are the incidence functions *from transitions to places and from places to transitions, respectively. The initial state* $\mathbf{m}_0$ *is a function* $P \to \mathbb{N}$.

The state can thus be written both as a vector in $\mathbb{N}^n$ and as a function $\mathbf{m} : P \to \mathbb{N}$. Analogously, transition vectors $\boldsymbol{\delta}$ can be considered as vectors in Δ or as functions $\boldsymbol{\delta} : T \to \mathbb{N}$ with $\boldsymbol{\delta}(t) \neq 0$ for at least one $t \in T$. For state vectors as well as for transition vectors both notations are used interchangeably. A function $P \to \mathbb{N}$, respectively $T \to \mathbb{N}$, together with a linear order of elements in P, respectively T, completely determines the corresponding vector. To simplify the notation, the initial state $\mathbf{m}_0$ is often not mentioned in the definition of a Petri net. The definition of the incidence matrices is deduced straightforwardly from the incidence functions f^+ and f^-:

$$F^+_{p,t} := f^+(t,p) \quad \text{and} \quad F^-_{p,t} := f^-(p,t) \qquad (p \in P, t \in T).$$

In this definition, $F^+_{p,t}$ and $F^-_{p,t}$ are the elements of $\mathbf{F}^+$ and $\mathbf{F}^-$, respectively, which correspond to p and t. Analogously, $\mathbf{m}(p)$ represents the component of $\mathbf{m}$ which corresponds to the place p and $\boldsymbol{\delta}(t)$ represents the component of $\boldsymbol{\delta}$ which corresponds to the transition t.

Transition vectors can also be seen as *multi-sets*. A multi-set is an extension of the notion of set: it can contain an element more than once. We say that transition t *belongs* to $\boldsymbol{\delta}$ if $\boldsymbol{\delta}(t) > 0$. We say that t is *enabled* under $\mathbf{m}$ if $\mathbf{m} \geq \mathbf{F}^-(t)$, where $\mathbf{F}^-(t)$ is the column of $\mathbf{F}^-$ which corresponds to t. Consequently, t is enabled under $\mathbf{m}$ if and only if $\boldsymbol{\delta}_t \in \Delta_{\mathbf{m}}$, where $\boldsymbol{\delta}_t(t) = 1$ and $\boldsymbol{\delta}_t(t') = 0$ for $t' \neq t$.

In the rest of this section we introduce a number of definitions originating from the graphical representation of a Petri net.

Definition 1.6. *For* $p \in P$ *define*

$${}^\bullet p := \{t \in T \mid f^+(t,p) > 0\} \quad \textit{and} \quad p^\bullet := \{t \in T \mid f^-(p,t) > 0\}$$

as the set of input *and* output transitions of p, *respectively. Analogously, for* $t \in T$ *define*

$${}^\bullet t := \{p \in P \mid f^-(p,t) > 0\} \quad \textit{and} \quad t^\bullet := \{p \in P \mid f^+(t,p) > 0\}$$

as the set of input *and* output places of t, *respectively. Define for* $\bar{P} \subseteq P$ *and* $\bar{T} \subseteq T$:

$${}^\bullet\bar{P} := \bigcup_{p \in \bar{P}} {}^\bullet p, \qquad \bar{P}^\bullet := \bigcup_{p \in \bar{P}} p^\bullet,$$

$${}^\bullet\bar{T} := \bigcup_{t \in \bar{T}} {}^\bullet t, \qquad \bar{T}^\bullet := \bigcup_{t \in \bar{T}} t^\bullet.$$

For each $p \in P$, ${}^\bullet p$ is the set of transitions t for which there exists an arc from t to p in the graph. On the other hand, $p^\bullet$ is the set of transitions t for which the graph contains an arc from p to t. Analogously for ${}^\bullet t$ and $t^\bullet$ with $t \in T$. Note that

$$p \in {}^\bullet t \Leftrightarrow t \in p^\bullet \quad \text{and} \quad t \in {}^\bullet p \Leftrightarrow p \in t^\bullet.$$

With Definition 1.6 the enabling condition $\mathbf{m} \geq \mathbf{F}^- \boldsymbol{\delta}$ can be written as:

$$\mathbf{m}(p) \geq \sum_{t \in p^\bullet} f^-(p,t)\boldsymbol{\delta}(t) \qquad (p \in P). \tag{1.11}$$

This is because $f^-(p,t) = 0$ for all $t \in P \setminus p^\bullet$. With a similar argument the state equation $\mathbf{m}' = \mathbf{m} + \mathbf{F}\boldsymbol{\delta}$ can be brought into the form

$$\mathbf{m}'(p) = \mathbf{m}(p) + \sum_{t \in {}^\bullet p} f^+(t,p)\boldsymbol{\delta}(t) - \sum_{t \in p^\bullet} f^-(p,t)\boldsymbol{\delta}(t) \qquad (p \in P).$$

Both of these alternative formulations are useful.

Definition 1.7. *A Petri net $\mathcal{N}_1 = (P_1, T_1, f_1^+, f_1^-, \mathbf{m}_1)$ is a* sub-Petri net *of a Petri net $\mathcal{N}_2 = (P_2, T_2, f_2^+, f_2^-, \mathbf{m}_2)$ if*

(i) $P_1 \subseteq P_2$,
(ii) $T_1 \subseteq T_2$,
(iii) $f_1^+ = f_2^+|_{T_1 \times P_1}$,
(iv) $f_1^- = f_2^-|_{P_1 \times T_1}$,
(v) $\mathbf{m}_1 = \mathbf{m}_2|_{P_1}$.

Condition (iii) states that f_1^+ is the restriction of the function f_2^+ to the subset $T_1 \times P_1$ of its domain $T_2 \times P_2$. Making abstraction of condition (v), $\mathcal{N}_1$ is a sub-Petri net of $\mathcal{N}_2$ if the graph of $\mathcal{N}_1$ is a subgraph of $\mathcal{N}_2$.

Definition 1.8. *A* path *in a Petri net $\mathcal{N} = (P, T, f^+, f^-)$ is a sequence $\pi = (a_1, \ldots, a_l)$ with $a_i \in P \cup T$ for $i = 1, \ldots, l$ and $a_{i+1} \in a_i^\bullet$ for $i = 1, \ldots, l-1$. We refer to l as the* length *of π. The first and the last element of π are denoted ${}^*\pi$ and π^*, respectively. A* cycle *in $\mathcal{N}$ is a path $(a_1, \ldots, a_l)$ in $\mathcal{N}$ with $a_1 \in a_l^\bullet$. A* self-loop *is a cycle with length two. A Petri net is* acyclic *if it is free of cycles.*

Algebraically, there exists a self-loop (p,t) if and only if $f^+(t,p) \neq 0$ and $f^-(p,t) \neq 0$.

Definition 1.9. *A* source transition *is a transition which does not have any input places. A* sink transition *is a transition which does not have any output places. A* choice place *is a place which has at least two output transitions. A* synchronising transition *is a transition which has at least two input places.*

The sets of source and sink transitions are denoted

$$T^{\mathrm{so}} := \{t \in T \mid {}^{\bullet}t = \emptyset\} \quad \text{and} \quad T^{\mathrm{si}} := \{t \in T \mid t^{\bullet} = \emptyset\},$$

respectively. Because $\mathbf{F}^{-}(t) = \mathbf{0}$ holds for each $t \in T^{\mathrm{so}}$, each source transition is enabled under any state.

Example 1.7. We illustrate the above definitions with the graph in Figure 1.2, the Petri net of the railway example. We find that ${}^{\bullet}p_1 = \{t_1\}$, $t_6^{\bullet} = \{p_5, p_6\}$ and $t_9^{\bullet} = \emptyset$. The sequence $(t_5, p_4, t_6, p_5, t_9)$ is a path and (p_7, t_8, p_6, t_7) is a cycle. There are two source transitions and there is one sink transition, namely $T^{\mathrm{so}} = \{t_1, t_4\}$ and $T^{\mathrm{si}} = \{t_9\}$. The place p_6 is a choice place because it has two output transitions, namely t_3 and t_7. The transition t_3 is a synchronising transition because it has two input places, namely p_2 and p_6. The transitions t_4, t_5 and t_6, the places p_3, p_4 and p_7, and the arcs interconnecting them form an acyclic sub-Petri net as it contains no cycle. △

8. Reachability via subsets of transitions

Sets of states which can be reached via transitions in a subset T' of T are important tools to develop a supervisory control theory for Petri nets. In this section, we define such sets and consider some of their properties.

Definition 1.10. *Define for all $T' \subseteq T$, $\mathbf{m} \in \mathbb{N}^n$ and $k \in \mathbb{N}$ the set of states*

$$\mathcal{R}_k^{T'}(\mathbf{m}) := \{\mathbf{m}' \in \mathbb{N}^n \mid \exists \mathbf{m}'' \in \mathcal{R}_{k-1}^{T'}(\mathbf{m}), \\ \exists \boldsymbol{\delta} \in \Delta_{\mathbf{m}''}, \boldsymbol{\delta}|_{T\setminus T'} = \mathbf{0} : \mathbf{m}' = \mathbf{m}'' + \mathbf{F}\boldsymbol{\delta}\} \cup \mathcal{R}_{k-1}^{T'}(\mathbf{m}),$$

where $\mathcal{R}_0^{T'}(\mathbf{m}) := \{\mathbf{m}\}$.

In comparison with the original definition of reachable sets (see Definition 1.3, p. 12), the only difference is the condition $\boldsymbol{\delta}|_{T\setminus T'} = \mathbf{0}$, i.e. the transition vectors which are used to reach states should be zero outside T'. The state of the Petri net can only change via transitions in T'. It is clear that $\mathcal{R}_k^{T'}(\mathbf{m}) \subseteq \mathcal{R}_k(\mathbf{m})$ for all $\mathbf{m} \in \mathbb{N}^n$, $k \in \mathbb{N}$ and $T' \subseteq T$. As in Section 6, because of Propositions 1.1 and 1.2 the set

$$\mathcal{R}_\infty^{T'}(\mathbf{m}) := \lim_{k\to\infty} \mathcal{R}_k^{T'}(\mathbf{m}) \tag{1.12}$$

exists for all $T' \subseteq T$ and $\mathbf{m} \in \mathbb{N}^n$. One immediately deduces that

$$\mathcal{R}_k^{T}(\mathbf{m}) = \mathcal{R}_k(\mathbf{m}) \quad \text{for all } \mathbf{m} \in \mathbb{N}^n \text{ and } k \in \mathbb{N} \cup \{\infty\}.$$

The larger the set T' in Definition 1.10, the larger the set of states that can be reached.

Proposition 1.10. *Let* $\mathbf{m} \in \mathbb{N}^n$ *and* $k \in \mathbb{N} \cup \{\infty\}$. *If* $T_1 \subseteq T_2 \subseteq T$, *then it holds that* $\mathcal{R}_k^{T_1}(\mathbf{m}) \subseteq \mathcal{R}_k^{T_2}(\mathbf{m})$.

Proof. This follows directly from Definition 1.10. □

If $T' = \{t\}$ with $t \in T$, then we write $\mathcal{R}_\cdot^t(\cdot)$ instead of $\mathcal{R}_\cdot^{\{t\}}(\cdot)$ for the set of states which can be reached via t only.

Proposition 1.11. *Let* $\mathbf{m} \in \mathbb{N}^n$ *and* $\boldsymbol{\delta} \in \Delta$ *with* $\boldsymbol{\delta}|_{T \setminus T'} = \mathbf{0}$. *Then* $\boldsymbol{\delta} \in \Delta_{\mathbf{m}}$ *if and only if*

$$\mathbf{m}(p) \geq \sum_{t \in p^\bullet \cap T'} f^-(p,t)\boldsymbol{\delta}(t) \quad \text{for } p \in {}^\bullet T'.$$

Proof. As shown in (1.11), $\boldsymbol{\delta} \in \Delta_{\mathbf{m}}$ is equivalent to

$$\mathbf{m}(p) \geq \sum_{t \in p^\bullet} f^-(p,t)\boldsymbol{\delta}(t) \quad \text{for } p \in P. \tag{1.13}$$

As $\boldsymbol{\delta}|_{T \setminus T'} = \mathbf{0}$, this condition can be rewritten as

$$\mathbf{m}(p) \geq \sum_{t \in p^\bullet \cap T'} f^-(p,t)\boldsymbol{\delta}(t) \quad \text{for } p \in P. \tag{1.14}$$

This can be further simplified to

$$\mathbf{m}(p) \geq \sum_{t \in p^\bullet \cap T'} f^-(p,t)\boldsymbol{\delta}(t) \quad \text{for } p \in {}^\bullet T', \tag{1.15}$$

because $\mathbf{m} \geq \mathbf{0}$, and $p^\bullet \cap T' \neq \emptyset$ if and only if $p \in {}^\bullet T'$. □

The next property is used in Chapter 4.

Proposition 1.12. *Let* $\mathbf{m}, \mathbf{m}' \in \mathbb{N}^n, T' \subseteq T$ *and* $\boldsymbol{\delta}_i \in \Delta$ $(i = 1, \ldots, l)$. *If* $(\boldsymbol{\delta}_1, \ldots, \boldsymbol{\delta}_l) \in \bar{\Delta}_{\mathbf{m}}$, $\boldsymbol{\delta}_i|_{T \setminus T'} = \mathbf{0}$ $(i = 1, \ldots, l)$ *and* $\mathbf{m}'|_{{}^\bullet T'} \geq \mathbf{m}|_{{}^\bullet T'}$, *then* $(\boldsymbol{\delta}_1, \ldots, \boldsymbol{\delta}_l) \in \bar{\Delta}_{\mathbf{m}'}$.

In words, an enabled sequence under $\mathbf{m}$ that only contains transitions in T', is also enabled under $\mathbf{m}'$ if the state of all input places of transitions in T' is greater than or equal to the corresponding component of $\mathbf{m}$.

Proof. We know that

$$\mathbf{m} \equiv \mathbf{m}_0 \xrightarrow{\boldsymbol{\delta}_1} \cdots \xrightarrow{\boldsymbol{\delta}_l} \mathbf{m}_l. \tag{1.16}$$

We are going to show that

$$\mathbf{m}' \equiv \mathbf{m}'_0 \xrightarrow{\boldsymbol{\delta}_1} \cdots \xrightarrow{\boldsymbol{\delta}_l} \mathbf{m}'_l. \tag{1.17}$$

From $\boldsymbol{\delta}_1 \in \Delta_{\mathbf{m}}$, $\mathbf{m}'|_{^\bullet T'} \geq \mathbf{m}|_{^\bullet T'}$, $\boldsymbol{\delta}_1|_{T \setminus T'}$ and Proposition 1.11 we find that $\boldsymbol{\delta}_1 \in \Delta_{\mathbf{m}'}$. The other steps in (1.17) are proven by induction. Suppose that $(\boldsymbol{\delta}_1, \ldots, \boldsymbol{\delta}_{i-1}) \in \bar{\Delta}_{\mathbf{m}'}$, i.e. that

$$\mathbf{m}' \equiv \mathbf{m}'_0 \xrightarrow{\boldsymbol{\delta}_1} \cdots \xrightarrow{\boldsymbol{\delta}_{i-1}} \mathbf{m}'_{i-1}. \tag{1.18}$$

Next, we prove that $\boldsymbol{\delta}_i \in \Delta_{\mathbf{m}'_{i-1}}$. From (1.16) and the induction hypothesis (1.18) we find that

$$\mathbf{m}'_{i-1}(p) - \mathbf{m}_{i-1}(p) = \mathbf{m}'_0(p) - \mathbf{m}_0(p) \quad \text{for } p \in {}^\bullet T'.$$

So, with $\mathbf{m}'|_{^\bullet T'} \geq \mathbf{m}|_{^\bullet T'}$, we have for $p \in {}^\bullet T'$ that (recall that $\mathbf{m} = \mathbf{m}_0$ and $\mathbf{m}' = \mathbf{m}'_0$)

$$\mathbf{m}'_{i-1}(p) \geq \mathbf{m}_{i-1}(p) \geq \sum_{t \in p^\bullet \cap T'} f^-(p,t)\boldsymbol{\delta}_i(t). \tag{1.19}$$

The last inequality follows from $\boldsymbol{\delta}_i \in \Delta_{\mathbf{m}_{i-1}}$. Consequently, we obtain with Proposition 1.11 that $\boldsymbol{\delta}_i \in \Delta_{\mathbf{m}_i}$. □

9. Other concurrency assumptions

So far we have used the transition bag assumption. The collections of events which can occur simultaneously can contain any number of events any number of times. Mathematically, this was expressed by the set $\Delta = \mathbb{N}^m \setminus \{\mathbf{0}\}$ in (1.2). In many applications the transition bag assumption is not realistic. It is not always true that any collection of events can occur together. In the railway example, two trains that are in the same section cannot move to the next one simultaneously (under the assumption that each state change or event is observed 'immediately'). Therefore, other concurrency assumptions, or *model speeds*, have their importance and need to be studied. In this section we discuss two of them: the *no concurrency assumption* and the *concurrency assumption.* They both have great practical importance. The acquired insight of this section, allows us to propose a unifying definition of a Petri net in Section 10 which comprises all concurrency assumptions.

9.1. The no concurrency assumption

Often, instead of Δ in (1.2), the set of possible transition vectors is limited to

$$\Delta_{\mathrm{N}} := \left\{ \boldsymbol{\delta} \in \Delta \middle| \sum_{j=1}^{m} \delta_j = 1 \right\}. \tag{1.20}$$

Apart from one element, which is equal to one, all elements of a transition vector in Δ_{N} are equal to zero. There exists a bijective relation between Δ_{N} and the set of subsets of T which contain precisely one element, i.e. the set of singletons in 2^T, the power set of T. The cardinality of Δ_{N} is equal to m. When the transition vectors should belong to (1.20) in the Petri net state equation, we say that the *no concurrency assumption* is used.

Example 1.8. For $m = 3$ the set in (1.20) is equal to

$$\Delta_{\mathrm{N}} = \left\{ \begin{bmatrix} 1 \\ 0 \\ 0 \end{bmatrix}, \begin{bmatrix} 0 \\ 1 \\ 0 \end{bmatrix}, \begin{bmatrix} 0 \\ 0 \\ 1 \end{bmatrix} \right\}$$

$\triangle$

The state evolution of a Petri net under the no concurrency assumption is as follows. If $\mathbf{m}_k$ is the present state, all possible next states $\mathbf{m}_{k+1}$ are given by $\mathbf{m}_{k+1} = \mathbf{m}_k + \mathbf{F}\boldsymbol{\delta}$ with $\boldsymbol{\delta} \in \Delta_{\mathrm{N}}$ and $\mathbf{m}_k \geq \mathbf{F}^{-}\boldsymbol{\delta}$. In contrast to the transition bag assumption, in each step at most one transition can be executed once. In other words, for a discrete event system, no two events can occur together and no event more than once at the same time.

Note that Δ_{N} is, with respect to the componentwise partial order, an antichain for all m. There are no two different comparable elements.

Under the no concurrency assumption reachable sets are defined in a similar manner as the sets in Section 6. As one would expect, $\Delta_{\mathbf{m}}$ is replaced in their definition by

$$\Delta_{\mathrm{N},\mathbf{m}} := \{\boldsymbol{\delta} \in \Delta_{\mathrm{N}} \mid \mathbf{m} \geq \mathbf{F}^{-}\boldsymbol{\delta}\}. \tag{1.21}$$

These reachable sets are denoted $\mathcal{R}_{\cdot,\mathrm{N}}(\cdot)$ to distinguish them from the reachable sets of Section 6. The definition of reachability via subsets of transitions (Section 8) is generalised in the same straightforward manner. Analogously to Proposition 1.6, $\Delta_{\mathrm{N},\mathbf{m}}$ is a down-set in $(\Delta_{\mathrm{N}}, \leq)$, and consequently in $(\Delta, \leq)$, for all states $\mathbf{m}$.

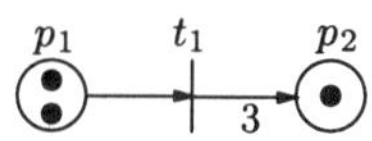

Figure 1.3. The initial state is $\mathbf{m}_0 = (2,1)$. One finds that $\mathcal{R}_1(\mathbf{m}_0) = \{(2,1),(1,4),(0,7)\}$ is a superset of $\mathcal{R}_{1,\mathrm{N}}(\mathbf{m}_0) = \{(2,1),(1,4)\}$.

For a given $\mathbf{m}$, the set of enabled transition vectors $\Delta_{\mathrm{N},\mathbf{m}}$ can equivalently be represented by the set of transitions

$$\{t \in T \mid \mathbf{m} \geq \mathbf{F}^{-}\boldsymbol{\delta} \text{ with } \boldsymbol{\delta}(t) = 1 \text{ and } \boldsymbol{\delta}|_{T\setminus\{t\}} = \mathbf{0}\}.$$

In contrast to a great part of the Petri net literature, this representation is not used in this book, because it hides the algebraic nature of Petri net dynamics.

As $\Delta_{\mathrm{N}} \subseteq \Delta$, reachable sets under the no concurrency assumption are subsets of the corresponding sets under the transition bag assumption.

Proposition 1.13. *For each $\mathbf{m} \in \mathbb{N}^n$ and each $k \in \mathbb{N}$ it holds that*

$$\mathcal{R}_{k,\mathrm{N}}(\mathbf{m}) \subseteq \mathcal{R}_k(\mathbf{m}).$$

Proof. This follows from $\Delta_{\mathrm{N},\mathbf{m}'} \subseteq \Delta_{\mathbf{m}'}$ for $\mathbf{m}' \in \mathbb{N}^n$. □

The subset relation in Proposition 1.13 can also be proper as illustrated in Figure 1.3.

Despite Proposition 1.13, the set of states which are reachable in any number of steps does not depend on the used concurrency assumption.

Proposition 1.14. *For each $\mathbf{m} \in \mathbb{N}^n$ it is true that*

$$\mathcal{R}_{\infty,\mathrm{N}}(\mathbf{m}) = \mathcal{R}_{\infty}(\mathbf{m}).$$

Proof. From Proposition 1.13 we have that $\mathcal{R}_{\infty,\mathrm{N}}(\mathbf{m}) \subseteq \mathcal{R}_{\infty}(\mathbf{m})$. On the other hand, each sequence of transition vectors $(\boldsymbol{\delta}_1, \ldots, \boldsymbol{\delta}_k)$ with $\boldsymbol{\delta}_i \in \Delta$ $(i = 1, \ldots, k)$ which is enabled under $\mathbf{m}$, can be decomposed with the aid of Proposition 1.7 until a sequence $(\boldsymbol{\delta}'_1, \ldots, \boldsymbol{\delta}'_{k'})$ with $\boldsymbol{\delta}'_i \in \Delta_{\mathrm{N}}$ $(i = 1, \ldots, k')$ is obtained. This sequence is also enabled under $\mathbf{m}$. It is clear that

$$\sum_{i=1}^{k} \boldsymbol{\delta}_i = \sum_{i=1}^{k'} \boldsymbol{\delta}'_i.$$

□

When constructing sets of states which are reachable in any number of steps, we know from Proposition 1.14 that the model speed one uses does not make any difference. It needs no proof to see that this result also holds for the reachable sets of Section 8.

When a Petri net is free of self-loops (Definition 1.8, p. 17), then the enabling condition $\mathbf{m} \geq \mathbf{F}^-\boldsymbol{\delta}$ can be rewritten as follows.

Proposition 1.15. *If a Petri net does not contain any self-loops, then for each $\boldsymbol{\delta} \in \Delta_\mathrm{N}$ the enabling condition $\mathbf{m} \geq \mathbf{F}^-\boldsymbol{\delta}$ is equivalent to $\mathbf{m}+\mathbf{F}\boldsymbol{\delta} \geq \mathbf{0}$.*

Proof. ($\Rightarrow$) Because of $\mathbf{F}^+\boldsymbol{\delta} \geq \mathbf{0}$ for each $\boldsymbol{\delta} \in \Delta_\mathrm{N}$ it is true that

$$\mathbf{m} \geq \mathbf{F}^-\boldsymbol{\delta} \quad \Rightarrow \quad \mathbf{m} + (\mathbf{F}^+ - \mathbf{F}^-)\boldsymbol{\delta} = \mathbf{m} + \mathbf{F}\boldsymbol{\delta} \geq \mathbf{0}.$$

($\Leftarrow$) Let t be the transition which corresponds to the element of $\boldsymbol{\delta}$ which is equal to one. The condition $\mathbf{m} + \mathbf{F}\boldsymbol{\delta} \geq \mathbf{0}$ can then be written as

$$\mathbf{m}(p) + \mathbf{F}^+(p,t) \geq \mathbf{F}^-(p,t) \qquad (p \in P). \tag{1.22}$$

If $\mathbf{F}^-(p,t) > 0$ (the case $\mathbf{F}^-(p,t) = 0$ is trivial) then it follows from the self-loop freeness of the Petri net that $\mathbf{F}^+(p,t) = 0$ and thus that $\mathbf{m}(p) \geq \mathbf{F}^-(p,t)$. □

Consequently, if a Petri net does not contain any self-loops and if the no concurrency assumption is used, then the evolution of that Petri net is completely determined by $\mathbf{F}$ and not by the pair $(\mathbf{F}^+, \mathbf{F}^-)$. Algebraically, a Petri net is free of self-loops if and only if for all $p \in P$ and $t \in T$:

$$f^-(p,t) \neq 0 \quad \Rightarrow \quad f^+(t,p) = 0,$$

or if and only if

$$f^+(t,p) \neq 0 \quad \Rightarrow \quad f^-(p,t) = 0.$$

9.2. The concurrency assumption

Another model speed is the *concurrency assumption* which is situated in between the no concurrency and the transition bag assumption. Events can occur simultaneously, but an event cannot occur more than once at the same time.

Transition vectors should belong to

$$\Delta_\mathrm{C} := \{0,1\}^m \setminus \{\mathbf{0}\}, \tag{1.23}$$

the set of m-dimensional vectors with elements equal to zero or one, minus the zero vector. There is a bijective relation between vectors in Δ_C and nonempty subsets of the set of transitions T. To $\boldsymbol{\delta} \in \Delta_\mathrm{C}$ corresponds the set of transitions

$$\{t \in T \mid \boldsymbol{\delta}(t) = 1\}.$$

Note that Δ_C is a down-set in $(\Delta, \leq)$. Its cardinality is equal to $2^m - 1$.

Example 1.9. For $m = 3$ the set in (1.23) is equal to

$$\Delta_C = \left\{ \begin{bmatrix}1\\1\\1\end{bmatrix}, \begin{bmatrix}1\\1\\0\end{bmatrix}, \begin{bmatrix}1\\0\\1\end{bmatrix}, \begin{bmatrix}0\\1\\1\end{bmatrix}, \begin{bmatrix}1\\0\\0\end{bmatrix}, \begin{bmatrix}0\\1\\0\end{bmatrix}, \begin{bmatrix}0\\0\\1\end{bmatrix} \right\}.$$

$\triangle$

Reachable sets are again defined in a similar way as above and are denoted with a subindex 'C'. For all $\mathbf{m} \in \mathbb{N}^n$, it is clear that

$$\mathcal{R}_{k,\mathrm{N}}(\mathbf{m}) \subseteq \mathcal{R}_{k,\mathrm{C}}(\mathbf{m}) \subseteq \mathcal{R}_k(\mathbf{m}) \tag{1.24}$$

for all $k \in \mathbb{N}$, and that

$$\mathcal{R}_{\infty,\mathrm{N}}(\mathbf{m}) = \mathcal{R}_{\infty,\mathrm{C}}(\mathbf{m}) = \mathcal{R}_{\infty}(\mathbf{m}). \tag{1.25}$$

These relations also hold for reachability via subsets of transitions.

10. A general Petri net definition

The three concurrency assumptions that we have discussed so far have in common that the sets to which transition vectors should belong, are all down-sets in $(\Delta, \leq)$, i.e.

$$\{\Delta, \Delta_N, \Delta_C\} \subseteq \mathcal{O}^{\downarrow}(\Delta, \leq).$$

These three model speeds are not sufficient to model all discrete event systems. In general, the concurrency character of a particular system can be a mixture of these three speeds or even be more general. Some events have transition bag features (for example the arrival, on a certain time scale, of http packages at a server), some events cannot occur twice at the same time (e.g. the closure of a door). Events with different concurrency properties can belong to the same discrete event system. That is why the possible concurrency assumptions should not be limited to the three we have discussed above.

By looking at a Petri net as a model for a discrete event system as in Section 5, it is clear that to each down-set Ω in $(\Delta, \leq)$, corresponds a valid concurrency assumption. This leads us to a Petri net definition for each $\Omega \in \mathcal{O}^{\downarrow}(\Delta, \leq)$. Above we have discussed already three of them, namely Δ, Δ_N and Δ_C.

A natural additional condition for Ω is that $\Delta_N \subseteq \Omega$. If not, the transitions or events which correspond to transition vectors $\boldsymbol{\delta} \in \Delta_N \setminus \Omega$ can never occur and can therefore be omitted from the model.

To make the further theoretical discussions generally applicable, we often use a general concurrency assumption Ω. To be complete and to avoid confusion, we end this section by restating some of the above definitions and properties for general model speeds Ω. We start with the state evolution.

The state of a Petri net with speed $\Omega \in \mathcal{O}^{\downarrow}(\Delta, \leq)$ changes stepwise as follows. If $\mathbf{m}_k$ is the state after step $k \in \mathbb{N}$, then all possible next states $\mathbf{m}_{k+1}$ are given by

$$\mathbf{m}_{k+1} = \mathbf{m}_k + (\mathbf{F}^+ - \mathbf{F}^-)\boldsymbol{\delta}, \tag{1.26}$$

where

$$\boldsymbol{\delta} \in \Omega \tag{1.27}$$

and

$$\mathbf{m}_k \geq \mathbf{F}^-\boldsymbol{\delta}. \tag{1.28}$$

Define for all $\mathbf{m} \in \mathbb{N}^n$,

$$\Omega_{\mathbf{m}} := \{\boldsymbol{\delta} \in \Omega \mid \mathbf{m} \geq \mathbf{F}^-\boldsymbol{\delta}\} \tag{1.29}$$

as the set of enabled transition vectors under $\mathbf{m}$.

Proposition 1.16. *For each $\mathbf{m} \in \mathbb{N}^n$ it holds that $\Omega_{\mathbf{m}} \in \mathcal{O}^{\downarrow}(\Omega, \leq)$.*

Reachable sets are defined in the obvious way by replacing $\Delta_{\mathbf{m}''}$ by $\Omega_{\mathbf{m}''}$ in Definitions 1.3 and 1.10. To indicate that we are using a speed Ω, reachable sets get an extra subscript 'Ω'. We write $\mathcal{R}^{\cdot}_{\cdot,\Omega}(\cdot)$. As such, $\mathcal{R}^{\cdot}_{\cdot}(\cdot)$ (transition bag assumption), $\mathcal{R}^{\cdot}_{\cdot,\mathrm{C}}(\cdot)$ (concurrency assumption) and $\mathcal{R}^{\cdot}_{\cdot,\mathrm{N}}(\cdot)$ (no concurrency assumption) are shorthand notations for $\mathcal{R}^{\cdot}_{\cdot,\Delta}(\cdot)$, $\mathcal{R}^{\cdot}_{\cdot,\Delta_\mathrm{C}}(\cdot)$ and $\mathcal{R}^{\cdot}_{\cdot,\Delta_\mathrm{N}}(\cdot)$, respectively.

Proposition 1.17. *Let $\mathbf{m}_0, \mathbf{m}_\mathrm{f} \in \mathbb{N}^n$. If $\mathbf{m}_\mathrm{f} \in \mathcal{R}_{\infty,\Omega}(\mathbf{m}_0)$, then there exists a vector $\mathbf{y} \in \mathbb{N}^m$ such that $\mathbf{m}_\mathrm{f} = \mathbf{m}_0 + \mathbf{F}\mathbf{y}$.*

Note that $\mathbf{y}$ does not necessarily belong to Ω. The following proposition is the generalisation of the relation in (1.24).

Proposition 1.18. *Let $\Omega_1, \Omega_2 \in \mathcal{O}^{\downarrow}(\Delta, \leq)$, $k \in \mathbb{N}$, $\mathbf{m} \in \mathbb{N}^n$ and $T' \subseteq T$. Then,*

$$\Omega_1 \subseteq \Omega_2 \quad \Rightarrow \quad \mathcal{R}^{T'}_{k,\Omega_1}(\mathbf{m}) \subseteq \mathcal{R}^{T'}_{k,\Omega_2}(\mathbf{m}).$$

Analogously, Proposition 1.19 is the generalisation of (1.25).

Proposition 1.19. *Let* $\Omega_1, \Omega_2 \in \mathcal{O}^{\downarrow}(\Delta, \leq)$, $k \in \mathbb{N}$, $\mathbf{m} \in \mathbb{N}^n$ *and* $T' \subseteq T$. *Then,*

$$\Delta_{\mathrm{N}} \subseteq \Omega_1 \text{ and } \Delta_{\mathrm{N}} \subseteq \Omega_2 \quad \Rightarrow \quad \mathcal{R}^{T'}_{\infty,\Omega_1}(\mathbf{m}) = \mathcal{R}^{T'}_{\infty,\Omega_2}(\mathbf{m}).$$

Because of this property, from now on we suppress the subscript 'Ω' of reachable set $\mathcal{R}^{\cdot}_{\infty,\Omega}(\cdot)$, writing $\mathcal{R}^{\cdot}_{\infty}(\cdot)$.

Although model speeds for which Ω is a finite set, e.g. the concurrency and the no concurrency assumption, seem and are more natural and realistic than the transition bag assumption, the importance of the latter is a consequence of Proposition 1.19. This is made clear when constructing supervisory controllers in the course of the book.

11. Notes and references

Examples of excellent books where the order theoretical concepts of Section 3 are treated in more detail are [9] and [33]. A good introduction to discrete event systems can be found in [6]. Interesting introductory books on Petri nets are [30] and [32]. Standard survey papers on how to model discrete event systems with the aid of Petri nets and how they can be analysed are [10] and [28], respectively. The notions *no concurrency assumption* and *concurrency assumption* are used in [19]. The *transition bag assumption* appears in [30]. In Petri net theory, one usually refers to the state of the Petri net as its *marking*. To the best of our knowledge, the definition of the general concurrency assumptions in Section 10 is new.

Chapter 2

SUPERVISORY CONTROL

This chapter introduces supervisory control of Petri nets with a *legal set*, a given subset of the state space $\mathbb{N}^n$, as the control goal. The aim of the controller is to keep the state within this set by dynamically disabling transition vectors. The main subject of this book is control of Petri nets of which some transitions cannot be influenced by the supervisor, so-called *uncontrollable* transitions. As it gives more insight, we start with Petri nets of which all transitions are controllable. This chapter is entirely devoted to completely controllable Petri nets. Systems which contain uncontrollable transitions/events are studied in Chapter 3.

In the literature one finds several control paradigms. Together with the multiple possible concurrency assumptions of Chapter 1, this often leads to misunderstandings about the exact definition of the Petri net model under study and about the validity of specific results under different assumptions. Therefore, in an attempt to overcome this nuisance, a unifying approach to supervisory control is presented (Sections 1, 2 and 3) and a general definition of optimality, so-called *maximal permissiveness*, is proposed (Section 4). Attention is given to control issues for the different Petri net models of Chapter 1 and it is shown how existing control approaches fit into the present framework and how these results can be generalised (Sections 5, 6 and 7).

1. Control goal and architecture

Often a discrete event system can evolve in an unacceptable way. For the railway example of Chapter 1, in order to avoid collisions, two trains should never be in the same zone. For this reason, the behaviour of Petri nets, and thus of the underlying discrete event systems, needs to be influenced in such a way that the closed-loop system satisfies a number of given conditions. We express these conditions by means of the Petri net state. The control goal is that the state always belongs to a given set of states $\mathcal{A} \subseteq \mathbb{N}^n$. We refer to $\mathcal{A}$ as the *legal set*. States in $\mathcal{A}$ are called *legal* states. It is evident that $\mathcal{A}$ can be assumed to be nonempty and to be a proper subset of $\mathbb{N}^n$; $\mathcal{A} = \emptyset$ would imply that there does not even exist any legal initial state and $\mathcal{A} = \mathbb{N}^n$ would mean that each state is legal and that no control of the system is required.

Example 2.1. For the railway example in Figure 1.1 and the corresponding Petri net in Figure 1.2, the legal set is

$$\mathcal{A} = \bigcap_{i=1}^{5} \{\mathbf{m} \in \mathbb{N}^7 \mid \mathbf{m}(p_i) \leq 1\}.$$

This legal set guarantees that two trains are never in the same zone. △

To keep the state of the system within $\mathcal{A}$ unceasingly, the occurrence of transition vectors can be affected by a *supervisory controller*. This controller can only interfere with the Petri net model in a prescribed way. This is discussed next intuitively for a general concurrency assumption $\Omega \in \mathcal{O}^{\downarrow}(\Delta, \leq)$; a formal definition is given in Section 2.

The behaviour of the system is influenced by reducing, for each state $\mathbf{m}$, the set of enabled transition vectors $\Omega_{\mathbf{m}}$. The supervisory controller limits the set of transition vectors via which the system can evolve, to a subset of $\Omega_{\mathbf{m}}$. However, the supervisory controller cannot choose each subset of $\Omega_{\mathbf{m}}$.

As discussed in Chapter 1, a Petri net model defines the maximal collections of events that can occur in a given state. Each subcollection of one of these maximal collections can occur as well. Therefore, in a Petri net it is impossible to model that an event is forced to occur. We want the closed-loop system of a Petri net and a supervisory controller also to have this property. Therefore, a supervisory controller cannot 'push' the behaviour of the Petri net in a given direction. A control action does not consist of forcing the occurrence of an event. What can be done, is that the occurrence of an event is preempted. The controller can influence the behaviour of the system by limiting the set of possible next states and by limiting the set of possible next events, instead of by determining the next state. Thus, a controller uses its influence in an indirect manner. This can be seen as control of the discrete event system on a higher level. That is why the term *supervisory control* is used.

Example 2.2. Consider again the railway example in Figure 1.1 (p. 3). At the entrance of each of the five zones there is a traffic light. When the phase of a traffic light is red, trains are not allowed to enter that zone. When green, they are. This way of controlling traffic complies with the above definition of supervisory control. The events 'a train enters Zone i' can be preempted from occurring (when the phase of the corresponding traffic light is red), but cannot be forced (when green the train is allowed but not obliged to enter the next zone). $\triangle$

Supervisory control of a Petri net is done in a dynamic way, i.e. via a feedback controller. Each time the controller observes an event, the control action is adjusted to the new state of the system. In first instance, we assume that all events are observable and that the state is known by the controller. The case in which some events are unobservable, is discussed in Chapter 8.

As mentioned above and illustrated by the example, supervisory control is control on a higher level. One could say that the controller does not decide what is going to happen, but merely what is not allowed to happen. The decisions to make events occur are taken at a lower level. The controller at this lower level can pick transition vectors from the set of transition vectors which are enabled by the supervisory controller and make them happen. It is also possible that the environment generates an event. For the railway example, the supervisory controller can influence the behaviour of the system via the traffic lights. The decisions whether or not to enter the next zone are taken by the train driver in compliance with the guidelines of the supervisory controller.

The legal set $\mathcal{A}$ can express various conditions: safety conditions (minimum supply, maximal waiting queue, collision avoidance between vehicles, ...), deadlock avoidance, the language of the controlled Petri net is a subset of a given language, ... Therefore, a thorough study of supervisory control for Petri nets with a legal set as the control goal is justified.

2. Formal definition

In this section, the discussion in Section 1 on the architecture of supervisory control is formalised. The intuitive description is cast into a mathematical formalism.

As for the Petri net model in Chapter 1, we give a general and unifying definition. Instead of adopting one of the three concurrency assumptions that have been discussed in detail in Chapter 1, we work with a general concurrency assumption $\Omega \in \mathcal{O}^{\downarrow}(\Delta, \leq)$. In general, as is made clear later, for each Ω a large number of controller definitions is possible. A global supervisory control theory is therefore not superfluous. Its aim is not only to generalise and abstract existing ideas, but to be widely applicable and to provide better insights into the subject. The essence of supervisory control is the following.

Let the set $\Omega \in \mathcal{O}^{\downarrow}(\Delta, \leq)$ be the concurrency assumption. From Proposition 1.16 we know that for all $\mathbf{m} \in \mathbb{N}^n$ the set of enabled transition vectors $\Omega_{\mathbf{m}}$ is a down-set in $(\Omega, \leq)$. Mathematically, the fact that the supervisory controller cannot make events occur, but only disable them, is expressed by stating that the subsets of $\Omega_{\mathbf{m}}$ to which transitions vectors are limited, are also down-sets in $(\Omega, \leq)$.

Definition 2.1. *Let $\Omega \in \mathcal{O}^{\downarrow}(\Delta, \leq)$ be the concurrency assumption of the Petri net under study. A* control set $\mathcal{U}$ *is a set which satisfies*

$$\mathcal{U} \subseteq \mathcal{O}^{\downarrow}(\Omega, \leq) \quad \text{and} \quad \emptyset, \Omega \in \mathcal{U}. \tag{2.1}$$

A control set $\mathcal{U}$ is a set of down-sets in $(\Omega, \leq)$ which contains both the empty set and Ω. Elements of $\mathcal{U}$ are referred to as *control values.* The set $\mathcal{U}$ consists of all possible control actions which can be selected by the supervisory controller. Note that $\mathcal{U}$ is a finite set if Ω is a finite set.

Definition 2.2. *Consider a Petri net with concurrency assumption Ω and control set $\mathcal{U}$. A* supervisory control law, *or* control law *for short, is a function*

$$u : \mathbb{N}^n \to \mathcal{U} : \mathbf{m} \mapsto u(\mathbf{m}). \tag{2.2}$$

The set of all control laws is denoted $\mathfrak{U}$.

A control law u influences the behaviour of the Petri net in the following way. When $\mathbf{m}$ is the Petri net state, the enabling condition $\mathbf{m} \geq \mathbf{F}^{-}\boldsymbol{\delta}$ in (1.28) is replaced by the stronger condition

$$\mathbf{m} \geq \mathbf{F}^{-}\boldsymbol{\delta} \quad \wedge \quad \boldsymbol{\delta} \in u(\mathbf{m}). \tag{2.3}$$

Consequently, the set of all possible next states is no longer determined by transition vectors in $\Omega_{\mathbf{m}}$, but by transition vectors in

$$\Omega_{\mathbf{m}} \cap u(\mathbf{m}). \tag{2.4}$$

From Proposition 1.16 we know that $\Omega_{\mathbf{m}}$ is a down-set in $(\Omega, \leq)$. As $u(\mathbf{m})$ is also a down-set in $(\Omega, \leq)$ and as the intersection of down-sets is again a down-set (Proposition 1.3, p. 7), it follows that

$$\Omega_{\mathbf{m}} \cap u(\mathbf{m}) \in \mathcal{O}^{\downarrow}(\Omega, \leq) \quad \text{for all } \mathbf{m} \in \mathbb{N}^n \text{ and all } u \in \mathfrak{U}.$$

If $\boldsymbol{\delta} \in \Omega_{\mathbf{m}} \cap u(\mathbf{m})$, then each $\boldsymbol{\delta}' \leq \boldsymbol{\delta}$ (with $\boldsymbol{\delta}' \in \Omega$) belongs to $\Omega_{\mathbf{m}} \cap u(\mathbf{m})$. Consequently, as required in Section 1 and with the intuitive interpretation of Chapter 1 (Section 5), the resulting enabling condition specifies maximal collections of events.

The significance of $\emptyset \in \mathcal{U}$ in (2.1) is that the possibility to disable all events simultaneously always exists. (The case in which not all events can be disabled by the controller is discussed in Chapter 3.) If $u(\mathbf{m}) = \emptyset$, then $\Omega_{\mathbf{m}} \cap u(\mathbf{m}) = \emptyset$ and the controller disables all further state changes.

Definition 2.3. *Define u_0 as the control law for which $u_0(\mathbf{m}) = \emptyset$ for all $\mathbf{m} \in \mathbb{N}^n$. We call this constant control law the* zero control law.

The following property is a trivial consequence of the above definition.

Proposition 2.1. *For all $u \in \mathfrak{U}$ it holds that*

$$u_0(\mathbf{m}) \subseteq u(\mathbf{m}) \quad \textit{for all } \mathbf{m} \in \mathbb{N}^n.$$

The significance of $\Omega \in \mathcal{U}$ in (2.1) is that the controller always can opt not to influence the behaviour of the system. If $u(\mathbf{m}) = \Omega$, then $\Omega_{\mathbf{m}} \cap u(\mathbf{m}) = \Omega_{\mathbf{m}}$ and the enabling condition under supervision (2.3) is equal to the enabling condition without supervision. The assumption $\Omega \in \mathcal{U}$ does not restrain the applicability of the theory that follows, but it is a natural one to make in order to separate clearly plant and controller models.

3. Reachable sets under supervision

In this section we define reachable sets under supervisory control. We consider two cases: reachable sets under a control law (Definition 2.4) and reachable sets under a control value (Definition 2.5). Again, a general concurrency assumption Ω is adopted.

Definition 2.4. *Define for $u \in \mathfrak{U}$, $\mathbf{m} \in \mathbb{N}^n$ and $k \in \mathbb{N}$, the set of states which are reachable in at most k steps under the control law u and starting from an initial state $\mathbf{m}$ as*

$$\begin{aligned}\mathcal{R}_{k,\Omega}(u,\mathbf{m}) := \{\mathbf{m}' \in \mathbb{N}^n \mid \exists \mathbf{m}'' \in \mathcal{R}_{k-1,\Omega}(u,\mathbf{m}),\\ \exists \boldsymbol{\delta} \in \Omega_{\mathbf{m}''} \cap u(\mathbf{m}'') : \mathbf{m}' = \mathbf{m}'' + \mathbf{F}\boldsymbol{\delta}\} \cup \mathcal{R}_{k-1,\Omega}(u,\mathbf{m}),\end{aligned}$$

where $\mathcal{R}_{0,\Omega}(u,\mathbf{m}) := \{\mathbf{m}\}$. Further define

$$\mathcal{R}_{\infty,\Omega}(u,\mathbf{m}) := \lim_{k\to\infty} \mathcal{R}_{k,\Omega}(u,\mathbf{m}).$$

Definition 2.5. *Define for $\zeta \in \mathcal{U}, \mathbf{m} \in \mathbb{N}^n$ and $k \in \mathbb{N}$, the set of states which are reachable in at most k steps under the control value ζ and starting from an initial state $\mathbf{m}$ as*

$$\begin{aligned}\mathcal{R}_{k,\Omega}(\zeta,\mathbf{m}) := \{\mathbf{m}' \in \mathbb{N}^n \mid \exists \mathbf{m}'' \in \mathcal{R}_{k-1,\Omega}(\zeta,\mathbf{m}),\\ \exists \boldsymbol{\delta} \in \Omega_{\mathbf{m}''} \cap \zeta : \mathbf{m}' = \mathbf{m}'' + \mathbf{F}\boldsymbol{\delta}\} \cup \mathcal{R}_{k-1,\Omega}(\zeta,\mathbf{m}),\end{aligned}$$

where $\mathcal{R}_{0,\Omega}(\zeta, \mathbf{m}) := \{\mathbf{m}\}$. *Further define*

$$\mathcal{R}_{\infty,\Omega}(\zeta, \mathbf{m}) := \lim_{k\to\infty} \mathcal{R}_{k,\Omega}(\zeta, \mathbf{m}).$$

The reachable sets of Definition 2.5 can be seen as reachable sets under a constant control law $u \in \mathfrak{U}$ for which $u(\mathbf{m}) = \zeta$ for all $\mathbf{m} \in \mathbb{N}^n$.

Proposition 2.2. *Let* $u \in \mathfrak{U}, \mathbf{m} \in \mathbb{N}^n$ *and* $k \in \mathbb{N}$. *Then* $\mathcal{R}_{k,\Omega}(u, \mathbf{m}) \subseteq \mathcal{R}_{k,\Omega}(\mathbf{m})$.

Reachable sets under whichever control law are subsets of the corresponding reachable sets without control.

Proposition 2.3. *Let* $u_1, u_2 \in \mathfrak{U}$ *and* $\mathbf{m} \in \mathbb{N}^n$. *If* $u_1(\mathbf{m}) \subseteq u_2(\mathbf{m})$ *for all states* $\mathbf{m}$, *then* $\mathcal{R}_{k,\Omega}(u_1, \mathbf{m}) \subseteq \mathcal{R}_{k,\Omega}(u_2, \mathbf{m})$.

The smaller the control value, the smaller the reachable sets.

Proposition 2.4. *Let* $u \in \mathfrak{U}$ *and* $\mathbf{m} \in \mathbb{N}^n$. *Then* $\mathcal{R}_{k,\Omega}(u_0, \mathbf{m}) = \{\mathbf{m}\} \subseteq \mathcal{R}_{k,\Omega}(u, \mathbf{m})$.

In words, a Petri net cannot evolve under the zero control law. Propositions 2.2 to 2.4 also hold for $k \to \infty$.

We extend the notation $\mathbf{m} \xrightarrow{\delta} \mathbf{m}'$ in (1.6) on page 9 to supervisory control. For $\zeta \in \mathcal{U}$, we write

$$\mathbf{m} \xrightarrow{\delta,\zeta} \mathbf{m}' \tag{2.5}$$

if $\mathbf{m} \xrightarrow{\delta} \mathbf{m}'$ and $\delta \in \zeta$.

4. Maximally permissive control laws

4.1. Introduction

As already mentioned in Section 1, the control goal we consider is that all possible future states belong to a given legal set $\mathcal{A} \subseteq \mathbb{N}^n$. This is done by selecting an appropriate control law $u \in \mathfrak{U}$. If $\mathbf{m} \in \mathcal{A}$ is the state of the Petri net, then u should satisfy

$$\mathcal{R}_\infty(u, \mathbf{m}) \subseteq \mathcal{A}. \tag{2.6}$$

Starting from a state $\mathbf{m} \in \mathcal{A}$, all states that can be reached under the control law u should belong to $\mathcal{A}$. As $\mathbf{m}$ is in $\mathcal{A}$, it follows from Proposition 2.4 that the zero control law u_0 always is a possible choice. As $\mathcal{R}_\infty(u_0, \mathbf{m}) = \{\mathbf{m}\}$, this would imply that each further state evolution is impossible. For this reason, there is a need for an additional optimality criterion. One has a lot of options: state or event based, small or large 'time' window, deadlock-free closed-loop system, ... In this book we opt for an event based objective function with a small 'time' window. For each $\mathbf{m} \in \mathbb{N}^n$, we want the control value $u(\mathbf{m})$ to be such that the enabled set $\Omega_{\mathbf{m}} \cap u(\mathbf{m})$ is as large as possible with respect to the $\subseteq$-relation.

If $\mathbf{m}$ is the state of the Petri net, then the chosen control value should satisfy two conditions. On the one hand, for all states which are reachable in one step from $\mathbf{m}$ a control law must exist which guarantees that all future states belong to $\mathcal{A}$. On the other hand, the chosen control value should limit the behaviour of the system as little as possible. These conditions are expressed mathematically in the following subsection.

4.2. Definition

We define in first instance 'good' control laws, i.e. those that keep the state within $\mathcal{A}$, and then the 'best' control laws, those that, in addition, restrain the set of enabled transition vectors as little as possible.

4.2.1 Permissive control laws

Let $\mathbf{m}$ be the Petri net state. The control value $u(\mathbf{m})$ should be such that, for all states in

$$\mathcal{R}_{1,\Omega}(u(\mathbf{m}), \mathbf{m}),$$

there exists a control law under which all future states remain in $\mathcal{A}$. Expressed formally, $u(\mathbf{m})$ should belong to

$$\mathcal{U}_{\mathbf{m}}^{\mathcal{A}} := \{\zeta \in \mathcal{U} \mid \forall \mathbf{m}' \in \mathcal{R}_{1,\Omega}(\zeta, \mathbf{m}) : \exists \bar{u} \in \mathfrak{U} : \mathcal{R}_{\infty}(\bar{u}, \mathbf{m}') \subseteq \mathcal{A}\}. \quad (2.7)$$

Herein, because of Proposition 2.4, $\bar{u}$ can always be chosen equal to the zero control law u_0. Therefore we obtain from the definition of $\mathcal{U}_{\mathbf{m}}^{\mathcal{A}}$ that

$$\mathcal{U}_{\mathbf{m}}^{\mathcal{A}} = \{\zeta \in \mathcal{U} \mid \mathcal{R}_{1,\Omega}(\zeta, \mathbf{m}) \subseteq \mathcal{A}\}. \quad (2.8)$$

Because of $\emptyset \in \mathcal{U}$ in (2.1) and $\mathcal{R}_{1,\Omega}(\emptyset, \mathbf{m}) = \{\mathbf{m}\}$, one has that $\mathcal{U}_{\mathbf{m}}^{\mathcal{A}} \neq \emptyset$ if and only if $\mathbf{m} \in \mathcal{A}$. Thus, if the present state $\mathbf{m}$ is legal, then the supervisor always has a nonempty set of control values, possibly only the singleton $\{\emptyset\}$, to choose from.

Definition 2.6. *Consider a legal set $\mathcal{A}$. A control value $\zeta \in \mathcal{U}$ is* permissive *for $\mathbf{m} \in \mathcal{A}$ if $\zeta \in \mathcal{U}_{\mathbf{m}}^{\mathcal{A}}$. A control law $u \in \mathfrak{U}$ is* permissive *if*

$$u(\mathbf{m}) \in \mathcal{U}_{\mathbf{m}}^{\mathcal{A}} \quad \textit{for all } \mathbf{m} \in \mathcal{A}.$$

For each permissive control law u and each $\mathbf{m} \in \mathcal{A}$, it is true that $\mathcal{R}_{\infty}(u, \mathbf{m}) \subseteq \mathcal{A}$, and vice versa. With Definition 2.5 it follows from (2.8) that

$$\mathcal{U}_{\mathbf{m}}^{\mathcal{A}} = \{\zeta \in \mathcal{U} \mid \forall \boldsymbol{\delta} \in \Omega_{\mathbf{m}} \cap \zeta : \mathbf{m} + \mathbf{F}\boldsymbol{\delta} \in \mathcal{A}\}. \quad (2.9)$$

Both (2.8) and (2.9) show that, if ζ belongs to $\mathcal{U}^{\mathcal{A}}_{\mathbf{m}}$, each subset ζ' of ζ which belongs to $\mathcal{U}$ also belongs to $\mathcal{U}^{\mathcal{A}}_{\mathbf{m}}$.

Proposition 2.5. *Let* $\mathcal{A} \subseteq \mathbb{N}^n$ *and* $\mathbf{m} \in \mathcal{A}$. *Then,* $\mathcal{U}^{\mathcal{A}}_{\mathbf{m}} \in \mathcal{O}^{\downarrow}(\mathcal{U}, \subseteq)$.

Note that for permissive control laws u only their value on $\mathcal{A}$ is important. As $\mathcal{R}_\infty(u, \mathbf{m}) \subseteq \mathcal{A}$, the image of u on $\mathbb{N}^n \setminus \mathcal{A}$ has no influence on the set $\mathcal{R}_\infty(u, \mathbf{m})$.

4.2.2 Optimality

We look for permissive control laws $u \in \mathfrak{U}$ which in addition limit the behaviour of the Petri net as little as possible. To express this formally, we introduce a number of definitions. For each $\mathbf{m}$ in $\mathbb{N}^n$, Definition 2.7 introduces an equivalence relation in $\mathcal{U}$.

Definition 2.7. *Let* $\mathbf{m} \in \mathbb{N}^n$ *and* $\zeta_1, \zeta_2 \in \mathcal{U}$. *Then* $\zeta_1 \sim_{\mathbf{m}} \zeta_2$ *if*

$$\Omega_{\mathbf{m}} \cap \zeta_1 = \Omega_{\mathbf{m}} \cap \zeta_2.$$

Define

$$[\zeta]_{\mathbf{m}} := \{\zeta' \in \mathcal{U} \mid \zeta' \sim_{\mathbf{m}} \zeta\}$$

as the set of control values which are $\sim_{\mathbf{m}}$-equivalent to $\zeta \in \mathcal{U}$. From Definition 2.5 it follows that $\mathcal{R}_1(\zeta_1, \mathbf{m}) = \mathcal{R}_1(\zeta_2, \mathbf{m})$ if $\zeta_1 \sim_{\mathbf{m}} \zeta_2$, and thus that $[\zeta]_{\mathbf{m}} \subseteq \mathcal{U}^{\mathcal{A}}_{\mathbf{m}}$ for all $\zeta \in \mathcal{U}^{\mathcal{A}}_{\mathbf{m}}$. This property can also be found by inspecting (2.9). The quotient set of $\mathcal{U}^{\mathcal{A}}_{\mathbf{m}}$ modulo $\sim_{\mathbf{m}}$ is denoted $\mathcal{E}^{\mathcal{A}}_{\mathbf{m}}$. Next, we introduce a partial order relation in this quotient set $\mathcal{E}^{\mathcal{A}}_{\mathbf{m}}$.

Definition 2.8. *Let* $\mathbf{m} \in \mathcal{A}$ *and* $\xi_1, \xi_2 \in \mathcal{E}^{\mathcal{A}}_{\mathbf{m}}$ *with* $\zeta_1 \in \xi_1$ *and* $\zeta_2 \in \xi_2$. *Then,* $\xi_1 \leq_{\mathbf{m}} \xi_2$ *if*

$$\Omega_{\mathbf{m}} \cap \zeta_1 \subseteq \Omega_{\mathbf{m}} \cap \zeta_2.$$

From Definition 2.5 it follows that

$$[\zeta_1]_{\mathbf{m}} \leq_{\mathbf{m}} [\zeta_2]_{\mathbf{m}} \quad \Rightarrow \quad \mathcal{R}_{1,\Omega}(\zeta_1, \mathbf{m}) \subseteq \mathcal{R}_{1,\Omega}(\zeta_2, \mathbf{m}).$$

If in addition $[\zeta_1]_{\mathbf{m}} \neq [\zeta_2]_{\mathbf{m}}$, then the control value ζ_2 is preferred to ζ_1. This is because ζ_2 satisfies the control goal and allows more transition vectors than ζ_1, i.e. $\Omega_{\mathbf{m}} \cap \zeta_1 \subset \Omega_{\mathbf{m}} \cap \zeta_2$. Optimal control laws have control values for which the $\sim_{\mathbf{m}}$-equivalence classes are maximal in $(\mathcal{E}^{\mathcal{A}}_{\mathbf{m}}, \leq_{\mathbf{m}})$.

Definition 2.9. *Consider a legal set* $\mathcal{A}$. *A control value* $\zeta \in \mathcal{U}$ *is* maximally permissive *for a state* $\mathbf{m} \in \mathcal{A}$ *if* $[\zeta]_{\mathbf{m}} \in \max(\mathcal{E}^{\mathcal{A}}_{\mathbf{m}}, \leq_{\mathbf{m}})$. *A control law* $u \in \mathfrak{U}$ *is* maximally permissive *if*

$$[u(\mathbf{m})]_{\mathbf{m}} \in \max(\mathcal{E}^{\mathcal{A}}_{\mathbf{m}}, \leq_{\mathbf{m}}) \quad \textit{for all } \mathbf{m} \in \mathcal{A}.$$

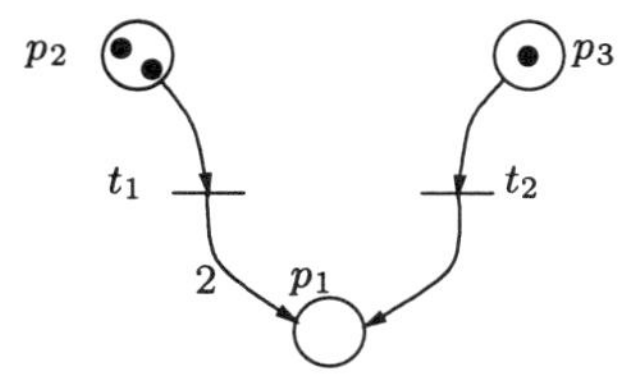

Figure 2.1. The Petri net of Example 2.3.

In general, the set $\max(\mathcal{E}_{\mathbf{m}}^{\mathcal{A}}, \leq_{\mathbf{m}})$ can contain more than one element, as the next example illustrates.

Example 2.3. Consider the Petri net in Figure 2.1. The concurrency assumption is used:

$$\Omega = \Delta_C = \{(1,0),(0,1),(1,1)\}.$$

For the indicated state $\mathbf{m} = (0,2,1)$, one finds that $\Omega_{\mathbf{m}} = \Omega$, i.e. all transition vectors in Ω are enabled under $\mathbf{m}$. The control set is:

$$\mathcal{U} = \{\{(1,0)\},\{(0,1)\},\Omega\}.$$

Consider two legal sets:

$$\mathcal{A}_1 = \{\mathbf{m} \in \mathbb{N}^n \mid \mathbf{m}(p_1) \leq 2\} \quad \text{and} \quad \mathcal{A}_2 = \{\mathbf{m} \in \mathbb{N}^n \mid \mathbf{m}(p_1) \leq 3\}.$$

It is easy to see that for $\mathcal{A}_1$

$$\mathcal{U}_{\mathbf{m}}^{\mathcal{A}_1} = \{\{(1,0)\},\{(0,1)\}\},$$

since $\boldsymbol{\delta} = (1,1)$ leads to $\mathbf{m}'$ with $\mathbf{m}'(p_1) = 3$. As $\{(1,0)\} \not\sim_{\mathbf{m}} \{(0,1)\}$ we find that $\mathcal{E}_{\mathbf{m}}^{\mathcal{A}_1} = \{\{\{(1,0)\}\},\{\{(0,1)\}\}\}$. As the equivalence classes $\{\{(1,0)\}\}$ and $\{\{(0,1)\}\}$ are not comparable with respect to $\leq_{\mathbf{m}}$, we have $\max(\mathcal{E}_{\mathbf{m}}^{\mathcal{A}_1}, \leq_{\mathbf{m}}) = \mathcal{E}_{\mathbf{m}}^{\mathcal{A}_1}$. Consequently, $\{(1,0)\}$ and $\{(0,1)\}$ both are maximally permissive control values.

On the other hand, we find for $\mathcal{A}_2$ that $\mathcal{U}_{\mathbf{m}}^{\mathcal{A}_2} = \mathcal{U}$ and

$$\mathcal{E}_{\mathbf{m}}^{\mathcal{A}_2} = \{\{\{(1,0)\}\},\{\{(0,1)\}\},\{\Omega\}\},$$

because no two control values are equivalent. Consequently, we find $\max(\mathcal{E}_{\mathbf{m}}^{\mathcal{A}_2}, \leq_{\mathbf{m}}) = \{\{\Omega\}\}$. For $\mathcal{A}_2$ there is a maximum control value, namely Ω.

Finally, we repeat this construction for the above set $\mathcal{A}_1$, the model speed

$$\Omega = \{(1,0),(2,0),(0,1),(0,2)\}$$

Figure 2.2. The Petri net of Example 2.4.

and the control set

$$\mathcal{U} = \{\{(1,0)\}, \{(1,0),(2,0)\}, \{(0,1)\}, \{(0,1),(0,2)\}\}.$$

In this case is $\mathcal{U}_{\mathbf{m}}^{\mathcal{A}_1} = \{\{(1,0)\}, \{(0,1)\}, \{(0,1),(0,2)\}\}$, $[\{(1,0)\}]_{\mathbf{m}} = \{\{(1,0)\}\}$ and $[\{(0,1)\}]_{\mathbf{m}} = \{\{(1,0)\}, \{(0,1),(0,2)\}\}$. As $[\{(1,0)\}]_{\mathbf{m}}$ and $[\{(0,1)\}]_{\mathbf{m}}$ are not comparable with respect to $\leq_{\mathbf{m}}$, the maximally permissive control values are $\{(1,0)\}$, $\{(0,1)\}$ and $\{(0,1),(0,2)\}$. △

Note that $\mathcal{E}_{\mathbf{m}}^{\mathcal{A}}$ can be an infinite set for which the property

$$\forall \xi \in \mathcal{E}_{\mathbf{m}}^{\mathcal{A}}, \exists \xi' \in \max(\mathcal{E}_{\mathbf{m}}^{\mathcal{A}}, \leq_{\mathbf{m}}) : \xi \leq_{\mathbf{m}} \xi' \tag{2.10}$$

does not hold for all $\mathcal{A}$ and $\mathbf{m} \in \mathcal{A}$. That is why we introduce an extra technical condition on $\mathcal{U}$ in Definition 2.1, namely that (2.10) holds for all nonempty, proper subsets $\mathcal{A}$ of $\mathbb{N}^n$ and for all $\mathbf{m} \in \mathcal{A}$. Note that a sufficient condition for (2.10) is that $\mathcal{U}$ is a finite set. In that case $\mathcal{E}_{\mathbf{m}}^{\mathcal{A}}$ is a finite set for all $\mathcal{A} \subseteq \mathbb{N}^n$ and all $\mathbf{m} \in \mathcal{A}$. To illustrate that this additional condition is very weak, in the next example we give a Petri net, a legal set and a control set which do not satisfy it.

Example 2.4. Consider the simple Petri net in Figure 2.2 with the transition bag assumption, i.e. $\Omega = \mathbb{N}$. Let $\zeta_k := \{n \in \mathbb{N} \mid n \leq k\}$ for $k \in \mathbb{N}$. The control set is $\mathcal{U} = \{\zeta_k \mid k \in \mathbb{N}\}$. Note that $\mathbb{N} \notin \mathcal{U}$. The legal set is $\mathcal{A} = \{\mathbf{m} \in \mathcal{A} \mid \mathbf{m}(p) \geq 1\}$. As the state $\mathbf{m}$ of Figure 2.2 belongs to $\mathcal{A}$ and as its value can only increase via the source transition t, we find that $\mathcal{U}_{\mathbf{m}}^{\mathcal{A}} = \mathcal{U}$. As $\zeta_k \not\sim_{\mathbf{m}} \zeta_{k'}$ if $k \neq k'$, we have $\mathcal{E}_{\mathbf{m}}^{\mathcal{A}} = \{\{\zeta_k\} \mid k \in \mathbb{N}\}$. Together with $\Omega_{\mathbf{m}} = \mathbb{N}$ we obtain $\{\zeta_k\} \leq_{\mathbf{m}} \{\zeta_{k'}\}$ if $k \leq k'$. Thus, as $\{\mathbb{N}\} \notin \mathcal{E}_{\mathbf{m}}^{\mathcal{A}}$, we find $\max(\mathcal{E}_{\mathbf{m}}^{\mathcal{A}}, \leq_{\mathbf{m}}) = \emptyset$. △

Definition 2.10. *Two control laws* $u_1, u_2 \in \mathfrak{U}$ *are* equivalent *if*

$$u_1(\mathbf{m}) \sim_{\mathbf{m}} u_2(\mathbf{m}) \quad \textit{for all } \mathbf{m} \in \mathcal{A}.$$

For equivalent control laws u_1 and u_2 it is true that

$$\Omega_{\mathbf{m}} \cap u_1(\mathbf{m}) = \Omega_{\mathbf{m}} \cap u_2(\mathbf{m}) \quad \text{for all } \mathbf{m} \in \mathcal{A}.$$

Thus, reachable sets under equivalent control laws are equal.

4.3. Construction

There is no general algorithm known yet which constructs maximally permissive control laws for all Petri nets and all legal sets. However, if $\Omega_{\mathbf{m}}$ is a finite set for all $\mathbf{m}$ in $\mathcal{R}_\infty(\mathbf{m}_0)$, $\mathbf{m}_0$ being the initial state, then $\mathcal{E}_{\mathbf{m}}^{\mathcal{A}}$ is a finite set for all $\mathbf{m}$ and optimal control laws can be constructed via enumeration. In particular, this is so when Ω is a finite set.

Proposition 2.6. *Consider a Petri net with concurrency assumption Ω. If the Petri net does not contain any source transitions, i.e. $T^{\mathrm{so}} = \emptyset$, then $\Omega_{\mathbf{m}}$ is a finite set for all $\mathbf{m}$ in $\mathbb{N}^n$.*

Proof. This is a direct consequence of the definition of source transitions (Definition 1.9, p. 17) and the Petri net enabling condition. □

For several classes of systems the construction of optimal control laws is feasible in a more intelligent manner than by brute enumeration. This is the subject of Sections 5 to 7 of this chapter. In the rest of this subsection, four useful properties for the construction of control laws are formulated.

Proposition 2.7. *Consider a Petri net with legal set $\mathcal{A}$, state $\mathbf{m} \in \mathcal{A}$ and control set $\mathcal{U}$. If $\zeta_1, \zeta_2 \in \mathcal{U}_{\mathbf{m}}^{\mathcal{A}}$ and $\zeta_1 \cup \zeta_2 \in \mathcal{U}$, then $\zeta_1 \cup \zeta_2 \in \mathcal{U}_{\mathbf{m}}^{\mathcal{A}}$. If, in addition, ζ_1 and ζ_2 are maximally permissive, then*

$$\zeta_1 \sim_{\mathbf{m}} \zeta_2 \sim_{\mathbf{m}} \zeta_1 \cup \zeta_2. \tag{2.11}$$

Proof. The fact that $\zeta_1 \cup \zeta_2 \in \mathcal{U}_{\mathbf{m}}^{\mathcal{A}}$ under the given conditions, immediately follows from (2.9). Next, we prove (2.11).

By definition, maximal permissiveness of ζ_1 and ζ_2 is equivalent to:

$$\forall \xi \in \mathcal{E}_{\mathbf{m}}^{\mathcal{A}} : \xi \geq_{\mathbf{m}} [\zeta_i]_{\mathbf{m}} \Rightarrow \xi = [\zeta_i]_{\mathbf{m}} \qquad (i = 1, 2). \tag{2.12}$$

For $i = 1, 2$ it is true that $[\zeta_1 \cup \zeta_2]_{\mathbf{m}} \geq_{\mathbf{m}} [\zeta_i]_{\mathbf{m}}$ because this is equivalent to $\Omega \cap (\zeta_1 \cup \zeta_2) \supseteq \Omega_{\mathbf{m}} \cap \zeta_1$. Consequently, by taking $\xi = [\zeta_1 \cup \zeta_2]_{\mathbf{m}}$ in (2.12) we obtain that $[\zeta_1 \cup \zeta_2]_{\mathbf{m}} = [\zeta_1]_{\mathbf{m}} = [\zeta_2]_{\mathbf{m}}$. □

The following property is a direct consequence of Proposition 2.7.

Proposition 2.8. *If $\mathcal{U}$ is closed under union, then all maximally permissive control values, and thus also all maximally permissive control laws, are equivalent.*

Proposition 2.9. *Consider ν sets of states $\mathcal{A}_i$ ($i = 1, \ldots, \nu$). If $\mathcal{A}$ is given by the intersection of these sets:*

$$\mathcal{A} = \bigcap_{i=1}^{\nu} \mathcal{A}_i,$$

then

$$\mathcal{U}_{\mathbf{m}}^{\mathcal{A}} = \bigcap_{i=1}^{\nu} \mathcal{U}_{\mathbf{m}}^{\mathcal{A}_i}.$$

Proof. Immediate from Equation (2.8). □

Proposition 2.10. *Consider ν sets of states $\mathcal{A}_i$ $(i = 1, \ldots, \nu)$. If $\mathcal{A}$ is given by the union of these sets:*

$$\mathcal{A} = \bigcup_{i=1}^{\nu} \mathcal{A}_i,$$

then

$$\mathcal{U}_{\mathbf{m}}^{\mathcal{A}} \supseteq \bigcup_{i=1}^{\nu} \mathcal{U}_{\mathbf{m}}^{\mathcal{A}_i}. \tag{2.13}$$

Proof. It holds that:

$$\begin{aligned}\bigcup_{i=1}^{\nu} \mathcal{U}_{\mathbf{m}}^{\mathcal{A}_i} &= \bigcup_{i=1}^{\nu} \{\zeta \in \mathcal{U} \mid \forall \boldsymbol{\delta} \in \Omega_{\mathbf{m}} \cap \zeta : \mathbf{m} + \mathbf{F}\boldsymbol{\delta} \in \mathcal{A}_i\}, \\ &\subseteq \bigcup_{i=1}^{\nu} \{\zeta \in \mathcal{U} \mid \forall \boldsymbol{\delta} \in \Omega_{\mathbf{m}} \cap \zeta : \mathbf{m} + \mathbf{F}\boldsymbol{\delta} \in \mathcal{A}\}, \\ &= \{\zeta \in \mathcal{U} \mid \forall \boldsymbol{\delta} \in \Omega_{\mathbf{m}} \cap \zeta : \mathbf{m} + \mathbf{F}\boldsymbol{\delta} \in \mathcal{A}\}, \\ &= \mathcal{U}_{\mathbf{m}}^{\mathcal{A}}.\end{aligned}$$

□

The subset relation in (2.13) can also be proper as is illustrated in Chapter 6, which is completely devoted to the decomposition of the control design.

5. Specific control sets

In this section we discuss two specific control sets, referred to as *fine control* and *on/off control*. A general concurrency assumption Ω is adopted.

5.1. Fine control

In this case, the control set is equal to

$$\mathcal{U} = \mathcal{O}^{\downarrow}(\Omega, \leq). \tag{2.14}$$

A control value can be any down-set in $(\Omega, \leq)$. That is the reason why the term *fine control* is used. As the union of down-sets is again a down-set (Proposition 1.3, p. 7), it follows from Proposition 2.8 that, under fine control, all maximally permissive control values are equivalent. The next proposition provides more information about these control values.

Proposition 2.11. *Consider a control set* $\mathcal{U} = \mathcal{O}^{\downarrow}(\Omega, \leq)$. *Then for each* $\mathcal{A} \subseteq \mathbb{N}^n$ *and for each* $\mathbf{m} \in \mathcal{A}$ *the set* $(\mathcal{E}^{\mathcal{A}}_{\mathbf{m}}, \leq_{\mathbf{m}})$ *has a maximum. This maximum is equal to* $\hat{\xi}^{\mathcal{A}}_{\mathbf{m}} := \left[\hat{\zeta}^{\mathcal{A}}_{\mathbf{m}}\right]_{\mathbf{m}}$ *with*

$$\hat{\zeta}^{\mathcal{A}}_{\mathbf{m}} := \{\boldsymbol{\delta} \in \Omega \mid \forall \boldsymbol{\delta}' \in \Omega, \boldsymbol{\delta}' \leq \boldsymbol{\delta} : \mathbf{m} + \mathbf{F}\boldsymbol{\delta}' \in \mathcal{A}\}. \tag{2.15}$$

Proof. The proof consists of two parts. First we show that $\hat{\zeta}^{\mathcal{A}}_{\mathbf{m}} \in \mathcal{U}$ and in the second part we show that $\hat{\xi}^{\mathcal{A}}_{\mathbf{m}}$ is the maximum of $(\mathcal{E}^{\mathcal{A}}_{\mathbf{m}}, \leq_{\mathbf{m}})$.

1. By definition, $\hat{\zeta}^{\mathcal{A}}_{\mathbf{m}}$ belongs to $\mathcal{U}$ or, equivalently, is a down-set in $(\Omega, \leq)$ if and only if

$$\forall \boldsymbol{\delta} \in \hat{\zeta}^{\mathcal{A}}_{\mathbf{m}} : \forall \boldsymbol{\delta}' \in \Omega, \boldsymbol{\delta}' \leq \boldsymbol{\delta} : \boldsymbol{\delta}' \in \hat{\zeta}^{\mathcal{A}}_{\mathbf{m}}. \tag{2.16}$$

This immediately follows from (2.15).

2. The equivalence class $\hat{\xi}^{\mathcal{A}}_{\mathbf{m}}$ is the maximum of $(\mathcal{E}^{\mathcal{A}}_{\mathbf{m}}, \leq_{\mathbf{m}})$ if and only if

$$\begin{aligned} & \forall \xi \in \mathcal{E}^{\mathcal{A}}_{\mathbf{m}} : \xi \leq_{\mathbf{m}} \hat{\xi}^{\mathcal{A}}_{\mathbf{m}}, \\ \Leftrightarrow\ & \forall \zeta \in \mathcal{U}^{\mathcal{A}}_{\mathbf{m}} : \zeta \cap \Omega_{\mathbf{m}} \subseteq \hat{\zeta}^{\mathcal{A}}_{\mathbf{m}} \cap \Omega_{\mathbf{m}}, \\ \Leftrightarrow\ & \forall \zeta \in \mathcal{U}^{\mathcal{A}}_{\mathbf{m}} : \zeta \cap \Omega_{\mathbf{m}} \subseteq \hat{\zeta}^{\mathcal{A}}_{\mathbf{m}}, \\ \Leftrightarrow\ & \forall \zeta \in \mathcal{U}^{\mathcal{A}}_{\mathbf{m}} : \zeta \cap \Omega_{\mathbf{m}} \subseteq \{\boldsymbol{\delta} \in \Omega_{\mathbf{m}} \mid \forall \boldsymbol{\delta}' \in \Omega, \boldsymbol{\delta}' \leq \boldsymbol{\delta} : \mathbf{m} + \mathbf{F}\boldsymbol{\delta}' \in \mathcal{A}\}. \end{aligned} \tag{2.17}$$

As ζ belongs to $\mathcal{U}^{\mathcal{A}}_{\mathbf{m}}$, it holds that

$$\begin{aligned} & \forall \boldsymbol{\delta} \in \zeta \cap \Omega_{\mathbf{m}} : \mathbf{m} + \mathbf{F}\boldsymbol{\delta} \in \mathcal{A} \\ \Leftrightarrow\ & \forall \boldsymbol{\delta} \in \zeta \cap \Omega_{\mathbf{m}} : \forall \boldsymbol{\delta}' \in \Omega, \boldsymbol{\delta}' \leq \boldsymbol{\delta} : \mathbf{m} + \mathbf{F}\boldsymbol{\delta}' \in \mathcal{A}\}. \end{aligned}$$

The second statement follows from the fact that $\zeta \cap \Omega_{\mathbf{m}}$ is a down-set. Substitution in (2.17) finishes the proof. □

From this property we can conclude that in the case of fine control all maximally permissive control values are equivalent and that all maximally permissive control laws are equivalent to the control law

$$\mathbf{m} \in \mathcal{A} \mapsto \hat{\zeta}^{\mathcal{A}}_{\mathbf{m}}, \tag{2.18}$$

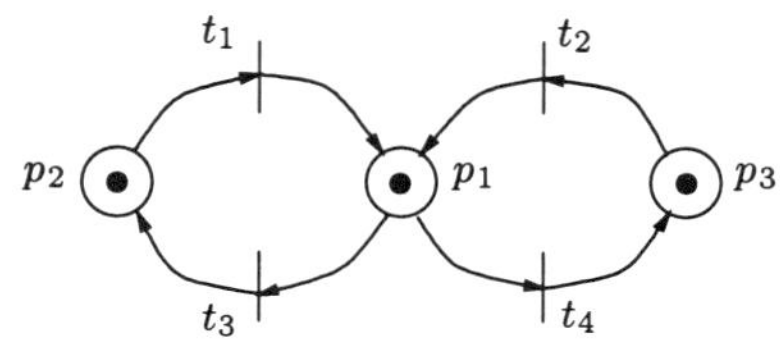

Figure 2.3. A Petri net with initial state $\mathbf{m} = (1, 1, 1)$.

where $\hat{\zeta}^{\mathcal{A}}_{\mathbf{m}}$ is given by (2.15). In Section 6 we show that for a legal set which is described by a finite conjunction of linear inequalities, this control law can be realised by a Petri net without modifying the Petri net plant model.

Example 2.5. For the Petri net in Figure 2.3 the initial state is $\mathbf{m} = (1, 1, 1)$. The transition bag assumption is used, i.e. $\Omega = \Delta = \mathbb{N}^4 \setminus \{\mathbf{0}\}$. The legal set is $\mathcal{A} = \{\mathbf{m} \in \mathbb{N}^3 \mid \mathbf{m}(p_1) \leq 2\}$. In the definition of $\hat{\zeta}^{\mathcal{A}}_{\mathbf{m}}$ we then have that $\mathbf{m} + \mathbf{F}\boldsymbol{\delta}' \in \mathcal{A}$ if and only if $\mathbf{A}(\mathbf{m} + \mathbf{F}\boldsymbol{\delta}') \leq b$, where $\mathbf{A} = (1, 0, 0)$, $b = 1$ and

$$\mathbf{F} = \begin{bmatrix} 1 & 1 & -1 & -1 \\ -1 & 0 & 1 & 0 \\ 0 & -1 & 0 & 1 \end{bmatrix}.$$

This condition is further equivalent to $\boldsymbol{\delta}'(t_1) + \boldsymbol{\delta}'(t_2) - \boldsymbol{\delta}'(t_3) - \boldsymbol{\delta}'(t_4) \leq 1$. Then following (2.15), $\boldsymbol{\delta} \in \hat{\zeta}^{\mathcal{A}}_{\mathbf{m}}$ if and only if

$$\boldsymbol{\delta}'(t_1) + \boldsymbol{\delta}'(t_2) - \boldsymbol{\delta}'(t_3) - \boldsymbol{\delta}'(t_4) \leq 1 \quad \text{for all } \boldsymbol{\delta}' \in \Omega \text{ with } \boldsymbol{\delta}' \leq \boldsymbol{\delta}.$$

This condition can be simplified to

$$\boldsymbol{\delta}'(t_1) + \boldsymbol{\delta}'(t_2) \leq 1 \quad \text{for all } \boldsymbol{\delta}' \in \Omega \text{ with } \boldsymbol{\delta}' \leq \boldsymbol{\delta},$$

or to $\boldsymbol{\delta}(t_1) + \boldsymbol{\delta}(t_2) \leq 1$. One concludes that

$$\hat{\zeta}^{\mathcal{A}}_{\mathbf{m}} = \{\boldsymbol{\delta} \in \mathbb{N}^4 \mid \boldsymbol{\delta}(t_1) + \boldsymbol{\delta}(t_2) \leq 1\}.$$

△

Proposition 2.12. *Consider ν sets of states $\mathcal{A}_i$ ($i = 1, \ldots, \nu$). If $\mathcal{A}$ is given by the intersection of these sets:*

$$\mathcal{A} = \bigcap_{i=1}^{\nu} \mathcal{A}_i,$$

then it holds under fine control:

$$\hat{\zeta}_{\mathbf{m}}^{\mathcal{A}} = \bigcap_{i=1}^{\nu} \hat{\zeta}_{\mathbf{m}}^{\mathcal{A}_i},$$

Proof. It holds that:

$$\begin{aligned}
\hat{\zeta}_{\mathbf{m}}^{\mathcal{A}} &= \{\boldsymbol{\delta} \in \Omega \mid \forall \boldsymbol{\delta}' \in \Omega, \boldsymbol{\delta}' \leq \boldsymbol{\delta} : \mathbf{m} + \mathbf{F}\boldsymbol{\delta}' \in \mathcal{A}\}, \\
&= \{\boldsymbol{\delta} \in \Omega \mid \forall \boldsymbol{\delta}' \in \Omega, \boldsymbol{\delta}' \leq \boldsymbol{\delta} : \mathbf{m} + \mathbf{F}\boldsymbol{\delta}' \in \cap_{i=1}^{\nu} \mathcal{A}_i\}, \\
&= \{\boldsymbol{\delta} \in \Omega \mid \wedge_{i=1}^{\nu} (\forall \boldsymbol{\delta}' \in \Omega, \boldsymbol{\delta}' \leq \boldsymbol{\delta} : \mathbf{m} + \mathbf{F}\boldsymbol{\delta}' \in \mathcal{A}_i)\}, \\
&= \bigcap_{i=1}^{\nu} \{\boldsymbol{\delta} \in \Omega \mid \forall \boldsymbol{\delta}' \in \Omega, \boldsymbol{\delta}' \leq \boldsymbol{\delta} : \mathbf{m} + \mathbf{F}\boldsymbol{\delta}' \in \mathcal{A}_i\}, \\
&= \bigcap_{i=1}^{\nu} \hat{\zeta}_{\mathbf{m}}^{\mathcal{A}_i}.
\end{aligned}$$

□

Proposition 2.13. *Consider ν sets of states $\mathcal{A}_i$ ($i = 1, \ldots, \nu$). If $\mathcal{A}$ is given by the union of these sets:*

$$\mathcal{A} = \bigcup_{i=1}^{\nu} \mathcal{A}_i,$$

then it holds for fine control:

$$\hat{\zeta}_{\mathbf{m}}^{\mathcal{A}} \supseteq \bigcup_{i=1}^{\nu} \hat{\zeta}_{\mathbf{m}}^{\mathcal{A}_i},$$

Proof. It holds that:

$$\begin{aligned}
\hat{\zeta}_{\mathbf{m}}^{\mathcal{A}} &= \{\boldsymbol{\delta} \in \Omega \mid \forall \boldsymbol{\delta}' \in \Omega, \boldsymbol{\delta}' \leq \boldsymbol{\delta} : \mathbf{m} + \mathbf{F}\boldsymbol{\delta}' \in \mathcal{A}\}, \\
&= \{\boldsymbol{\delta} \in \Omega \mid \forall \boldsymbol{\delta}' \in \Omega, \boldsymbol{\delta}' \leq \boldsymbol{\delta} : \mathbf{m} + \mathbf{F}\boldsymbol{\delta}' \in \cup_{i=1}^{\nu} \mathcal{A}_i\}, \\
&\supseteq \{\boldsymbol{\delta} \in \Omega \mid \vee_{i=1}^{\nu} (\forall \boldsymbol{\delta}' \in \Omega, \boldsymbol{\delta}' \leq \boldsymbol{\delta} : \mathbf{m} + \mathbf{F}\boldsymbol{\delta}' \in \mathcal{A}_i)\}, \\
&= \bigcup_{i=1}^{\nu} \{\boldsymbol{\delta} \in \Omega \mid \forall \boldsymbol{\delta}' \in \Omega, \boldsymbol{\delta}' \leq \boldsymbol{\delta} : \mathbf{m} + \mathbf{F}\boldsymbol{\delta}' \in \mathcal{A}_i\}, \\
&= \bigcup_{i=1}^{\nu} \hat{\zeta}_{\mathbf{m}}^{\mathcal{A}_i}.
\end{aligned}$$

□

5.2. On/off control

In this case control values are parameterised by sets of transitions. Define for each subset T' of the set of transitions T, a set of transition vectors

$$\zeta_{T'} := \{\boldsymbol{\delta} \in \Omega \mid \boldsymbol{\delta}|_{T'} = \mathbf{0}\}. \tag{2.19}$$

It is clear that $\zeta_\emptyset = \Omega$, $\zeta_T = \emptyset$ and that $\zeta_{T'}$ is a down-set in $(\Omega, \leq)$ for each $T' \subseteq T$. Here, the control set is equal to

$$\mathcal{U} = \{\zeta_{T'} \mid T' \subseteq T\}. \tag{2.20}$$

The control law decides which components of the transition vectors should be zero. For each transition vector component there is a choice between forcing to zero and imposing no additional constraint. That is why this type of supervisory control is called *on/off control.* As $\mathcal{U}$ is a finite set of sets $\zeta_{T'}$, the technical condition (2.10) is satisfied.

In general, the union of two control values $\zeta_{T_1}, \zeta_{T_2} \in \mathcal{U}$ with $T_1, T_2 \subseteq T$ does not belong to $\mathcal{U}$, i.e. there does not exist a set $T_3 \subseteq T$ such that $\zeta_{T_3} = \zeta_{T_1} \cup \zeta_{T_2}$. This implies that for on/off control there can be two or more maximally permissive control values which are not equivalent. We illustrate this with an example.

Example 2.6. Revisit Example 2.5 and the Petri net in Figure 2.3 for on/off control and the transition bag assumption. For the indicated state $\mathbf{m}$, the control values $\zeta_{\{t_1\}}$ and $\zeta_{\{t_2\}}$ are maximally permissive. Under $\zeta_{\{t_1\}}$ transition t_1 is disabled and under $\zeta_{\{t_2\}}$ transition t_2 is disabled. There is no unique maximally permissive control law, because $\zeta_{\{t_1\}} \not\sim_{\mathbf{m}} \zeta_{\{t_2\}}$, i.e. $\zeta_{\{t_1\}}$ and $\zeta_{\{t_2\}}$ are not $\sim_{\mathbf{m}}$-equivalent. △

Because of the parameterisation in (2.19) and (2.20), an on/off control law can also be represented by a function

$$\mathbb{N}^n \to 2^T. \tag{2.21}$$

In the literature one often uses this definition. A control value is then a set of transitions T'. The set of transition vectors which are enabled both by the plant model and by the control law, is found by setting the components of the transition vectors in $\Omega_{\mathbf{m}}$ corresponding to transitions in T', to zero. Dually, the function (2.21) can be interpreted as a function which decides which transitions are not affected by the supervisor. In our control model, these interpretations correspond to the sets of transition vectors $\Omega_{\mathbf{m}} \cap \zeta_{T'}$ and $\Omega_{\mathbf{m}} \cap \zeta_{T \setminus T'}$, respectively. On/off control for the no concurrency assumption is probably the most widely used combination in supervisory control theory of Petri nets. It is discussed in detail in Section 7 of this chapter.

6. Linear inequalities as a legal set

Often the legal set is given by a finite conjunction of linear inequalities with integer coefficients, see for instance the railway example in Example 2.1 (p. 28).

In this section we show that for these legal sets, if fine control for a general concurrency assumption Ω is used (see Section 5.1), then the maximally permissive control law in (2.18) can be realised by extending the Petri net $\mathcal{N}$ with extra places. These places are connected in an appropriate manner to the transitions of the Petri net under control. The fact that the closed-loop system is a Petri net, is important for analysis purposes. For in that case, theoretical results and algorithms from Petri net theory can be used to study its behaviour.

Before solving the general case in Section 6.2, we first consider legal sets which are given by one inequality.

6.1. One linear inequality

Consider a legal set which is characterised by one linear inequality with integer coefficients:

$$\mathcal{A} = \{\mathbf{m} \in \mathbb{N}^n \mid \mathbf{A}^T\mathbf{m} \leq b\} \quad \text{with } \mathbf{A} \in \mathbb{Z}^n \text{ and } b \in \mathbb{Z}. \tag{2.22}$$

Let $\mathbf{m} \in \mathcal{A}$. Then, following (2.15), $\boldsymbol{\delta} \in \Omega$ belongs to $\hat{\zeta}_{\mathbf{m}}^{\mathcal{A}}$ if and only if

$$\forall \boldsymbol{\delta}' \in \Omega, \boldsymbol{\delta}' \leq \boldsymbol{\delta} : \mathbf{m} + \mathbf{F}\boldsymbol{\delta}' \in \mathcal{A},$$

or, with (2.22), if and only if

$$\forall \boldsymbol{\delta}' \in \Omega, \boldsymbol{\delta}' \leq \boldsymbol{\delta} : \mathbf{A}^T\mathbf{F}\boldsymbol{\delta}' \leq b - \mathbf{A}^T\mathbf{m}. \tag{2.23}$$

This condition can be simplified further by employing the following lemma.

Lemma 2.1. *Let* $\mathbf{C} \in \mathbb{Z}^v, \mathbf{x} \in \mathbb{N}^v$ *and* $d \in \mathbb{Z}$. *Then*

$$\forall \mathbf{x}' \in \mathbb{N}^v, \mathbf{x}' \leq \mathbf{x} : \mathbf{C}^T\mathbf{x}' \leq d$$

is equivalent to

$$\bar{\mathbf{C}}^T\mathbf{x} \leq d.$$

Herein, $\bar{\mathbf{C}} \in \mathbb{N}^v$ *with* $\bar{C}_i = \max(C_i, 0)$ *for* $i = 1, \ldots, v$.

Proof. ($\Leftarrow$) From the definition of $\bar{\mathbf{C}}$ it follows that $\bar{C}_i \geq C_i$ and $\bar{C}_i \geq 0$ for $i = 1, \ldots, v$. As $\mathbf{x} \geq \mathbf{0}$, i.e. $x_i \geq 0$ for $i = 1, \ldots, v$, we obtain that

$$\bar{C}_i x_i \geq C_i x_i \qquad (i = 1, \ldots, v). \tag{2.24}$$

If $C_i \geq 0$, then it follows from $\mathbf{x} \geq \mathbf{x}' \geq \mathbf{0}$, i.e. $x_i \geq x_i'$ $(i = 1, \ldots, v)$, that $C_i x_i \geq C_i x_i'$ or, with (2.24), that

$$\bar{C}_i x_i \geq C_i x_i' \qquad (i = 1, \ldots, v; C_i \geq 0). \tag{2.25}$$

If $C_i < 0$, then is $\bar{C}_i = 0$ and we have, since $\mathbf{x}' \in \mathbb{N}^v$, that $C_i x_i' \leq 0$ or, because of $\bar{C}_i x_i = 0$, that

$$\bar{C}_i x_i \geq C_i x_i' \qquad (i = 1, \ldots, v; C_i < 0). \tag{2.26}$$

From $\bar{\mathbf{C}}^T \mathbf{x} \leq d$, (2.25) and (2.26) we obtain $\mathbf{C}^T \mathbf{x}' \leq d$.

($\Rightarrow$) We show that for each $\mathbf{x} \in \mathbb{N}^v$ there exists a $\bar{\mathbf{x}}' \in \mathbb{N}^v$ with $\bar{\mathbf{x}}' \leq \mathbf{x}$ such that $\bar{\mathbf{C}}^T \mathbf{x} = \mathbf{C}^T \bar{\mathbf{x}}'$. Define $J_1 := \{i \in \{1, \ldots, v\} \mid C_i \geq 0\}$ and $J_2 := \{i \in \{1, \ldots, v\} \mid C_i < 0\}$. From the definition of $\bar{\mathbf{C}}$ it follows that $\bar{C}_i = C_i$ if $i \in J_1$ and that $\bar{C}_i = 0$ if $i \in J_2$. Define $\bar{\mathbf{x}}'$ as follows:

$$\bar{x}_i' := \begin{cases} x_i & i \in J_1 \\ 0 & i \in J_2 \end{cases} \qquad (i = 1, \ldots, v).$$

Thus $\bar{\mathbf{x}}' \in \mathbb{N}^v$ and $\bar{\mathbf{x}}' \leq \mathbf{x}$. Consequently, it holds for each $i \in J_1$ that $C_i \bar{x}_i' = \bar{C}_i x_i$ and for each $i \in J_2$ that $C_i \bar{x}_i' = \bar{C}_i x_i = 0$. This implies that $\bar{\mathbf{C}}^T \mathbf{x} = \mathbf{C}^T \bar{\mathbf{x}}'$. □

As $\mathbf{0} \notin \Omega$, for Lemma 2.1 to be applicable to (2.23) we first need to show that $\mathbf{A}^T \mathbf{F} \boldsymbol{\delta}' \leq b - \mathbf{A}^T \mathbf{m}$ holds for $\boldsymbol{\delta}' = \mathbf{0}$. This is true because $\mathbf{m} \in \mathcal{A}$ is equivalent to $b - \mathbf{A}^T \mathbf{m} \geq 0$. Therefore, with Lemma 2.1, the condition in (2.23) is equivalent to

$$\mathbf{B}^{-,T} \boldsymbol{\delta} \leq b - \mathbf{A}^T \mathbf{m},$$

where $\mathbf{B}^- \in \mathbb{N}^m$ with

$$B_i^- = \max((\mathbf{A}^T \mathbf{F})_i, 0) \quad \text{for } i = 1, \ldots, m. \tag{2.27}$$

This implies that the maximum control value of Proposition 2.11 can be described by a single linear inequality:

$$\hat{\zeta}_{\mathbf{m}}^{\mathcal{A}} = \{\boldsymbol{\delta} \in \Omega \mid \mathbf{B}^{-,T} \boldsymbol{\delta} \leq b - \mathbf{A}^T \mathbf{m}\}.$$

We know that under fine control all maximally permissive control laws are equivalent to the control law

$$\mathcal{A} \to \mathcal{U} : \mathbf{m} \mapsto \hat{\zeta}_{\mathbf{m}}^{\mathcal{A}}. \tag{2.28}$$

Condition (2.3) for such a control law, the conjunction of the plant and the supervisor enabling condition for a transition vector $\boldsymbol{\delta}$, can then be written as

$$\mathbf{m} \geq \mathbf{F}^{-}\boldsymbol{\delta} \quad \wedge \quad \mathbf{b} - \mathbf{A}^T\mathbf{m} \geq \mathbf{B}^{-,T}\boldsymbol{\delta}. \tag{2.29}$$

If we extend the state $\mathbf{m}$ with one component to

$$\bar{\mathbf{m}} := \begin{bmatrix} \mathbf{m} \\ b - \mathbf{A}^T\mathbf{m} \end{bmatrix} \tag{2.30}$$

and the incidence matrix $\mathbf{F}^-$ with one row to

$$\bar{\mathbf{F}}^- := \begin{bmatrix} \mathbf{F}^- \\ \mathbf{B}^{-,T} \end{bmatrix},$$

then the condition in (2.29) can be rewritten as

$$\bar{\mathbf{m}} \geq \bar{\mathbf{F}}^-\boldsymbol{\delta}.$$

As $b - \mathbf{A}^T\mathbf{m} \in \mathbb{N}$ (because of $\mathbf{m} \in \mathcal{A}$) and $\mathbf{B}^- \in \mathbb{N}^m$ this condition has the same form as the enabling condition in (1.28). The only difference is that the number of state components has increased with one from n to $n+1$.

Next, we study the evolution of the extended state $\bar{\mathbf{m}}$ in (2.30) as the state $\mathbf{m}$ evolves. The above reasoning can be repeated for each $\mathbf{m}' \in \mathcal{R}_{1,\Omega}(\hat{\zeta}_{\mathbf{m}}, \mathbf{m})$. With $\mathbf{m}' = \mathbf{m} + \mathbf{F}\boldsymbol{\delta}$ we obtain for the last component of the new extended state $\bar{\mathbf{m}}'$ that

$$\bar{m}'_{n+1} = b - \mathbf{A}^T\mathbf{m}' = b - \mathbf{A}^T(\mathbf{m} + \mathbf{F}\boldsymbol{\delta}) = (b - \mathbf{A}^T\mathbf{m}) - \mathbf{A}^T\mathbf{F}\boldsymbol{\delta}.$$

Thus the reached extended state equals

$$\bar{\mathbf{m}}' = \begin{bmatrix} \mathbf{m}' \\ (b - \mathbf{A}^T\mathbf{m}) - \mathbf{A}^T\mathbf{F}\boldsymbol{\delta} \end{bmatrix} = \bar{\mathbf{m}} + \begin{bmatrix} \mathbf{F} \\ -\mathbf{A}^T\mathbf{F} \end{bmatrix}\boldsymbol{\delta}. \tag{2.31}$$

Define the vector $\mathbf{B}^+$ such that

$$\mathbf{B}^{+,T} := \mathbf{B}^{-,T} - \mathbf{A}^T\mathbf{F}. \tag{2.32}$$

It follows from (2.27) that

$$B_i^+ = \max(-(\mathbf{A}^T\mathbf{F})_i, 0) \quad \text{for } i = 1, \ldots, m.$$

Hence, $\mathbf{B}^+ \in \mathbb{N}^m$. Consequently, the closed-loop system of the Petri net and the supervisory control law in (2.28) can be represented by a Petri net with incidence matrices

$$\bar{\mathbf{F}}^+ := \begin{bmatrix} \mathbf{F}^+ \\ \mathbf{B}^{+,T} \end{bmatrix} \quad \text{and} \quad \bar{\mathbf{F}}^- = \begin{bmatrix} \mathbf{F}^- \\ \mathbf{B}^{-,T} \end{bmatrix}.$$

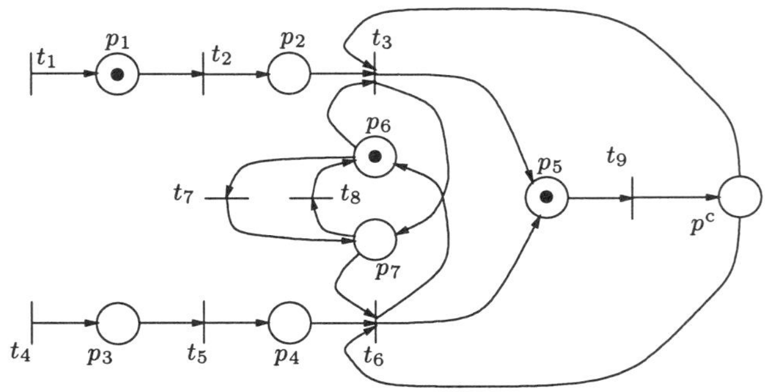

Figure 2.4. Closed-loop system of the Petri net model of the railway example together with a maximally permissive supervisor which guarantees that the number of trains within Zone 5 is not greater than one.

This extended Petri net has $n+1$ places and m transitions: in comparison with the original Petri net, there is one extra place. We refer to this place as the *control place*. It is denoted p^c.

The vectors $\mathbf{B}^+$ and $\mathbf{B}^-$ contain the weights of arcs from transitions in T to p^c and of arcs from p^c to transitions in T, respectively. Notice that $B_i^+ > 0$ implies $B_i^- = 0$, and that $B_i^- > 0$ implies $B_i^+ = 0$ for all $i = 1, \ldots, m$. Therefore, in the closed-loop Petri net there do not exist self-loops which contain the control place.

If $\mathbf{m}$ is the state of the original Petri net, then the state of p^c always equals $b - \mathbf{A}^T\mathbf{m}$ (see (2.30)).

Example 2.7. Revisit the railway example. We use the no concurrency assumption and fine control. (In Section 7 it is shown that under the no concurrency assumption on/off control and fine control are equivalent.) Thus we assume that events cannot occur simultaneously and each event can be controlled individually. Supervisory control is done via the seven traffic lights at the extremities of the zones and the possibility to preempt the switch from being moved manually (transitions t_7 and t_8). Note that if we would assume the concurrency assumption, i.e. $\Omega = \Delta_C$, then on/off control, i.e. with the aid of traffic lights, would not be equivalent to fine control.

The legal set which guarantees that there are never two or more trains in the same zone was given in Example 2.1 (p. 28). Here, we enforce this condition for Zone 5 only, i.e. the legal set is

$$\mathcal{A} = \{\mathbf{m} \in \mathbb{N}^n \mid \mathbf{A}^T\mathbf{m} \leq b\} \quad \text{with } \mathbf{A} = (0,0,0,0,1,0,0) \text{ and } b = 1.$$

For the incidence matrices $\mathbf{F}^+$ and $\mathbf{F}^-$ of the Petri net model we refer to Example 1.4 (p. 11). One finds that

$$-\mathbf{A}^T\mathbf{F} = (0,0,-1,0,0,-1,0,0,1).$$

Hence,

$$\mathbf{B}^+ = (0,0,0,0,0,0,0,0,1) \quad \text{and} \quad \mathbf{B}^- = (0,0,1,0,0,1,0,0,0).$$

Thus the supervisory controller consists of one control place p^{c} and the arcs (t_9, p^{c}), (p^{c}, t_3) and (p^{c}, t_6) with weights equal to one. For the state of Figure 1.2, the state of p^{c} equals $b - \mathbf{A}^T\mathbf{m} = 0$. Intuitively, this is because in Figures 1.1 and 1.2 there already is a train in Zone 5. The discussed closed-loop system is depicted in Figure 2.4. △

The above result has an instructive intuitive explanation. When $\mathbf{m}$ is the state of the original Petri net, then the state of the control place p^{c} equals $b - \mathbf{A}^T\mathbf{m}$. This value can be seen as a security margin with respect to the legal set constraint $\mathbf{A}^T\mathbf{m} \leq b$. Execution of a transition t leads —see the definitions of $\mathbf{B}^+$ and $\mathbf{B}^-$— to an increase of this margin with $B^+(t) - B^-(t) = -\mathbf{A}^T\mathbf{F}(t)$ ($\mathbf{F}(t)$ is the column of $\mathbf{F}$ which corresponds to t). One verifies that this value is positive if $t \in {}^\bullet p^{\mathrm{c}}$, negative if $t \in (p^{\mathrm{c}})^\bullet$ and zero in all other cases. If the margin, and thus the state of the control place, is zero, then none of the output transitions of the control place is enabled. This avoids that the margin decreases further and thus becomes negative, violating the constraint $\mathbf{A}^T\mathbf{m} \leq b$.

If $\mathbf{B}^- = \mathbf{0}$ or, equivalently, $\mathbf{A}^T\mathbf{F} \leq \mathbf{0}$, then p^{c} has no output transitions and, consequently, its state does not influence the behaviour of the Petri net. This property can also be explained by using the necessary condition for reachability of Proposition 1.9 (p. 13). If $\mathbf{m} \in \mathcal{A}$ is the initial state, then in this case it holds for each $\mathbf{m}' \in \mathcal{R}_\infty(\mathbf{m})$ that (recall that $\mathbf{y} \geq \mathbf{0}$):

$$\mathbf{A}^T\mathbf{m}' = \mathbf{A}^T(\mathbf{m} + \mathbf{F}\mathbf{y}) = \mathbf{A}^T\mathbf{m} + \mathbf{A}^T\mathbf{F}\mathbf{y} \leq \mathbf{A}^T\mathbf{m} \leq b.$$

The weighted sum $\mathbf{A}^T\mathbf{m}'$ is smaller than b for all reachable states $\mathbf{m}'$. As all reachable states are legal, control of the Petri net is not needed.

6.2. Conjunctions of linear inequalities

Next we show that if the legal set is described by a finite conjunction of linear inequalities, then under fine control the maximally permissive control law of Proposition 2.11 can also be realised by a Petri net. One merely needs to apply the design of the previous section repeatedly.

Consider for $i = 1, \ldots, \nu$ the sets:

$$\mathcal{A}_i = \{\mathbf{m} \in \mathbb{N}^n \mid \mathbf{A}_i^T\mathbf{m} \leq b_i\} \quad \text{with } \mathbf{A}_i \in \mathbb{Z}^n \text{ and } b_i \in \mathbb{Z}.$$

By applying the result of Section 6.1 for each $\mathcal{A}_i$, we find that

$$\hat{\zeta}_{\mathbf{m}}^{\mathcal{A}_i} = \{\boldsymbol{\delta} \in \Omega \mid \mathbf{B}_i^{-,T}\boldsymbol{\delta} \leq b_i - \mathbf{A}_i^T\mathbf{m}\} \qquad (i = 1, \ldots, \nu).$$

Herein, $\mathbf{B}_i^-$ is defined from $\mathbf{A}_i$ as is $\mathbf{B}^-$ from $\mathbf{A}$ in (2.27). If $\mathcal{A} = \cap_{i=1}^{\nu}\mathcal{A}_i$, then we find with Proposition 2.12 (p. 40) that

$$\hat{\zeta}_{\mathbf{m}}^{\mathcal{A}} = \bigcap_{i=1}^{\nu}\{\boldsymbol{\delta} \in \Omega \mid \mathbf{B}_i^{-,T}\boldsymbol{\delta} \leq b_i - \mathbf{A}_i^T\mathbf{m}\}.$$

The enabling condition of the closed-loop system is then:

$$\bar{\mathbf{m}} \geq \bar{\mathbf{F}}^-\boldsymbol{\delta},$$

where

$$\bar{\mathbf{m}} = \begin{bmatrix} \mathbf{m} \\ b_1 - \mathbf{A}_1^T\mathbf{m} \\ \vdots \\ b_\nu - \mathbf{A}_\nu^T\mathbf{m} \end{bmatrix} \quad \text{and} \quad \bar{\mathbf{F}}^- = \begin{bmatrix} \mathbf{F}^- \\ \mathbf{B}_1^{-,T} \\ \vdots \\ \mathbf{B}_\nu^{-,T} \end{bmatrix}.$$

The state equation for the extended state is:

$$\bar{\mathbf{m}}' = \bar{\mathbf{m}} + (\bar{\mathbf{F}}^+ - \bar{\mathbf{F}}^-)\boldsymbol{\delta},$$

with

$$\bar{\mathbf{F}}^+ = \begin{bmatrix} \mathbf{F}^+ \\ \mathbf{B}_1^{+,T} \\ \vdots \\ \mathbf{B}_\nu^{+,T} \end{bmatrix}$$

and $\mathbf{B}_i^{+,T} = \mathbf{B}_i^{-,T} - \mathbf{A}_i^T\mathbf{F}$ for $i = 1, \ldots, \nu$.

We conclude that the discrete event system which consists of a Petri net plus a supervisor can be realised by a Petri net without modifying the Petri net plant model. It has $n + \nu$ places and m transitions. The closed-loop system has ν control places p_i^{c} $(i = 1, \ldots, \nu)$ with respective states $b_i - \mathbf{A}_i^T\mathbf{m}$.

Here we have enforced all ν inequalities simultaneously. No great effort is needed to see that one would find the same closed-loop Petri net by enforcing the inequality constraints successively one by one. After having added a constraint, one obtains a Petri net and one can therefore add the next one. One obtains the maximally permissive supervisor for $\mathcal{A}$ by applying the result of the previous subsection ν times.

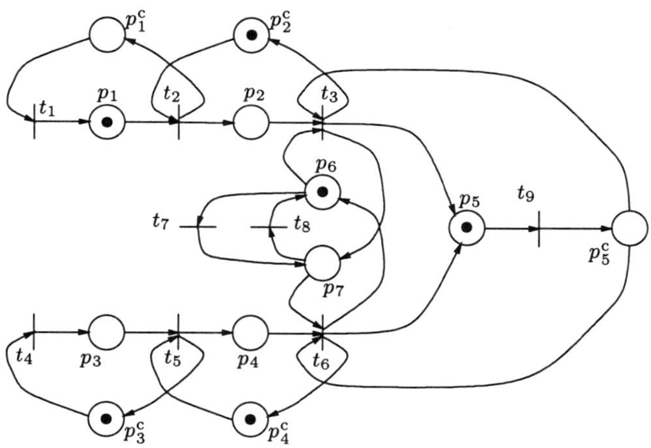

Figure 2.5. Closed-loop system of the Petri net model of the railway example together with a maximally permissive supervisor which guarantees that the number of trains within each of the Zones 1 to 5 is not greater than one.

Example 2.8. The closed-loop system for the railway example and the legal set

$$\mathcal{A} = \bigcap_{i=1}^{5} \{\mathbf{m} \in \mathbb{N}^n \mid \mathbf{m}(p_i) \leq 1\}$$

of Example 2.1 is depicted in Figure 2.5. There are five control places $p_1^c, \ldots, p_5^c$, which correspond to Zones 1 to 5, respectively. △

Note that the above constructed maximally permissive control law is not always the simplest one. If $\mathcal{R}_\infty(\mathbf{m}) \subseteq \mathcal{A}$, i.e. all reachable states of the Petri net plant are legal, then the control places do not influence the behaviour of the Petri net. Consequently, they can be removed. We formulate a sufficient condition to check the redundancy of control places. It is based on the necessary condition for reachability in Proposition 1.9 (p. 13). For the above legal set $\mathcal{A} = \cap_{i=1}^{\nu} \mathcal{A}_i$ the condition consists of ν integer linear programming inequalities:

$$\max_{\substack{\mathbf{y} \in \mathbb{N}^m \\ \mathbf{m}+\mathbf{F}\mathbf{y} \geq \mathbf{0}}} \left[\mathbf{A}_i^T(\mathbf{m} + \mathbf{F}\mathbf{y})\right] \leq b_i \quad \text{for } i = 1, \ldots, \nu,$$

or their linear programming relaxations:

$$\max_{\substack{\mathbf{x} \in \mathbb{Q}_+^m \\ \mathbf{m}+\mathbf{F}\mathbf{x} \geq \mathbf{0}}} \left[\mathbf{A}_i^T(\mathbf{m} + \mathbf{F}\mathbf{x})\right] \leq b_i \quad \text{for } i = 1, \ldots, \nu.$$

7. Control design under the no concurrency assumption

Control design under the no concurrency assumption, i.e. $\Omega = \Delta_\mathrm{N}$, is an important issue in Petri net theory. This section is devoted to it. All possible control sets $\mathcal{U}$ are given by (cfr. (2.1) on page 30):

$$\mathcal{U} \subseteq \mathcal{O}^{\downarrow}(\Delta_\mathrm{N}, \leq) \quad \text{with } \emptyset, \Delta_\mathrm{N} \in \mathcal{U}. \tag{2.33}$$

Example 2.9. For $m = 4$ we have that

$$\Delta_\mathrm{N} = \left\{ \begin{bmatrix}1\\0\\0\\0\end{bmatrix}, \begin{bmatrix}0\\1\\0\\0\end{bmatrix}, \begin{bmatrix}0\\0\\1\\0\end{bmatrix}, \begin{bmatrix}0\\0\\0\\1\end{bmatrix} \right\}.$$

An example of a control set, i.e. a set that satisfies (2.33), is then:

$$\mathcal{U} = \left\{ \left\{ \begin{bmatrix}1\\0\\0\\0\end{bmatrix}, \begin{bmatrix}0\\0\\1\\0\end{bmatrix} \right\}, \left\{ \begin{bmatrix}0\\0\\0\\1\end{bmatrix} \right\}, \emptyset, \Delta_\mathrm{N} \right\}.$$

$\triangle$

As $(\Delta_\mathrm{N}, \leq)$ is an antichain (see Section 3 of Chapter 1), each subset of Δ_N is a down-set in $(\Delta_\mathrm{N}, \leq)$. This can be verified in the above example. Consequently, the condition in (2.33) is equivalent to

$$\mathcal{U} \subseteq 2^{\Delta_\mathrm{N}} \quad \text{with } \emptyset, \Delta_\mathrm{N} \in \mathcal{U}.$$

Because 2^{Δ_N} is a finite set for all Petri nets, the technical condition in (2.10) is satisfied.

Next we study fine and on/off control (see Section 5) under the no concurrency assumption. For fine control the control set is equal to

$$\mathcal{U} = \mathcal{O}^{\downarrow}(\Delta_\mathrm{N}, \leq) = 2^{\Delta_\mathrm{N}}, \tag{2.34}$$

and for on/off control equal to

$$\mathcal{U} = \{\zeta_{T'} \mid T' \subseteq T\}, \tag{2.35}$$

where (take $\Omega = \Delta_\mathrm{N}$ in (2.19)):

$$\zeta_{T'} = \{\boldsymbol{\delta} \in \Delta_\mathrm{N} \mid \boldsymbol{\delta}|_{T'} = \mathbf{0}\} \quad \text{with } T' \subseteq T. \tag{2.36}$$

From the definition of Δ_N it is clear that the control sets in (2.34) and (2.35) are equal. In particular,

$$\zeta_\emptyset = \Delta_\mathrm{N} \quad \text{and} \quad \zeta_T = \emptyset.$$

Consequently, fine and on/off control are equivalent under the no concurrency assumption.

This implies that all results in Sections 5.1 and 6 for fine control are recovered for the no concurrency assumption and on/off control. From Proposition 2.11 we find that for all legal sets $\mathcal{A}$, all maximally permissive control laws are equivalent to:

$$\mathbf{m} \in \mathcal{A} \mapsto \hat{\zeta}^{\mathcal{A}}_{\mathbf{m}} = \{\boldsymbol{\delta} \in \Delta_{\mathrm{N}} \mid \forall \boldsymbol{\delta}' \in \Delta_{\mathrm{N}}, \boldsymbol{\delta}' \leq \boldsymbol{\delta} : \mathbf{m} + \mathbf{F}\boldsymbol{\delta}' \in \mathcal{A}\}. \quad (2.37)$$

As $(\Delta_{\mathrm{N}}, \leq)$ is an antichain, $\hat{\zeta}^{\mathcal{A}}_{\mathbf{m}}$ can be written as

$$\hat{\zeta}^{\mathcal{A}}_{\mathbf{m}} = \{\boldsymbol{\delta} \in \Delta_{\mathrm{N}} \mid \mathbf{m} + \mathbf{F}\boldsymbol{\delta} \in \mathcal{A}\}.$$

In particular, when the legal set is equal to

$$\mathcal{A} = \bigcap_{i=1}^{\nu} \bigcup_{j=1}^{\nu_i} \{\mathbf{m} \in \mathbb{N}^n \mid \mathbf{A}_{i,j}^T \mathbf{m} \leq b_{i,j}\} \quad (2.38)$$

with

$$\mathbf{A}_{i,j} \in \mathbb{Z}^n \quad \text{and} \quad b_{i,j} \in \mathbb{Z} \qquad (i = 1, \ldots, \nu; j = 1, \ldots, \nu_i),$$

then we have for all $\mathbf{m} \in \mathcal{A}$ that

$$\hat{\zeta}^{\mathcal{A}}_{\mathbf{m}} = \left\{ \boldsymbol{\delta} \in \Delta_{\mathrm{N}} \,\middle|\, \bigwedge_{i=1}^{\nu} \bigvee_{j=1}^{\nu_i} \mathbf{A}_{i,j}^T(\mathbf{m} + \mathbf{F}\boldsymbol{\delta}) \leq b_{i,j} \right\}. \quad (2.39)$$

This result allows the easy construction of maximally permissive control laws.

Because of the disjunctions in (2.39), in general the closed-loop system for $\mathcal{A}$ in (2.38) cannot be represented by a Petri net without modifying the original model. However, from Section 6 we know that this is always possible if $\nu_i = 1$ for $i = 1, \ldots, \nu$ in (2.38). Next, we give a much shorter proof of this result when the no concurrency and fine control assumptions are used. The price we pay for the shorter proof is the additional condition that the Petri net is free of self-loops.

If $\nu_i = 1$ for $i = 1, \ldots, \nu$ in (2.38), then the condition in (2.39) can be rewritten as (suppressing the second index):

$$\bigwedge_{i=1}^{\nu} 0 \leq (b_i - \mathbf{A}_i^T \mathbf{m}) - \mathbf{A}_i^T \mathbf{F}\boldsymbol{\delta}. \quad (2.40)$$

If the Petri net does not contain any self-loops (see Definition 1.8, p. 17), then, following Proposition 1.15 (p. 23), the enabling condition $\mathbf{m} \geq \mathbf{F}^{-}\boldsymbol{\delta}$

is equivalent to $\mathbf{m} + \mathbf{F}\delta \geq \mathbf{0}$. As (2.40) has the form of the latter condition, the closed-loop system can be represented by a Petri net. The incidence matrix is

$$\bar{\mathbf{F}} = \bar{\mathbf{F}}^{+} - \bar{\mathbf{F}}^{-} = \begin{bmatrix} \mathbf{F} \\ -\mathbf{A}_1^T\mathbf{F} \\ \vdots \\ -\mathbf{A}_\nu^T\mathbf{F} \end{bmatrix}.$$

There are ν control places with respective states $b_i - \mathbf{A}_i^T\mathbf{m}$ $(i = 1, \ldots, \nu)$. The approach of this alternative proof is used in Chapter 8.

In contrast to the no concurrency assumption, for the concurrency assumption on/off control is not equivalent to fine control. If $m = 2$, then

$$\Delta_{\mathrm{C}} = \{(1,0), (0,1), (1,1)\}.$$

For fine control we find that

$$\mathcal{U} = \{\{(1,0)\}, \{(0,1)\}, \{(1,0), (0,1)\}, \Delta_{\mathrm{C}}, \emptyset\}$$

and for on/off control that

$$\mathcal{U} = \{\{(1,0)\}, \{(0,1)\}, \Delta_{\mathrm{C}}, \emptyset\}.$$

8. Notes and references

The problem we have formulated in this chapter is often referred to as the *forbidden state supervisory control problem.* In that case one uses a *forbidden set* $\mathcal{F} := \mathbb{N}^n \setminus \mathcal{A}$, the complement of the legal set. The condition in (2.6) can then be written as $\mathcal{R}_\infty(u, \mathbf{m}) \cap \mathcal{F} = \emptyset$. A survey paper on supervisory control for Petri nets is [19].

The detailed discussion of different control laws for general concurrency assumptions has not been done before. The result of Section 6 on Petri net supervisors is deduced in [42] for the case where two additional conditions are satisfied, namely the no concurrency assumption and the self-loop freeness of the studied Petri nets. It is also presented (without proof) in [23] and appears in [14, 15] for specific concurrency and control assumptions. The result of the second part of Section 7 also appears in [24]. Supervisory of Petri nets under the concurrency assumption is studied in [25].

Another name for a control place is a *monitor place.*

Chapter 3

UNCONTROLLABLE EVENTS AND TRANSITIONS

In this chapter, the supervisory control problem for Petri nets of which a number of transitions/events are uncontrollable is defined. Also the control problems which form the subject of the remainder of the book are described.

In Section 1, uncontrollable events and transitions are introduced intuitively. These ideas are formalised in Section 2, where supervisory control laws are redefined for general concurrency assumptions. Section 3 discusses these control laws for some specific concurrency and control assumptions. Section 4 defines (sub)optimality of control laws. In Section 5, the design of optimal control laws is studied and it is shown that control design for a Petri net with uncontrollable transitions and a legal set $\mathcal{A}$ is equivalent to control design for another Petri net without uncontrollable transitions and a legal set $\mathcal{A}^* \subseteq \mathcal{A}$. This new legal set, the *supremal controllable subset* of $\mathcal{A}$, is discussed in Section 6.

1. Introduction

In practice the supervisory controller cannot influence all events. For example it is not always possible to disable each event in the discrete event system. Expressed by means of the terminology of Chapter 2, the empty set does not necessarily belong to the control set $\mathcal{U}$. Events which cannot be disabled by the supervisory controller are called *uncontrollable*. Examples of inherently uncontrollable events are a machine breakdown or a request from outside the model. An event can also be uncontrollable because of other reasons, e.g. actuator limitations (the supervisor cannot access each part of the system under control) and control strategy (to give priority to that event at any time) [6, p. 138].

For the railway example an uncontrollable event is a train that leaves Zone 1 and enters Zone 2 when this train is at such a high speed that it cannot stop within Zone 1. Even when the corresponding traffic light phase is red, the supervisor knows in advance that it cannot preempt this event from occurring (assuming that the supervisor has access to information about the speed of the train). To represent this kind of behaviour it is necessary to extend the railway model of Chapter 1. This is done at the end of this section.

The presence of uncontrollable events makes control design more difficult. In Chapter 2, the supervisory controller had the possibility to choose the empty set as a control value, implying that the state of the Petri net cannot evolve any further. However, due to uncontrollable events, the controller no longer has full control over the system. For control design, as they cannot be affected, the consequence is that the controller needs to foresee the occurrence of uncontrollable events.

Transitions which correspond to uncontrollable events are *uncontrollable.* Define $T^{\mathrm{uc}} \subseteq T$ as the set of uncontrollable transitions. Its complement is the set of *controllable* transitions $T^{\mathrm{c}} := T \setminus T^{\mathrm{uc}}$. To make clear that the set of transitions is partitioned in a controllable and an uncontrollable part, i.e. $T = T^{\mathrm{uc}} \dot{\cup} T^{\mathrm{c}}$, we extend the definition of a Petri net to a six-tuple

$$\mathcal{N} = (P, T, f^+, f^-, \mathbf{m}_0, T^{\mathrm{uc}}).$$

Herein, $(P, T, f^+, f^-, \mathbf{m}_0)$ is a Petri net as defined in Definition 1.5. In the graphical representation of a Petri net, controllable transitions are indicated with a label 'c'. All other transitions are uncontrollable. A Petri net is called *uncontrollable* if $T^{\mathrm{uc}} = T$, i.e. if all transitions are uncontrollable. A transition vector $\boldsymbol{\delta} \in \Omega$ is *uncontrollable* if $\boldsymbol{\delta}|_{T^{\mathrm{c}}} = \mathbf{0}$, i.e. if all components which correspond to controllable transitions are equal to zero.

Example 3.1. We revisit the railway example of Figure 1.1 in Chapter 1. So far, it was assumed that, when necessary, trains can be stopped within the zone they are in. If the length of a zone is small with respect to a typical braking distance of a train, then this model is not complete enough for supervisory control design. Therefore, we represent the inertia of trains as follows.

The four state components which model the states of Zones 1 to 4, are each split up into two new states. One state component indicates the number of trains at 'High speed' in the corresponding zone, the other state component gives the number of trains at 'Low speed'. A train at high speed cannot always stop within its present zone. A train at low speed can always stop.

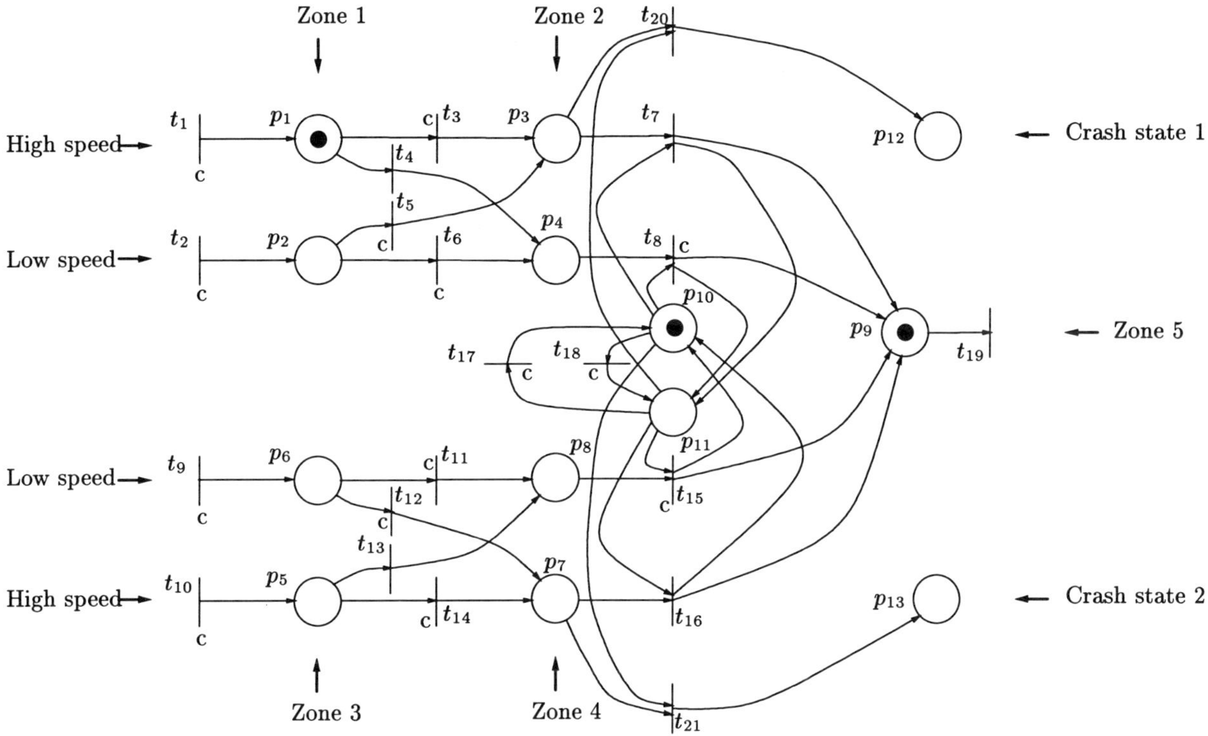

Figure 3.1. The extended Petri net model of Example 3.1. There is one train at high speed in Zone 1 and one train in Zone 5. The switch is in Position 0, allowing a train to pass from Zone 2 to 5.

The Petri net of this more detailed model is depicted in Figure 3.1. The places p_1 and p_2 correspond to high and low speed in Zone 1, respectively. Analogously, p_3 and p_4 correspond to Zone 2, p_5 and p_6 to Zone 3, and p_7 and p_8 to Zone 4. Zone 5 is modelled by one place only, namely p_9, because the evolution of the railway system downstream from Zone 5 is of no importance for the control design. As can be seen in Figure 3.1, trains can change speed when going from Zone 1 to 2 (transitions t_3 to t_6), or from Zone 3 to 4 (transitions t_{11} to t_{14}). Like in the first railway model, the position of the switch is represented by two places, namely p_{10} and p_{11}. Transitions t_{20} and t_{21}, and places p_{12} and p_{13} model the possible crashing of trains. Crashes occur when trains arrive at high speed at a switch which is not in the correct position. For example, a train at high speed in Zone 2 (place p_3) can cause a crash (via transition t_{20} to place p_{12}) when the switch is in Position 1 (place p_{11}). One of the control goals of the supervisor is to avoid crashes.

So far, supervisory control of the railway section was done via traffic lights with two phases (red and green) at the extremities of the zones. Here, supervisory control is more sophisticated. Disabling t_3 means that a train that leaves Zone 1 at high speed is not allowed to enter Zone 2 at high speed. Disabling t_5 means that a train that leaves Zone 1 at low speed is not allowed to accelerate and enter Zone 2 at high speed. Transition t_4 (train leaves Zone 1 at high speed and enters Zone 2 at low speed) is uncontrollable, because a breakdown or a passenger caused emergency stop of the train when at high speed in Zone 1 makes the corresponding event occur. Because of similar reasonings, transitions t_7, t_{13}, t_{16} and t_{19} are also uncontrollable. Finally, the 'crash transitions' t_{20} and t_{21} are also uncontrollable in order to express that the supervisor cannot make trains at high speed in Zones 2 and 4 stop for an incorrectly positioned switch. Consequently, in this example is

$$
\begin{aligned}
T^{\text{uc}} &= \{t_4, t_7, t_{13}, t_{16}, t_{19}, t_{20}, t_{21}\}, \\
T^{\text{c}} &= \{t_1, t_2, t_3, t_5, t_6, t_8, t_9, t_{10}, t_{11}, t_{12}, t_{14}, t_{15}, t_{17}, t_{18}\}.
\end{aligned}
$$

The control objective is, as in Example 2.1, to avoid that there are ever two or more trains in the same zone. In addition, crashes should be avoided. Thus, the legal set equals

$$
\begin{aligned}
\mathcal{A} = \{\mathbf{m} \in \mathbb{N}^{13} \mid \mathbf{m}(p_1) + \mathbf{m}(p_2) &\leq 1 \wedge \\
\mathbf{m}(p_3) + \mathbf{m}(p_4) &\leq 1 \wedge \\
\mathbf{m}(p_5) + \mathbf{m}(p_6) &\leq 1 \wedge \\
\mathbf{m}(p_7) + \mathbf{m}(p_8) &\leq 1 \wedge \\
\mathbf{m}(p_9) &\leq 1 \wedge
\end{aligned}
$$

$$\mathbf{m}(p_{12}) \leq 0 \wedge$$
$$\mathbf{m}(p_{13}) \leq 0\}.$$

△

2. Supervisory control laws

The fact that a number of events are uncontrollable, implies that the control set $\mathcal{U}$ is, in comparison to Chapter 2 (Definition 2.1), defined differently. Before formulating this new definition we introduce a number of order relations in Δ.

Definition 3.1. *Let $\boldsymbol{\delta}_1, \boldsymbol{\delta}_2 \in \Delta$ and $t \in T$. Then $\boldsymbol{\delta}_1 \leq_t \boldsymbol{\delta}_2$ if*

$$\boldsymbol{\delta}_1(t) \leq \boldsymbol{\delta}_2(t) \quad \textit{and} \quad \boldsymbol{\delta}_1|_{T\setminus\{t\}} = \boldsymbol{\delta}_2|_{T\setminus\{t\}}.$$

Proposition 3.1. *For each $t \in T$, $(\Delta, \leq_t)$ is a partially ordered set.*

A set is an up-set (down-set) in $(\Delta, \leq_t)$ if this set is closed when going up (down) in the t-direction.

Proposition 3.2. *Let $\Delta' \subseteq \Delta$. Then*

$$\Delta' \in \mathcal{O}^{\downarrow}(\Delta, \leq) \quad \Leftrightarrow \quad \Delta' \in \mathcal{O}^{\downarrow}(\Delta, \leq_t) \quad \textit{for all } t \in T,$$

and

$$\Delta' \in \mathcal{O}^{\uparrow}(\Delta, \leq) \quad \Leftrightarrow \quad \Delta' \in \mathcal{O}^{\uparrow}(\Delta, \leq_t) \quad \textit{for all } t \in T.$$

Proof. Straightforward. □

Consider a Petri net with a general concurrency assumption $\Omega \in \mathcal{O}^{\downarrow}(\Delta, \leq)$. Define

$$\zeta_0 = \{\boldsymbol{\delta} \in \Omega \mid \boldsymbol{\delta}|_{T^c} = \mathbf{0}\} \tag{3.1}$$

as the set of transition vectors in Ω of which the elements corresponding to controllable transitions are zero. To take the presence of T^{uc}, the set of uncontrollable transitions, into account, the definition of a supervisory controller is extended as follows.

Definition 3.2. *Let $\Omega \in \mathcal{O}^{\downarrow}(\Delta, \leq)$ be the concurrency assumption. For a Petri net $\mathcal{N} = (P, T, f^+, f^-, T^{uc})$ a* control set *$\mathcal{U}$ is a set which satisfies the four following conditions:*

$$\mathcal{U} \subseteq \mathcal{O}^{\downarrow}(\Omega, \leq), \tag{3.2}$$
$$\Omega \in \mathcal{U}, \tag{3.3}$$

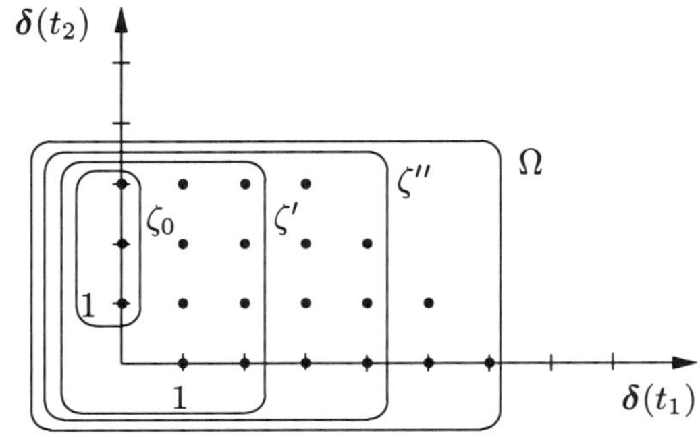

Figure 3.2. The control set $\mathcal{U} = \{\zeta_0, \zeta', \zeta'', \Omega\}$ for a Petri net with two transitions, namely $t_1 \in T^c$ and $t_2 \in T^{uc}$. Elements of Ω are indicated with black dots.

$$\zeta \cup \{\mathbf{0}\} \in \mathcal{O}^{\uparrow}(\Omega \cup \{\mathbf{0}\}, \leq_t) \quad \textit{for all } \zeta \in \mathcal{U} \textit{ and all } t \in T^{uc}, \tag{3.4}$$

$$\zeta_0 \in \mathcal{U}. \tag{3.5}$$

As in Section 2 of Chapter 2, a control set is a set of down-sets in $(\Omega, \leq)$, its elements are called *control values*, a *control law* is a function $\mathbb{N}^n \to \mathcal{U}$ and $\mathfrak{U}$ is the set of all control laws. Condition (3.4) states that the union of each control value ζ and the singleton $\{\mathbf{0}\}$ is an up-set in $(\Omega \cup \{\mathbf{0}\}, \leq_t)$ for all $t \in T^{uc}$. One easily finds that $\zeta_0 \in \mathcal{O}^{\downarrow}(\Omega, \leq)$ and that $\zeta \cup \{\mathbf{0}\} \in \mathcal{O}^{\uparrow}(\Omega \cup \{\mathbf{0}\}, \leq_t)$ for all $t \in T^{uc}$.

If the Petri net does not contain any uncontrollable transitions, i.e. $T^{uc} = \emptyset$ and thus $T = T^c$, then the condition (3.4) is always satisfied and in (3.1) we find that

$$\zeta_0 = \{\boldsymbol{\delta} \in \Omega \mid \boldsymbol{\delta}|_T = \mathbf{0}\} = \emptyset \tag{3.6}$$

because $\mathbf{0} \notin \Omega \subseteq \Delta$. Thus, if $T^{uc} = \emptyset$, one recovers Definition 2.1 (p. 30) as one would expect.

Example 3.2. Consider a Petri net with two transitions, one is controllable and one is uncontrollable: $T^c = \{t_1\}$, $T^{uc} = \{t_2\}$. The concurrency assumption is Ω and is depicted in Figure 3.2. The elements of Ω are indicated with black dots. Note that, in accordance with its definition, $\mathbf{0} \notin \Omega$. Following (3.1), we have here that $\zeta_0 = \{(0,1),(0,2),(0,3)\}$. An example of a control set is then (verify the four conditions of Definition 3.2):

$$\mathcal{U} = \{\zeta_0, \zeta', \zeta'', \Omega\}.$$

These control values are indicated in Figure 3.2. Notice that the control freedom of the supervisor is limited to the t_1-dimension. $\triangle$

Proposition 3.3. *Let $t \in T^{\mathrm{uc}}$ and $\Omega \in \mathcal{O}^{\downarrow}(\Delta, \leq)$. Then,*

$$\zeta \cup \{\mathbf{0}\} \in \mathcal{O}^{\uparrow}(\Omega \cup \{\mathbf{0}\}, \leq_t) \quad \Rightarrow \quad \zeta \in \mathcal{O}^{\uparrow}(\Omega, \leq_t).$$

Proof. By definition, $\zeta \cup \{\mathbf{0}\} \in \mathcal{O}^{\uparrow}(\Omega \cup \{\mathbf{0}\}, \leq_t)$ is equivalent to

$$\begin{aligned} & \forall \boldsymbol{\delta} \in \zeta \cup \{\mathbf{0}\}, \forall \boldsymbol{\delta}' \in \Omega \cup \{\mathbf{0}\} : \boldsymbol{\delta} \leq \boldsymbol{\delta}' \Rightarrow \boldsymbol{\delta}' \in \zeta \cup \{\mathbf{0}\}, \\ \Rightarrow \quad & \forall \boldsymbol{\delta} \in \zeta, \forall \boldsymbol{\delta}' \in \Omega : \boldsymbol{\delta} \leq \boldsymbol{\delta}' \Rightarrow \boldsymbol{\delta}' \in \zeta \cup \{\mathbf{0}\}. \end{aligned} \tag{3.7}$$

As $\boldsymbol{\delta} \in \zeta$, it is true that $\boldsymbol{\delta} \geq 0$ and $\boldsymbol{\delta} \neq \mathbf{0}$. Therefore, one obtains $\boldsymbol{\delta}' \neq \mathbf{0}$. Consequently, (3.7) is equivalent to $\zeta \in \mathcal{O}^{\uparrow}(\Omega, \leq_t)$. □

On the other hand, we obtain from (3.2) and Proposition 3.2 that each control value ζ is a down-set in $(\Omega, \leq_t)$. Consequently for each control value $\zeta \in \mathcal{U}$ and for each uncontrollable transition t the predicate '$\in \zeta$' is invariant in Ω under both going up and down in the t-direction. This is how the uncontrollable nature of a transition is expressed. Proposition 3.4 states this formally.

Proposition 3.4. *Let $\zeta \in \mathcal{U}$, $\boldsymbol{\delta} \in \zeta$ and $\boldsymbol{\delta}' \in \Omega$. If $\boldsymbol{\delta}'|_{T^{\mathrm{c}}} = \boldsymbol{\delta}|_{T^{\mathrm{c}}}$, then $\boldsymbol{\delta}' \in \zeta$.*

Note that the value of $\boldsymbol{\delta}'|_{T^{\mathrm{uc}}}$ is of no importance in this proposition.

Proof. It suffices to consider the case $T^{\mathrm{uc}} \neq \emptyset$. As discussed above, we have that $\zeta \in \mathcal{O}^{\downarrow}(\Omega, \leq_t) \cap \mathcal{O}^{\uparrow}(\Omega, \leq_t)$ for each $t \in T^{\mathrm{uc}}$. By using the definitions of up- and down-set (Definition 1.1, p. 7) this is equivalent to

$$\forall \bar{\boldsymbol{\delta}}_1 \in \zeta, \forall \bar{\boldsymbol{\delta}}_2 \in \Omega : (\bar{\boldsymbol{\delta}}_1 \leq_t \bar{\boldsymbol{\delta}}_2 \Rightarrow \bar{\boldsymbol{\delta}}_2 \in \zeta) \wedge (\bar{\boldsymbol{\delta}}_1 \geq_t \bar{\boldsymbol{\delta}}_2 \Rightarrow \bar{\boldsymbol{\delta}}_2 \in \zeta).$$

With Definition 3.1 this is equivalent to

$$\begin{aligned} \forall \bar{\boldsymbol{\delta}}_1 \in \zeta, \forall \bar{\boldsymbol{\delta}}_2 \in \Omega : [(& \bar{\boldsymbol{\delta}}_1(t) \leq \bar{\boldsymbol{\delta}}_2(t) \wedge \bar{\boldsymbol{\delta}}_1|_{T\setminus\{t\}} = \bar{\boldsymbol{\delta}}_2|_{T\setminus\{t\}}) \\ & \vee (\bar{\boldsymbol{\delta}}_1(t) \geq \bar{\boldsymbol{\delta}}_2(t) \wedge \bar{\boldsymbol{\delta}}_1|_{T\setminus\{t\}} = \bar{\boldsymbol{\delta}}_2|_{T\setminus\{t\}})] \Rightarrow \bar{\boldsymbol{\delta}}_2 \in \zeta. \end{aligned} \tag{3.8}$$

Finally, this can be rewritten as:

$$\forall \bar{\boldsymbol{\delta}}_1 \in \zeta, \forall \bar{\boldsymbol{\delta}}_2 \in \Omega : \bar{\boldsymbol{\delta}}_1|_{T\setminus\{t\}} = \bar{\boldsymbol{\delta}}_2|_{T\setminus\{t\}} \Rightarrow \bar{\boldsymbol{\delta}}_2 \in \zeta. \tag{3.9}$$

Introduce the notation $T^{\mathrm{uc}} = \{t_1, \ldots, t_k\}$. Define for $\boldsymbol{\delta}_1 \in \zeta$ and $\boldsymbol{\delta}_2 \in \Omega$ the transition vectors $\boldsymbol{\delta}'_i$ $(i = 1, \ldots, k+1)$ such that $\boldsymbol{\delta}'_1 = \boldsymbol{\delta}_1$ and

$$\boldsymbol{\delta}'_i|_{T\setminus\{t_{i-1}\}} = \boldsymbol{\delta}'_{i-1}|_{T\setminus\{t_{i-1}\}}, \tag{3.10}$$

$$\boldsymbol{\delta}'_i(t_{i-1}) = \boldsymbol{\delta}_2(t_{i-1}), \tag{3.11}$$

for $i = 2, \ldots, k+1$. Because of $\boldsymbol{\delta}_1|_{T^c} = \boldsymbol{\delta}_2|_{T^c}$, this construction is such that $\boldsymbol{\delta}'_{k+1} = \boldsymbol{\delta}_2$. For each pair of transition vectors $\boldsymbol{\delta}'_{i-1}$ and $\boldsymbol{\delta}'_i$ $(i = 2, \ldots, k+1)$ we can use (3.9) for t_{i-1} (see (3.10)–(3.11)). In k steps we obtain from $\boldsymbol{\delta}_1 = \boldsymbol{\delta}'_1 \in \zeta$ that $\boldsymbol{\delta}_2 = \boldsymbol{\delta}'_{k+1} \in \zeta$. □

Next, we show that ζ_0 is the minimum of $(\mathcal{U}, \subseteq)$.

Proposition 3.5. *Consider a control set $\mathcal{U}$. Then $\zeta_0 \subseteq \zeta$ for all $\zeta \in \mathcal{U}$.*

Proof. From $\zeta \in \mathcal{O}^{\downarrow}(\Omega, \leq)$ it follows straightforwardly that $\zeta \cup \{\mathbf{0}\} \in \mathcal{O}^{\downarrow}(\Omega \cup \{\mathbf{0}\}, \leq_t)$ for all $t \in T^{uc}$. As in the proof of Proposition 3.4 one can show that

$$\forall \zeta \in \mathcal{U}, \forall \boldsymbol{\delta} \in \zeta \cup \{\mathbf{0}\}, \forall \boldsymbol{\delta}' \in \Omega \cup \{\mathbf{0}\} : \boldsymbol{\delta}'|_{T^c} = \boldsymbol{\delta}|_{T^c} \Rightarrow \boldsymbol{\delta}' \in \zeta \cup \{\mathbf{0}\}.$$

By taking $\boldsymbol{\delta} = \mathbf{0}$ herein, one obtains the statement of the proposition. □

Definition 3.3. *Consider a Petri net $\mathcal{N} = (P, T, f^+, f^-, T^{uc})$. Define u_0 as the control law for which $u_0(\mathbf{m}) = \zeta_0$ for all $\mathbf{m} \in \mathbb{N}^n$. We call this constant control law the* zero control law.

Proposition 3.6 follows from Propositions 2.3 and 3.5.

Proposition 3.6. *Let $u \in \mathfrak{U}$ and $\mathbf{m} \in \mathbb{N}^n$. It holds that $\mathcal{R}_\infty(u_0, \mathbf{m}) \subseteq \mathcal{R}_\infty(u, \mathbf{m})$.*

If ζ_0 is used as a control value, then all controllable transitions are disabled. That is why, in contradiction to the result of Proposition 2.4 for fully controlled Petri nets, it is now possible that the state evolves under the zero control law.

Proposition 3.7. *For each $\mathbf{m} \in \mathbb{N}^n$ it holds that*

$$\mathcal{R}_{k,\Omega}(u_0, \mathbf{m}) = \mathcal{R}_{k,\Omega}(\zeta_0, \mathbf{m}) = \mathcal{R}^{T^{uc}}_{k,\Omega}(\mathbf{m}) \quad \textit{for each } k \in \mathbb{N} \cup \{\infty\}.$$

Proof. Compare the definition of $\mathcal{R}_{k,\Omega}(\zeta, \mathbf{m})$ (Definition 2.5, p. 31) for $\zeta = \zeta_0$ with the definition of $\mathcal{R}^{T'}_{k,\Omega}(\mathbf{m})$ (Definition 1.10, p. 18) for $T' = T^{uc}$. □

The reachable set under the zero control law is equal to the set of states which are reachable via uncontrollable transitions only.

3. Specific concurrency and control assumptions

Transition bag assumption. Next we look at the definitions of Section 2 for the transition bag assumption, i.e. $\Omega = \Delta$. If $T^{uc} \neq \emptyset$, then for each control value ζ it is, with the aid of Proposition 3.4, true that:

$$\zeta \cup \{\mathbf{0}\} = \underbrace{\zeta'}_{t \in T^c} \times \underbrace{\mathbb{N} \times \cdots \times \mathbb{N}}_{t \in T^{uc}},$$

where $\zeta' \in \mathcal{O}^{\downarrow}(\mathbb{N}^{\sharp T^c}, \leq)$. For the minimum control value ζ_0 one finds:

$$\zeta_0 \cup \{\mathbf{0}\} = \underbrace{\{0\} \times \cdots \times \{0\}}_{t \in T^c} \times \underbrace{\mathbb{N} \times \cdots \times \mathbb{N}}_{t \in T^{uc}}.$$

This expression is only valid when the set of controllable transitions T^c is a nonempty subset of T, i.e. if there is at least one uncontrollable transition. If $T^{uc} = \emptyset$, then $\zeta_0 = \emptyset$.

Fine control. As in Chapter 2, fine control refers to the fact that all control values which are allowed by the definition of the control set can be chosen by the supervisory controller. Thus, in this case one can verify that the following set is a control set:

$$\mathcal{U} = \{\zeta \in \mathcal{O}^{\downarrow}(\Omega, \leq) \mid \zeta \cup \{\mathbf{0}\} \in \mathcal{O}^{\uparrow}(\Omega \cup \{\mathbf{0}\}, \leq_t) \text{ for all } t \in T^{uc}\}. \quad (3.12)$$

Note that both Ω and ζ_0 belong to this set. For the Petri net of Example 3.2 (p. 58), $\mathcal{U}$ consists of seven control values in the case of fine control (see Figure 3.2).

On/off control. Recall from Chapter 2 (Section 5.2) that in the case of on/off control, each of the components of a transition vector can be chosen not to be influenced or to be put to zero. As the transitions in T^{uc} are uncontrollable, the control set is in this case (compare with (2.20) on page 42):

$$\mathcal{U} = \{\zeta_{T'} \mid T' \subseteq T^c\}, \quad (3.13)$$

where $\zeta_{T'} = \{\boldsymbol{\delta} \in \Omega \mid \boldsymbol{\delta}|_{T'} = \mathbf{0}\}$. The zero control value is $\zeta_0 = \zeta_{T^c}$. An on/off control law can be represented as a function $u : \mathbb{N}^n \to 2^{T^c}$.

4. Maximally permissive control laws

For Petri nets with uncontrollable transitions we now search, as in Section 4 of Chapter 2, for the most liberal (maximally permissive) control laws $u : \mathbb{N}^n \to \mathcal{U}$ which keep the state within $\mathcal{A}$. The definitions of Chapter 2 are reused. However, due to the presence of uncontrollable transitions, their consequences are different. Given an initial state $\mathbf{m}$ we are looking for the most liberal control laws u which satisfy $\mathcal{R}_\infty(u, \mathbf{m}) \subseteq \mathcal{A}$. First we define permissiveness, thereafter we define maximal permissiveness. A general concurrency assumption Ω is used.

4.1. Permissive control laws

As in Section 4 of Chapter 2, the control value $u(\mathbf{m}) \in \mathcal{U}$ should be such that, for all states in $\mathcal{R}_{1,\Omega}(u(\mathbf{m}), \mathbf{m})$, there exists a control law under which all possible future states belong to $\mathcal{A}$. Therefore, as in (2.7), $u(\mathbf{m})$ should belong to

$$\mathcal{U}^{\mathcal{A}}_{\mathbf{m}} = \{\zeta \in \mathcal{U} \mid \forall \mathbf{m}' \in \mathcal{R}_{1,\Omega}(\zeta, \mathbf{m}) : \exists \bar{u} \in \mathfrak{U} : \mathcal{R}_\infty(\bar{u}, \mathbf{m}') \subseteq \mathcal{A}\}. \quad (3.14)$$

Because of Propositions 3.6 and 3.7, we find, by taking $\bar{u} = u_0$, that

$$\mathcal{U}^{\mathcal{A}}_{\mathbf{m}} = \{\zeta \in \mathcal{U} \mid \mathcal{R}^{T^{\mathrm{uc}}}_\infty(\mathcal{R}_{1,\Omega}(\zeta, \mathbf{m})) \subseteq \mathcal{A}\}. \quad (3.15)$$

Definition 3.4. *Consider a Petri* $\mathcal{N} = (P, T, f^+, f^-, T^{\mathrm{uc}})$ *and a legal set* $\mathcal{A}$. *Introduce the set of states*

$$\mathcal{A}^* := \{\mathbf{m} \in \mathbb{N}^n \mid \mathcal{R}^{T^{\mathrm{uc}}}_\infty(\mathbf{m}) \subseteq \mathcal{A}\}. \quad (3.16)$$

This set is referred to as the supremal controllable subset of $\mathcal{A}$.

A state belongs to $\mathcal{A}^*$ if all the states which are reachable via uncontrollable transitions from this state belong to $\mathcal{A}$. It is clear that $\mathcal{A}^* \subseteq \mathcal{A}$. If $T^{\mathrm{uc}} = \emptyset$, then $\mathcal{A}^* = \mathcal{A}$. With the definition of $\mathcal{A}^*$, we obtain from (3.15) that

$$\mathcal{U}^{\mathcal{A}}_{\mathbf{m}} = \{\zeta \in \mathcal{U} \mid \mathcal{R}_{1,\Omega}(\zeta, \mathbf{m}) \subseteq \mathcal{A}^*\}. \quad (3.17)$$

Proposition 3.8. $\mathcal{U}^{\mathcal{A}}_{\mathbf{m}} \neq \emptyset$ *if and only if* $\mathbf{m} \in \mathcal{A}^*$.

Proof. Because of Propositions 2.3 and 3.5, we have that

$$\begin{aligned} \mathcal{U}^{\mathcal{A}}_{\mathbf{m}} \neq \emptyset \quad &\Leftrightarrow \quad \mathcal{R}_{1,\Omega}(\zeta_0, \mathbf{m}) \subseteq \mathcal{A}^* \\ &\Leftrightarrow \quad \mathcal{R}^{T^{\mathrm{uc}}}_{1,\Omega}(\mathbf{m}) \subseteq \mathcal{A}^*. \end{aligned} \quad (3.18)$$

The last equivalence is because of Proposition 3.7. A necessary condition for (3.18) is that $\mathbf{m} \in \mathcal{A}^*$. For the sufficiency part we refer to Proposition 3.11 on page 65. □

Definition 3.5. *Consider a Petri net* $\mathcal{N} = (P, T, f^+, f^-, T^{\mathrm{uc}})$ *and a legal set* $\mathcal{A}$. *A control value* $\zeta \in \mathcal{U}$ *is* permissive *for* $\mathbf{m} \in \mathbb{N}^n$ *if* $\zeta \in \mathcal{U}^{\mathcal{A}}_{\mathbf{m}}$. *A control law* $u \in \mathfrak{U}$ *is* permissive *if* $u(\mathbf{m})$ *is permissive for all* $\mathbf{m} \in \mathcal{A}^*$.

Starting from a state $\mathbf{m}$, there exists at least one permissive control law if and only if $\mathbf{m} \in \mathcal{A}^*$. If $\mathbf{m} \in \mathcal{A} \setminus \mathcal{A}^*$, then a state outside $\mathcal{A}$ can always be reached via uncontrollable transitions.

As for the case in which all transitions are controllable (see Proposition 2.5), it also holds here that, if ζ belongs to $\mathcal{U}^{\mathcal{A}}_{\mathbf{m}}$, then each subset ζ' of ζ which belongs to $\mathcal{U}$, also belongs to $\mathcal{U}^{\mathcal{A}}_{\mathbf{m}}$.

Proposition 3.9. *Consider a Petri net* $\mathcal{N} = (P, T, f^+, f^-, T^{\mathrm{uc}})$. *Let* $\mathcal{A} \subseteq \mathbb{N}^n$ *and* $\mathbf{m} \in \mathcal{A}$. *Then* $\mathcal{U}^{\mathcal{A}}_{\mathbf{m}} \in \mathcal{O}^{\downarrow}(\mathcal{U}, \subseteq)$.

4.2. Optimality

Define $\mathcal{E}^{\mathcal{A}}_{\mathbf{m}}$ as the quotient set of $\mathcal{U}^{\mathcal{A}}_{\mathbf{m}}$ modulo $\sim_{\mathbf{m}}$ (see Definition 2.7). Then, like in Definition 2.8, we can define a partial order relation $\leq_{\mathbf{m}}$ in $\mathcal{E}^{\mathcal{A}}_{\mathbf{m}}$ to characterise optimal control laws.

Definition 3.6. *Consider a Petri net $\mathcal{N} = (P, T, f^+, f^-, T^{\mathrm{uc}})$ and a legal set $\mathcal{A}$. A control value $\zeta \in \mathcal{U}$ is* maximally permissive *for the state $\mathbf{m} \in \mathbb{N}^n$ if $\zeta \in \max(\mathcal{E}^{\mathcal{A}}_{\mathbf{m}}, \leq_{\mathbf{m}})$. A control law $u \in \mathfrak{U}$ is* maximally permissive *if $u(\mathbf{m})$ is maximally permissive for all $\mathbf{m} \in \mathcal{A}^*$.*

5. Control design

In this section we study how one can design control laws for Petri nets with uncontrollable transitions. We show that this problem can be converted to that of Chapter 2, i.e to control design for Petri nets with controllable transitions only.

Proposition 3.10. *For all $\mathbf{m} \in \mathbb{N}^n$ and for all $\zeta \in \mathcal{O}^{\downarrow}(\Omega, \leq)$ it holds that*

$$\mathcal{R}^{T^{\mathrm{uc}}}_{\infty}(\mathcal{R}_{1,\Omega}(\zeta, \mathbf{m})) = \mathcal{R}^{T^{\mathrm{uc}}}_{\infty}(\mathcal{R}^{T^{\mathrm{c}}}_{1,\Omega}(\zeta, \mathbf{m})).$$

Proof. ($\supseteq$) This directly follows from Proposition 1.10.

($\subseteq$) By definition, for each state $\mathbf{m}' \in \mathcal{R}^{T^{\mathrm{uc}}}_{\infty}(\mathcal{R}_{1,\Omega}(\zeta, \mathbf{m}))$ there exist transition vectors $\boldsymbol{\delta}_1, \ldots, \boldsymbol{\delta}_k$ with $\boldsymbol{\delta}_i|_{T^{\mathrm{c}}} = \mathbf{0}$ $(i = 1, \ldots, k)$ such that

$$\mathbf{m} \xrightarrow{\boldsymbol{\delta}_0, \zeta} \mathbf{m}'' \xrightarrow{(\boldsymbol{\delta}_1, \ldots, \boldsymbol{\delta}_k)} \mathbf{m}'. \tag{3.19}$$

Decompose $\boldsymbol{\delta}_0$ in a controllable part $\boldsymbol{\delta}'$ and an uncontrollable part $\boldsymbol{\delta}''$:

$$\begin{aligned} \boldsymbol{\delta}'|_{T^{\mathrm{c}}} &= \boldsymbol{\delta}_0|_{T^{\mathrm{c}}}, & \boldsymbol{\delta}'|_{T^{\mathrm{uc}}} &= \mathbf{0}, \\ \boldsymbol{\delta}''|_{T^{\mathrm{c}}} &= \mathbf{0}, & \boldsymbol{\delta}''|_{T^{\mathrm{uc}}} &= \boldsymbol{\delta}_0|_{T^{\mathrm{uc}}}. \end{aligned}$$

As ζ is a down-set in $(\Omega, \leq)$, both $\boldsymbol{\delta}'$ and $\boldsymbol{\delta}''$ belong to $\zeta \cup \{\mathbf{0}\}$. (Recall that we allow the notation $\mathbf{m} \xrightarrow{\mathbf{0}} \mathbf{m}$.) From (3.19) we obtain, with Proposition 1.7 (p. 10), that

$$\mathbf{m} \xrightarrow{\boldsymbol{\delta}', \zeta} \cdot \xrightarrow{\boldsymbol{\delta}'', \zeta} \mathbf{m}'' \xrightarrow{(\boldsymbol{\delta}_1, \ldots, \boldsymbol{\delta}_k)} \mathbf{m}'. \tag{3.20}$$

Because of $\boldsymbol{\delta}''|_{T^{\mathrm{c}}} = \mathbf{0}$, it follows that

$$\mathbf{m} \xrightarrow{\boldsymbol{\delta}', \zeta} \cdot \xrightarrow{\boldsymbol{\delta}'', \zeta_0} \mathbf{m}'' \xrightarrow{(\boldsymbol{\delta}_1, \ldots, \boldsymbol{\delta}_k)} \mathbf{m}', \tag{3.21}$$

or that $\mathbf{m}' \in \mathcal{R}^{T^{\mathrm{uc}}}_{\infty}(\mathcal{R}^{T^{\mathrm{c}}}_{1,\Omega}(\zeta, \mathbf{m}))$. □

With this property we obtain from (3.15) that

$$\mathcal{U}_{\mathbf{m}}^{\mathcal{A}} = \{\zeta \in \mathcal{U} \mid \mathcal{R}_{\infty}^{T^{\mathrm{uc}}}(\mathcal{R}_{1,\Omega}^{T^{\mathrm{c}}}(\zeta, \mathbf{m})) \subseteq \mathcal{A}\}, \tag{3.22}$$

and, comparing with (3.17), that

$$\mathcal{U}_{\mathbf{m}}^{\mathcal{A}} = \{\zeta \in \mathcal{U} \mid \mathcal{R}_{1,\Omega}^{T^{\mathrm{c}}}(\zeta, \mathbf{m}) \subseteq \mathcal{A}^*\}. \tag{3.23}$$

When comparing (3.23) with (2.8) on page 33, one concludes that the problem of control design for a Petri net with uncontrollable transitions is reduced to the problem stated in Chapter 2. Instead of for a Petri net $\mathcal{N} = (P, T, f^+, f^-, T^{\mathrm{uc}})$ and legal set $\mathcal{A}$, control design should now be done for the sub-Petri net

$$(P, T^{\mathrm{c}}, f^+|_{T^{\mathrm{c}} \times P}, f^-|_{P \times T^{\mathrm{c}}}, \mathbf{m}_0, \emptyset),$$

of $\mathcal{N}$ with the supremal controllable subset $\mathcal{A}^*$ of $\mathcal{A}$ as the control objective. In this Petri net, T^{c} is the set of all transitions. This Petri net does not contain any uncontrollable transitions. The concurrency assumption can be found by projecting Ω on $\mathbb{N}^{\sharp T^{\mathrm{c}}}$. One obtains the control set by projecting its elements on $\mathbb{N}^{\sharp T^{\mathrm{c}}}$. For the control sets in (3.12) (fine control) and (3.13) (on/off control) the projections are of the same form as (2.14) and (2.20), respectively. Consequently, the results of Sections 5 to 7 in Chapter 2 remain valid if $\mathcal{A}$ is replaced by $\mathcal{A}^*$.

This shows that control design in the presence of uncontrollable transitions is in essence the same as control design for Petri nets of which all controllable transitions are controllable. The main difference is that the legal set is now $\mathcal{A}^*$ instead of $\mathcal{A}$. Once $\mathcal{A}^*$ has been constructed, the control design techniques which have been developed in Chapter 2 can be applied. Therefore, the construction of the supremal controllable subset of a given legal set is the key step in control design for Petri nets with uncontrollable transitions. Although $\mathcal{A}$ is given, $\mathcal{A}^*$ is not known directly in general. The classes of systems for which there exist algorithms that construct $\mathcal{A}^*$ is limited at present. In this book we extend this set of classes by deducing general algorithms. Some basic properties of $\mathcal{A}^*$ are discussed in the following section.

6. The supremal controllable subset

We end this chapter by formulating some properties of the supremal controllable subset $\mathcal{A}^*$ of a given legal set $\mathcal{A}$. Because of Proposition 1.19 (p. 25), the definition of $\mathcal{A}^*$ in Definition 3.4 does not depend on the speed of the Petri net model. This is important for the following reason. When one designs controllers for a Petri net, the construction of $\mathcal{A}^*$ can be done with a concurrency assumption which can be chosen independent

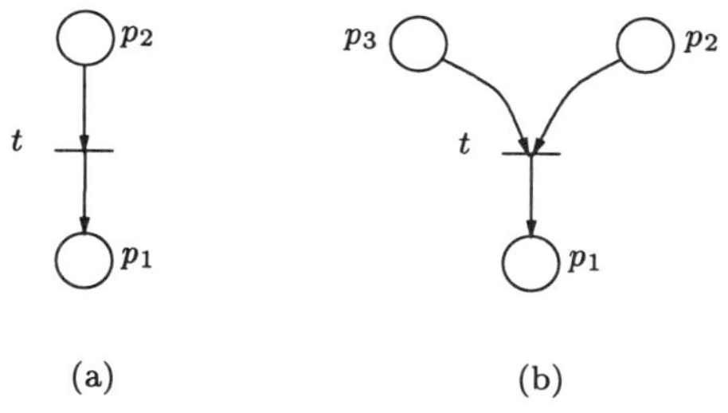

Figure 3.3. The two uncontrollable Petri nets of Example 3.3.

of the model speed of the Petri net under control. In this book, for the construction of the supremal controllable subset, we use the transition bag assumption in Chapters 6 and 7 and the no concurrency assumption in Chapter 8.

In the following proposition it is shown that $\mathcal{A}^*$ is invariant under the execution of uncontrollable transitions.

Proposition 3.11. *For each $\mathcal{A} \subseteq \mathbb{N}^n$ it is true that $\mathcal{R}_\infty^{T^{uc}}(\mathcal{A}^*) = \mathcal{A}^*$.*

Proof. Because of $\mathcal{R}_0(\mathcal{A}^*) = \mathcal{A}^*$ and Proposition 1.8 (p. 13), it suffices to prove that $\mathcal{R}_\infty^{T^{uc}}(\mathcal{A}^*) \subseteq \mathcal{A}^*$. From (3.16) we have that $\mathcal{R}_\infty^{T^{uc}}(\mathcal{A}^*) \subseteq \mathcal{A}$. Suppose there exists a state $\mathbf{m} \in \mathcal{R}_\infty^{T^{uc}}(\mathcal{A}^*)$ such that $\mathbf{m} \in \mathcal{A} \setminus \mathcal{A}^*$. Then there exists an $\mathbf{m}' \in \mathcal{A}^*$ for which $\mathbf{m} \in \mathcal{R}_\infty^{T^{uc}}(\mathbf{m}')$ and there exists an $\mathbf{m}'' \in \mathcal{R}_\infty^{T^{uc}}(\mathbf{m}) \setminus \mathcal{A}$. As $\mathbf{m}'' \notin \mathcal{A}$, $\mathbf{m}'' \in \mathcal{R}_\infty^{T^{uc}}(\mathbf{m}')$ contradicts $\mathbf{m}' \in \mathcal{A}^*$. □

The supremal controllable subset $\mathcal{A}^*$ is the largest subset of $\mathcal{A}$ for which this property holds.

Proposition 3.12. *Let $\mathcal{A} \subseteq \mathbb{N}^n$. Then each subset $\mathcal{A}_1$ of $\mathcal{A}$ which satisfies $\mathcal{R}_\infty^{T^{uc}}(\mathcal{A}_1^*) = \mathcal{A}_1^*$ is a subset of $\mathcal{A}^*$.*

Proof. Suppose that there exists a state $\mathbf{m}'$ in $\mathcal{A}_1 \setminus \mathcal{A}^*$. As $\mathbf{m}' \in \mathcal{A}_1$, we have that $\mathcal{R}_\infty^{T^{uc}}(\mathbf{m}') \subseteq \mathcal{A}_1 \subseteq \mathcal{A}$. By the definition of $\mathcal{A}^*$ in (3.16), it follows that $\mathbf{m}' \in \mathcal{A}^*$, which leads to a contradiction. □

Example 3.3. We illustrate the construction of the supremal controllable subset with the aid of two standard examples for which this construction can be done by hand. Consider the Petri net in Figure 3.3(a) with uncontrollable transition t and legal set $\mathcal{A} = \{\mathbf{m} \in \mathbb{N}^2 \mid \mathbf{m}(p_1) \leq k\}$ with $k \in \mathbb{N}$. One straightforwardly obtains that $\mathcal{A}^* = \{\mathbf{m} \in \mathbb{N}^2 \mid \mathbf{m}(p_1) + \mathbf{m}(p_2) \leq k\}$.

The Petri net in Figure 3.3(b) is also uncontrollable. For $\mathcal{A} = \{\mathbf{m} \in \mathbb{N}^3 \mid \mathbf{m}(p_1) \leq k\}$ with $k \in \mathbb{N}$, one finds

$$\begin{aligned}\mathcal{A}^* &= \{\mathbf{m} \in \mathbb{N}^3 \mid \mathbf{m}(p_1) + \min(\mathbf{m}(p_2), \mathbf{m}(p_3)) \leq k\}, \\ &= \{\mathbf{m} \in \mathbb{N}^3 \mid \mathbf{m}(p_1) + \mathbf{m}(p_2) \leq k\} \cup \{\mathbf{m} \in \mathbb{N}^3 \mid \mathbf{m}(p_1) + \mathbf{m}(p_3) \leq k\}.\end{aligned}$$

This example shows that, even for simple systems, the supremal controllable subset can be 'nonconvex'. △

The next proposition shows that the supremal controllable subset of the intersection equals the intersection of the supremal controllable subsets. This property is important when splitting the construction of $\mathcal{A}^*$ into smaller subproblems. In Chapter 6 we discuss this problem in more detail.

Proposition 3.13. *For each family $\{\mathcal{A}_i \subseteq \mathbb{N}^n \mid i \in I\}$ it is true that*

$$\left(\bigcap_{i \in I} \mathcal{A}_i\right)^* = \bigcap_{i \in I} \mathcal{A}_i^*.$$

Proof. This immediately follows from (3.16). □

To construct $\mathcal{A}^*$ one needs to study reachability via uncontrollable transitions. That is why it suffices to consider the Petri net

$$\mathcal{N}^{\mathrm{uc}} := (P, T^{\mathrm{uc}}, f^+|_{T^{\mathrm{uc}} \times P}, f^-|_{P \times T^{\mathrm{uc}}}, \mathbf{m}_0, T^{\mathrm{uc}}). \tag{3.24}$$

This is an uncontrollable sub-Petri net of $\mathcal{N}$. It is obtained by removing the controllable transitions and the corresponding arcs from $\mathcal{N}$. We often refer to $\mathcal{N}^{\mathrm{uc}}$ as the *uncontrollable part* or the *uncontrollable subnet* of $\mathcal{N}$.

Example 3.4. Consider again the Petri net $\mathcal{N}$ in Figure 3.1 that models the extended railway example. One obtains the Petri net $\mathcal{N}^{\mathrm{uc}}$ by removing the set of controllable transitions $T^{\mathrm{c}} = \{t_1, t_2, t_3, t_5, t_6, t_8, t_9, t_{10}, t_{11}, t_{12}, t_{14}, t_{15}, t_{17}, t_{18}\}$ and the corresponding arcs. It is depicted in Figure 3.4. △

Throughout the book, we assume that there is a bijective relation between transitions in the Petri net and events in the discrete event system. If there are two transitions in the Petri net which correspond to the same event, then care must be taken when defining the dynamics of the model with respect to the discrete event system. As studied in detail in automata theory, the possible problem of nondeterminism arises. For this reason, Petri nets without a bijective relation between transitions and events are not within the scope of this work. Nevertheless, we believe

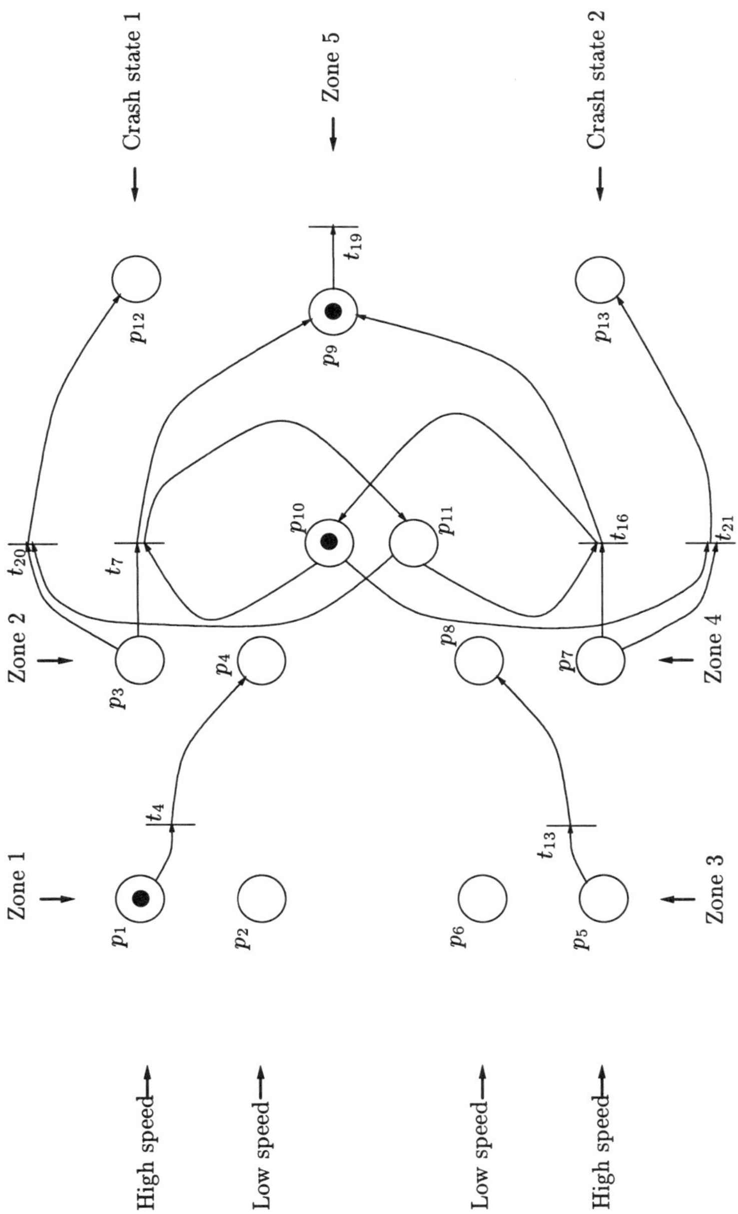

Figure 3.4. The uncontrollable part of the Petri net in Figure 3.1.

that the results for the supremal controllable subset presented hereafter remain valid for this more general class of systems. This is because in the definition of the supremal controllable subset reachability in any number of steps via whichever of transition vectors plays a role, cfr. $\mathcal{R}^{\cdot}_{\infty}(\cdot)$.

In the rest of this book we mainly focus on properties, the construction and the approximation of the supremal controllable subset of a given legal set. Its construction leads to maximal permissive supervisors, its approximation to permissive ones. Chapter 4 presents reduction theorems for the control design. In Chapter 6 decomposition of the control problem is studied. Chapters 7 and 8 provide algorithms that construct (subsets of) the supremal controllable subset of a given legal set.

7. Notes and references

Supervisory control of Petri nets with uncontrollable transitions has been the subject of many papers. An extensive survey is [19]. Other references are [5], [17], [18], [22], [24] and [25]. All control paradigms herein fit into the framework of this chapter. An extension of on/off control which allows to express dependence between events, is used in [18]. Translation of regular specification languages for Petri nets to legal sets is done in [22]. The form of these sets shows the need for the study of general legal sets. Examples of papers on deadlock avoidance and legal sets are [2], [3], [23] and [41].

Chapter 4

REDUCTION THEOREMS

The predicate '$\in \mathcal{A}$' in the state space $\mathbb{N}^n$ is often invariant under translation in a number of directions and the incidence matrices of a Petri net are mostly sparse. As a consequence, in many cases, the construction of the supremal controllable subset $\mathcal{A}^*$ of a legal set $\mathcal{A}$ can be done more efficiently than by directly applying its definition in (3.16). For the same reason, reduction of the design of (maximally) permissive control laws is also possible.

In this chapter, invariance properties of the legal set are studied and formalised and two reduction theorems are proven. We show that, for the construction of $\mathcal{A}^*$ and for the design of (maximally) permissive control laws, it suffices to do a reachability analysis of the sub-Petri nets $\mathcal{N}_{\mathcal{A}}^{\mathrm{uc}}$ and $\mathcal{N}_{\mathcal{A}}$ of $\mathcal{N}^{\mathrm{uc}}$ and $\mathcal{N}$, respectively. Both subnets are defined in this chapter. Moreover, we determine a set of transitions which does not need to be observed by the supervisory controller.

In Section 1, examples are given which make it intuitively clear how invariance properties of the legal set $\mathcal{A}$ can be used advantageously. Based on these invariance properties, Section 2 partitions the set of places into four mutually disjoint sets. For these sets, sets of places and sets of uncontrollable transitions, which are the building blocks of the uncontrollable Petri net $\mathcal{N}_{\mathcal{A}}^{\mathrm{uc}}$, are introduced in Section 3. Sections 4 and 5 provide a reduction result for $\mathcal{A}^*$ and for control design, respectively. Finally, in Section 6 it is studied how, for a given legal set, the above invariance properties can be determined.

1. Intuition for $\mathcal{A}^*$

From Section 6 in Chapter 3 we know that a reachability analysis of the uncontrollable Petri net $\mathcal{N}^{\mathrm{uc}}$ (see (3.24)) is needed to determine the

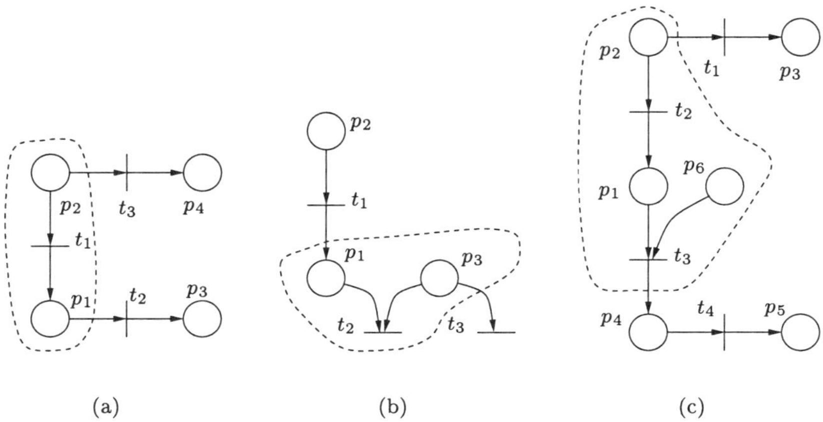

Figure 4.1. Three uncontrollable Petri nets, $\mathcal{N}_1^{\mathrm{uc}}$ in (a), $\mathcal{N}_2^{\mathrm{uc}}$ in (b) and $\mathcal{N}_3^{\mathrm{uc}}$ in (c), for which it suffices to study the part indicated with dotted lines to determine the supremal controllable subset of $\mathcal{A}_1$ in (4.1), $\mathcal{A}_2$ in (4.4) and $\mathcal{A}_3$ in (4.6), respectively.

supremal controllable subset $\mathcal{A}^*$ of a given legal set $\mathcal{A}$. In this section we illustrate by means of three examples how invariance properties of $\mathcal{A}$ can be used to simplify the construction of $\mathcal{A}^*$. As it is shown, it suffices to make a reachability analysis for a proper sub-Petri net of $\mathcal{N}^{\mathrm{uc}}$.

Consider the Petri net $\mathcal{N}_1^{\mathrm{uc}}$ in Figure 4.1(a). It is the uncontrollable part of a Petri net $\mathcal{N}_1$; all transitions in $\mathcal{N}_1^{\mathrm{uc}}$ are uncontrollable. The legal set is

$$\mathcal{A}_1 = \{\mathbf{m} \in \mathbb{N}^4 \mid \mathbf{m}(p_1) \leq 3\}. \tag{4.1}$$

The state of the place p_1 should always be less than or equal to three. Following Definition 3.4, $\mathcal{A}_1^*$ is the set of states for which all states reachable via uncontrollable transitions belong to $\mathcal{A}_1$. Here, $\{t_1, t_2, t_3\}$ is the set of uncontrollable transitions. Because of the structure of $\mathcal{A}_1$, for each state $\mathbf{m}$ we have the following optimisation problem:

$$\mathbf{m} \in \mathcal{A}_1^* \quad \Leftrightarrow \quad \max_{\mathbf{m}' \in \mathcal{R}_\infty^{T^{\mathrm{uc}}}(\mathbf{m})} \mathbf{m}'(p_1) \leq 3. \tag{4.2}$$

The maximum does exist since $\mathcal{R}_\infty^{T^{\mathrm{uc}}}(\mathbf{m})$ is a finite set for all $\mathbf{m}$. From Figure 4.1(a) one can deduce that in the search for a state in $\mathcal{R}_\infty^{T^{\mathrm{uc}}}(\mathbf{m})$ which maximises $\mathbf{m}'(p_1)$, transition t_2 can be assumed absent. This is because the execution of t_2 causes the objective function $\mathbf{m}'(p_1)$ to decrease. Then, the state of p_3 increases, but has no further influence on $\mathbf{m}'(p_1)$. Transition t_3 can also be assumed absent since it decreases the state of p_2 when executed. Consequently, the maximal number of times that t_1 can be executed decreases and this has a negative influence on $\mathbf{m}'(p_1)$. It is clear that the maximal value of $\mathbf{m}'(p_1)$ is independent of the state of p_3 and p_4. Therefore, it suffices to consider the indicated sub-Petri net (places p_1 and p_2, and transition t_1) when constructing $\mathcal{A}_1^*$. The above reasoning uses the following invariance property of $\mathcal{A}_1$:

$$\forall \mathbf{m}_1 \in \mathcal{A}_1, \forall \mathbf{m}_2 \in \mathbb{N}^4 : \mathbf{m}_2(p_1) \leq \mathbf{m}_1(p_1) \Rightarrow \mathbf{m}_2 \in \mathcal{A}_1. \tag{4.3}$$

As a second example, consider the uncontrollable Petri net $\mathcal{N}_2^{\mathrm{uc}}$ in Figure 4.1(b) and the legal set

$$\mathcal{A}_2 = \{\mathbf{m} \in \mathbb{N}^3 \mid \mathbf{m}(p_1) \geq 2\}. \tag{4.4}$$

The state of p_2 should be greater than or equal to two. Then,

$$\mathbf{m} \in \mathcal{A}_2^* \quad \Leftrightarrow \quad \min_{\mathbf{m}' \in \mathcal{R}_\infty^{T^{\mathrm{uc}}}(\mathbf{m})} \mathbf{m}'(p_1) \geq 2.$$

With a reasoning similar to the one above and the invariance property

$$\forall \mathbf{m}_1 \in \mathcal{A}_2, \forall \mathbf{m}_2 \in \mathbb{N}^3 : \mathbf{m}_2(p_1) \geq \mathbf{m}_1(p_1) \Rightarrow \mathbf{m}_2 \in \mathcal{A}_2, \tag{4.5}$$

transitions t_1 and t_3 can be assumed absent. Transition t_1 only increases the state of p_1. Hence, the state of p_2 is of no importance. On the other hand, transition t_3 decreases the state of p_3. This decreases the maximum number of times, to decrease the state of p_1 as much as possible, that t_2 can be executed. Therefore, it suffices to consider the indicated sub-Petri net in Figure 4.1(b).

Finally, consider the uncontrollable Petri net $\mathcal{N}_3^{\mathrm{uc}}$ in Figure 4.1(c) and the legal set

$$\mathcal{A}_3 = \{\mathbf{m} \in \mathbb{N}^6 \mid 3 \leq \mathbf{m}(p_1) \leq 7\}. \tag{4.6}$$

In this case we use the invariance property:

$$\forall \mathbf{m}_1 \in \mathcal{A}_3, \forall \mathbf{m}_2 \in \mathbb{N}^6 : \mathbf{m}_2(p_1) = \mathbf{m}_1(p_1) \Rightarrow \mathbf{m}_2 \in \mathcal{A}_3. \tag{4.7}$$

Then,

$$\mathbf{m} \in \mathcal{A}_3^* \quad \Leftrightarrow \quad \max_{\mathbf{m}' \in \mathcal{R}_\infty^{T^{\mathrm{uc}}}(\mathbf{m})} \mathbf{m}'(p_1) \leq 7 \wedge \min_{\mathbf{m}' \in \mathcal{R}_\infty^{T^{\mathrm{uc}}}(\mathbf{m})} \mathbf{m}'(p_1) \geq 3.$$

Therefore, we can combine the reasonings used in the first two examples. Because of $3 \leq \mathbf{m}(p_1)$, transitions in $\{t_1, t_3, t_4\}$ can be assumed disabled. Because of $\mathbf{m}(p_1) \leq 7$, transitions in $\{t_1, t_2, t_4\}$ can be assumed disabled. The set of transitions $\{t_1, t_4\}$ which is obtained by taking the intersection of both sets, can be removed from the model. The state components of p_3, p_4 and p_5 are of no importance. It suffices to study the indicated sub-Petri net.

Another way to come to the same conclusion for this third example is the following. If, in an uncontrollable sequence of transition vectors, transition t_1 or t_4 would be executed, then the same state of p_1 is reached by eliminating both transitions from this sequence. For this reason, t_1 and t_4 can be assumed absent.

In the next section, the invariance properties (4.3), (4.5) and (4.7) are defined formally for general legal sets.

2. Invariance properties of the legal set

The next definition introduces for each $p \in P$ the relation $\leq_p$ in $\mathbb{N}^n$.

Definition 4.1. *Let* $\mathbf{m}_1, \mathbf{m}_2 \in \mathbb{N}^n$ *and* $p \in P$. *Then,* $\mathbf{m}_1 \leq_p \mathbf{m}_2$ *if* $\mathbf{m}_1(p) \leq \mathbf{m}_2(p)$ *and* $\mathbf{m}_1|_{P\setminus\{p\}} = \mathbf{m}_2|_{P\setminus\{p\}}$.

Proposition 4.1. *For each* $p \in P$, $(\mathbb{N}^n, \leq_p)$ *is a partially ordered set.*

Note that similar partial order relations $\leq_t$ were defined in Δ in Definition 3.1 (p. 57) in the context of uncontrollable transitions. The sets of Definition 4.2 characterise invariance properties of the predicate '$\in \mathcal{A}$' in $\mathbb{N}^n$.

Definition 4.2. *Consider a Petri net* $\mathcal{N} = (P, T, f^+, f^-, T^{\text{uc}})$. *Define for* $\mathcal{A} \subseteq \mathbb{N}^n$ *the sets of places:*

$$\begin{aligned}
P^{\text{i}}(\mathcal{A}) &:= \{p \in P \mid \mathcal{A} \in \mathcal{O}^{\downarrow}(\mathbb{N}^n, \leq_p) \cap \mathcal{O}^{\uparrow}(\mathbb{N}^n, \leq_p)\}, \\
P^{\downarrow}(\mathcal{A}) &:= \{p \in P \mid \mathcal{A} \in \mathcal{O}^{\downarrow}(\mathbb{N}^n, \leq_p) \setminus \mathcal{O}^{\uparrow}(\mathbb{N}^n, \leq_p)\}, \\
P^{\uparrow}(\mathcal{A}) &:= \{p \in P \mid \mathcal{A} \in \mathcal{O}^{\uparrow}(\mathbb{N}^n, \leq_p) \setminus \mathcal{O}^{\downarrow}(\mathbb{N}^n, \leq_p)\}, \\
P^{\text{d}}(\mathcal{A}) &:= \{p \in P \mid \mathcal{A} \notin \mathcal{O}^{\downarrow}(\mathbb{N}^n, \leq_p) \cup \mathcal{O}^{\uparrow}(\mathbb{N}^n, \leq_p)\}.
\end{aligned}$$

A place p belongs to $P^{\text{i}}(\mathcal{A})$ if $\mathcal{A}$ is closed under both going down and up in the p-direction. A place p belongs to $P^{\downarrow}(\mathcal{A})$ if $\mathcal{A}$ is closed under going down in the p-direction, but not under going up. Analogous statements for $P^{\uparrow}(\mathcal{A})$ and $P^{\text{d}}(\mathcal{A})$ can be made. The sets $P^{\text{i}}(\mathcal{A})$, $P^{\downarrow}(\mathcal{A})$, $P^{\uparrow}(\mathcal{A})$ and $P^{\text{d}}(\mathcal{A})$ are mutually disjoint and their union is equal to P. The upper indices refer to *invariant*, *downward invariant*, *upward invariant* and *dependent*, respectively.

Proposition 4.2. *For each* $\mathcal{A} \subseteq \mathbb{N}^n$ *it is true that*

$$P = P^{\text{i}}(\mathcal{A}) \dot{\cup} P^{\downarrow}(\mathcal{A}) \dot{\cup} P^{\uparrow}(\mathcal{A}) \dot{\cup} P^{\text{d}}(\mathcal{A}).$$

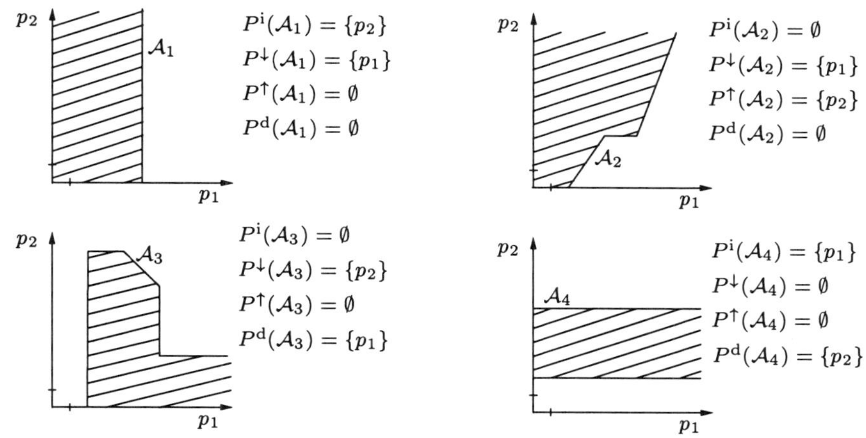

Figure 4.2. Illustration of the sets of Definition 4.2 for a Petri net with two places. The hatched regions indicate the legal sets.

Note that $P^{\mathrm{i}}(\emptyset) = P^{\mathrm{i}}(\mathbb{N}^n) = P$. The sets of Definition 4.2 are illustrated in Figure 4.2 for a Petri net with two places.

Example 4.1. For the three legal sets $\mathcal{A}_1$, $\mathcal{A}_2$ and $\mathcal{A}_3$ of Section 1 (see (4.1), (4.4) en (4.6)) it holds that

$$
\begin{array}{lll}
P^{\mathrm{i}}(\mathcal{A}_1) = \{p_2, p_3, p_4\}, & P^{\mathrm{i}}(\mathcal{A}_2) = \{p_2, p_3\}, & P^{\mathrm{i}}(\mathcal{A}_3) = \{p_2, p_3, p_4, p_5, p_6\}, \\
P^{\downarrow}(\mathcal{A}_1) = \{p_1\}, & P^{\downarrow}(\mathcal{A}_2) = \emptyset, & P^{\downarrow}(\mathcal{A}_3) = \emptyset, \\
P^{\uparrow}(\mathcal{A}_1) = \emptyset & P^{\uparrow}(\mathcal{A}_2) = \{p_1\}, & P^{\uparrow}(\mathcal{A}_3) = \emptyset, \\
P^{\mathrm{d}}(\mathcal{A}_1) = \emptyset, & P^{\mathrm{d}}(\mathcal{A}_2) = \emptyset, & P^{\mathrm{d}}(\mathcal{A}_3) = \{p_1\}.
\end{array}
$$

△

The result of this example can be extended to general inequalities. In Section 6 it is shown that if $\mathcal{A} = \{\mathbf{m} \in \mathbb{N}^n \mid \mathbf{A}^T\mathbf{m} \leq b\}$ with $\mathbf{A} \in \mathbb{Z}^n$ and $b \in \mathbb{Z}$, is a nonempty, proper subset of $\mathbb{N}^n$, then $P^{\mathrm{i}}(\mathcal{A}) = \{p \in P \mid \mathbf{A}(p) = 0\}$, $P^{\downarrow}(\mathcal{A}) = \{p \in P \mid \mathbf{A}(p) > 0\}$, $P^{\uparrow}(\mathcal{A}) = \{p \in P \mid \mathbf{A}(p) < 0\}$ and $P^{\mathrm{d}}(\mathcal{A}) = \emptyset$. With a proposition similar to Proposition 3.2, one verifies that $\mathcal{A} \in \mathcal{O}^{\downarrow}(\mathbb{N}^n, \leq)$ if and only if $P^{\uparrow}(\mathcal{A}) = \emptyset$ and $P^{\mathrm{d}}(\mathcal{A}) = \emptyset$, and that $\mathcal{A} \in \mathcal{O}^{\uparrow}(\mathbb{N}^n, \leq)$ if and only if $P^{\downarrow}(\mathcal{A}) = \emptyset$ and $P^{\mathrm{d}}(\mathcal{A}) = \emptyset$.

Proposition 4.3. *Let $\mathcal{A} \subseteq \mathbb{N}^n, \mathbf{m}_1 \in \mathcal{A}$ and $\mathbf{m}_2 \in \mathbb{N}^n$. The following three statements hold:*

(i) $\mathbf{m}_1|_{P \setminus P^{\mathrm{i}}(\mathcal{A})} = \mathbf{m}_2|_{P \setminus P^{\mathrm{i}}(\mathcal{A})} \Rightarrow \mathbf{m}_2 \in \mathcal{A}$,

(ii) $\mathbf{m}_1|_{P^{\downarrow}(\mathcal{A})} \geq \mathbf{m}_2|_{P^{\downarrow}(\mathcal{A})}$ *and* $\mathbf{m}_1|_{P \setminus P^{\downarrow}(\mathcal{A})} = \mathbf{m}_2|_{P \setminus P^{\downarrow}(\mathcal{A})} \Rightarrow \mathbf{m}_2 \in \mathcal{A}$,

(iii) $\mathbf{m}_1|_{P^{\uparrow}(\mathcal{A})} \leq \mathbf{m}_2|_{P^{\uparrow}(\mathcal{A})}$ *and* $\mathbf{m}_1|_{P \setminus P^{\uparrow}(\mathcal{A})} = \mathbf{m}_2|_{P \setminus P^{\uparrow}(\mathcal{A})} \Rightarrow \mathbf{m}_2 \in \mathcal{A}$.

The proof is similar to that of Proposition 3.4 (p. 59), but is given here for the sake of completeness.

Proof. (i) It suffices to consider the case $P^{\mathrm{i}}(\mathcal{A}) \neq \emptyset$. By definition, $p \in P^{\mathrm{i}}(\mathcal{A})$ if and only if $\mathcal{A} \in \mathcal{O}^{\uparrow}(\mathbb{N}^n, \leq_p) \cap \mathcal{O}^{\downarrow}(\mathbb{N}^n, \leq_p)$ or, with Definition 1.1:

$$\begin{aligned} &\forall \bar{\mathbf{m}}_1 \in \mathcal{A}, \forall \bar{\mathbf{m}}_2 \in \mathbb{N}^n : (\bar{\mathbf{m}}_1 \leq_p \bar{\mathbf{m}}_2 \Rightarrow \bar{\mathbf{m}}_2 \in \mathcal{A}) \\ &\qquad\qquad \wedge (\bar{\mathbf{m}}_1 \geq_p \bar{\mathbf{m}}_2 \Rightarrow \bar{\mathbf{m}}_2 \in \mathcal{A}), \\ \Leftrightarrow\ &\forall \bar{\mathbf{m}}_1 \in \mathcal{A}, \forall \bar{\mathbf{m}}_2 \in \mathbb{N}^n : (\bar{\mathbf{m}}_1 \leq_p \bar{\mathbf{m}}_2 \vee \bar{\mathbf{m}}_1 \geq_p \bar{\mathbf{m}}_2) \Rightarrow \bar{\mathbf{m}}_2 \in \mathcal{A}. \end{aligned}$$

With Definition 4.1 this is equivalent to

$$\begin{aligned} \forall \bar{\mathbf{m}}_1 \in \mathcal{A}, \forall \bar{\mathbf{m}}_2 \in \mathbb{N}^n : [(&\bar{\mathbf{m}}_1(p) \leq \bar{\mathbf{m}}_2(p) \wedge \bar{\mathbf{m}}_1|_{P\setminus\{p\}} = \bar{\mathbf{m}}_2|_{P\setminus\{p\}}) \vee \\ &(\bar{\mathbf{m}}_1(p) \geq \bar{\mathbf{m}}_2(p) \wedge \bar{\mathbf{m}}_1|_{P\setminus\{p\}} = \bar{\mathbf{m}}_2|_{P\setminus\{p\}})] \Rightarrow \bar{\mathbf{m}}_2 \in \mathcal{A}. \end{aligned}$$

Finally, this is equivalent to:

$$\forall \bar{\mathbf{m}}_1 \in \mathcal{A}, \forall \bar{\mathbf{m}}_2 \in \mathbb{N}^n : \bar{\mathbf{m}}_1|_{P\setminus\{p\}} = \bar{\mathbf{m}}_2|_{P\setminus\{p\}} \Rightarrow \bar{\mathbf{m}}_2 \in \mathcal{A}. \tag{4.8}$$

Introduce the notation $P^{\mathrm{i}}(\mathcal{A}) = \{p_1, \ldots, p_k\}$. Define for $\mathbf{m}_1 \in \mathcal{A}$ and $\mathbf{m}_2 \in \mathbb{N}^n$ with $\mathbf{m}_1|_{P\setminus P^{\mathrm{i}}(\mathcal{A})} = \mathbf{m}_2|_{P\setminus P^{\mathrm{i}}(\mathcal{A})}$ the states $\mathbf{m}'_i$ $(i = 1, \ldots, k+1)$ such that $\mathbf{m}'_1 = \mathbf{m}_1$ and that

$$\mathbf{m}'_i|_{P\setminus\{p_{i-1}\}} = \mathbf{m}'_{i-1}|_{P\setminus\{p_{i-1}\}}, \tag{4.9}$$

$$\mathbf{m}'_i(p_{i-1}) = \mathbf{m}_2(p_{i-1}), \tag{4.10}$$

for $i = 2, \ldots, k+1$. This construction is such that $\mathbf{m}'_{k+1} = \mathbf{m}_2$. For each pair of states $\mathbf{m}'_{i-1}$ and $\mathbf{m}'_i$ $(i = 2, \ldots, k+1)$ we can use (4.8) for p_{i-1} (see (4.9)–(4.10)). By induction, we obtain from $\mathbf{m}_1 = \mathbf{m}'_1 \in \mathcal{A}$ that $\mathbf{m}_2 = \mathbf{m}'_{k+1} \in \mathcal{A}$.

(ii) and (iii) can be proven in a similar way. □

Proposition 4.4 is essential for the reduction theorems of this chapter. Its proof is based on Proposition 4.3.

Proposition 4.4. *Let* $\mathcal{A} \subseteq \mathbb{N}^n$, $\mathbf{m}_1 \in \mathcal{A}$ *and* $\mathbf{m}_2 \in \mathbb{N}^n$. *If*

$$\begin{aligned} \mathbf{m}_1|_{P^{\downarrow}(\mathcal{A})} &\geq \mathbf{m}_2|_{P^{\downarrow}(\mathcal{A})}, \\ \mathbf{m}_1|_{P^{\uparrow}(\mathcal{A})} &\leq \mathbf{m}_2|_{P^{\uparrow}(\mathcal{A})}, \\ \mathbf{m}_1|_{P^{\mathrm{d}}(\mathcal{A})} &= \mathbf{m}_2|_{P^{\mathrm{d}}(\mathcal{A})}, \end{aligned}$$

then $\mathbf{m}_2 \in \mathcal{A}$.

Note that there is no condition on $\mathbf{m}_2|_{P^{\mathrm{i}}(\mathcal{A})}$.

Proof. Define $\mathbf{m}_3 \in \mathbb{N}^n$ such that

$$\mathbf{m}_3|_{P \setminus P^{\downarrow}(\mathcal{A})} = \mathbf{m}_1|_{P \setminus P^{\downarrow}(\mathcal{A})} \quad \text{and} \quad \mathbf{m}_3|_{P^{\downarrow}(\mathcal{A})} = \mathbf{m}_2|_{P^{\downarrow}(\mathcal{A})}.$$

The latter condition implies that $\mathbf{m}_3|_{P^{\downarrow}(\mathcal{A})} \leq \mathbf{m}_1|_{P^{\downarrow}(\mathcal{A})}$. After applying Proposition 4.3(ii) to $\mathbf{m}_1$ and $\mathbf{m}_3$, we obtain that $\mathbf{m}_3 \in \mathcal{A}$.

Next, define $\mathbf{m}_4 \in \mathbb{N}^n$ such that

$$\mathbf{m}_4|_{P \setminus P^{\uparrow}(\mathcal{A})} = \mathbf{m}_3|_{P \setminus P^{\uparrow}(\mathcal{A})} \quad \text{and} \quad \mathbf{m}_4|_{P^{\uparrow}(\mathcal{A})} = \mathbf{m}_2|_{P^{\uparrow}(\mathcal{A})}.$$

Consequently, it is true that $\mathbf{m}_4|_{P^{\uparrow}(\mathcal{A})} \geq \mathbf{m}_3|_{P^{\uparrow}(\mathcal{A})}$. If we apply Proposition 4.3(iii) to $\mathbf{m}_3$ and $\mathbf{m}_4$, we obtain that $\mathbf{m}_4 \in \mathcal{A}$.

Note that $\mathbf{m}_2|_{P \setminus P^{\mathrm{i}}(\mathcal{A})} = \mathbf{m}_4|_{P \setminus P^{\mathrm{i}}(\mathcal{A})}$, from which, together with Proposition 4.3(i), it follows that $\mathbf{m}_2 \in \mathcal{A}$. □

3. Sets of places and transitions

As already discussed in Section 1, we show in Section 4 that $\mathcal{A}^*$ can be constructed by studying a sub-Petri net $\mathcal{N}^{\mathrm{uc}}_{\mathcal{A}}$ of $\mathcal{N}^{\mathrm{uc}}$. In this section the building blocks of the sub-Petri net $\mathcal{N}^{\mathrm{uc}}_{\mathcal{A}}$ are defined and a number of their properties are formulated. These building blocks are sets of places and sets of uncontrollable transitions.

A path π in a Petri net $\mathcal{N}$ (Definition 1.8, p. 17) is *uncontrollable* if all transitions in π are uncontrollable, i.e. belong to T^{uc}. The set of uncontrollable paths in $\mathcal{N}$ is denoted Π^{uc}.

Definition 4.3. *Consider a Petri net $\mathcal{N} = (P, T, f^+, f^-, T^{\mathrm{uc}})$. Define for each $p \in P$ the sets*

$$\begin{aligned}
\alpha(p) &:= \{p' \in P \mid \exists \pi \in \Pi^{\mathrm{uc}} : {}^*\pi = p', \pi^* = p\},\\
\beta(p) &:= {}^{\bullet}\alpha(p) \cap T^{\mathrm{uc}},\\
\sigma(p) &:= {}^{\bullet}(p^{\bullet} \cap T^{\mathrm{uc}}) \cup \{p\},\\
\sigma'(p) &:= \sigma(p) \setminus \{p\}.
\end{aligned}$$

In words, $\alpha(p)$ is the set of places p' for which there exists an uncontrollable path in $\mathcal{N}$ which starts in p' and ends in p. Further, $\beta(p)$ is the set of uncontrollable input transitions of places in $\alpha(p)$ and $\sigma(p)$ is the union of the set of input places of uncontrollable output transitions of p and the singleton $\{p\}$. The singleton guarantees that $p \in \sigma(p)$ even if p has no uncontrollable output transitions. Analogous to the definition of $\alpha(p)$, it is clear that

$$\beta(p) = \{t \in T^{\mathrm{uc}} \mid \exists \pi \in \Pi^{\mathrm{uc}} : {}^*\pi = t, \pi^* = p\}.$$

For a place p, $\beta(p)$ is the set of uncontrollable transitions t for which there exists an uncontrollable path in $\mathcal{N}$ that starts in t and ends in p.

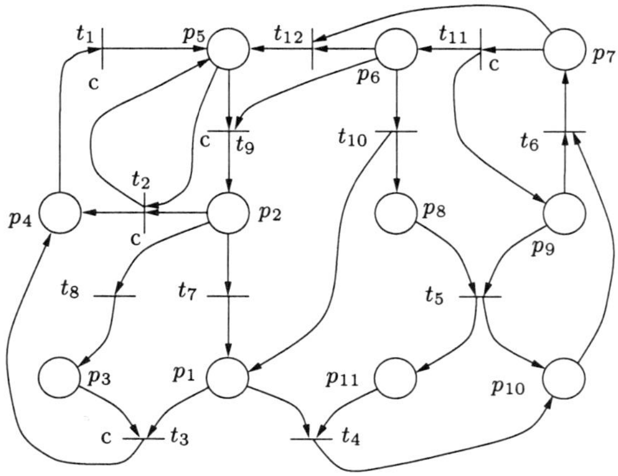

Figure 4.3. The Petri net of Example 4.2.

Example 4.2. For the Petri net in Figure 4.3 one verifies that

$$\begin{aligned} \alpha(p_1) &= \{p_1, p_2, p_6\}, & \alpha(p_{11}) &= \{p_6, p_8, p_9, p_{11}\}, \\ \beta(p_1) &= \{t_7, t_{10}\}, & \beta(p_{11}) &= \{t_5, t_{10}\}, \\ \sigma(p_1) &= \{p_1, p_{11}\}, & \sigma(p_{11}) &= \{p_1, p_{11}\}, \\ \sigma'(p_1) &= \{p_{11}\}, & \sigma'(p_{11}) &= \{p_1\}. \end{aligned}$$

Recall that, in the graphical representation of a Petri net, controllable transitions are indicated with a label 'c'. △

Proposition 4.5. *For each $p \in P$ it is true that*

(i) $p \in \alpha(p)$,

(ii) ${}^\bullet\beta(p) \cup \{p\} = \alpha(p)$.

Proof. (i) Each sequence which only consists of a place p, i.e. which does not contain any transitions, is an uncontrollable path in $\mathcal{N}$.

(ii) First, we show that ${}^\bullet\beta(p) \cup \{p\} \subseteq \alpha(p)$. Because of (i) it suffices to prove that ${}^\bullet\beta(p) \subseteq \alpha(p)$. By definition, for each $t \in \beta(p)$ there exists a $p' \in t^\bullet$ with $p' \in \alpha(p)$. For p' there exists an uncontrollable path $\pi \in \Pi^{uc}$ with ${}^*\pi = p'$ and $\pi^* = p$. Consequently, for each $p'' \in {}^\bullet t$ there exists an uncontrollable path $\pi' = (p'', t, \pi)$ for which it is true that ${}^*\pi' = p''$ and $(\pi')^* = p$.

Next, we show that $\alpha(p) \subseteq {}^\bullet\beta(p) \cup \{p\}$. For each $p_1 \in \alpha(p) \setminus \{p\}$ (the case $p_1 = p$ is trivial) there exists an uncontrollable path $\pi = (p_1, t_1, p_2, \ldots, p)$. The place p_2 belongs to $\alpha(p)$ (consider the uncontrollable path $(p_2, \ldots, p)$ which is a suffix of π). Then it follows from Definition 4.3 that $t_1 \in {}^\bullet p_2 \cap T^{uc} \subseteq \beta(p)$. Therefore, $p_1 \in {}^\bullet t_1 \subseteq {}^\bullet\beta(p)$. □

In the next definition $P^{\uparrow}(\mathcal{A})$ is partitioned into two disjoint sets.

Definition 4.4. *Define for a Petri net $\mathcal{N}$ and $\mathcal{A} \subseteq \mathbb{N}^n$ the sets of places:*

$$P^{\uparrow}_{\mathrm{u}}(\mathcal{A}) = \{p \in P^{\uparrow}(\mathcal{A}) \mid f^{-}(p,t) \leq 1, \forall t \in T^{\mathrm{uc}}\},$$
$$P^{\uparrow}_{\mathrm{d}}(\mathcal{A}) = P^{\uparrow}(\mathcal{A}) \setminus P^{\uparrow}_{\mathrm{u}}(\mathcal{A}).$$

By definition, $P^{\uparrow}(\mathcal{A}) = P^{\uparrow}_{\mathrm{u}}(\mathcal{A}) \dot{\cup} P^{\uparrow}_{\mathrm{d}}(\mathcal{A})$. A place $p \in P^{\uparrow}(\mathcal{A})$ belongs to $P^{\uparrow}_{\mathrm{u}}(\mathcal{A})$ if all arcs from p to uncontrollable output transitions of p have weight one. In contrast to $P^{\uparrow}(\mathcal{A})$, which only depends on the structure of the legal set $\mathcal{A}$, the sets $P^{\uparrow}_{\mathrm{u}}(\mathcal{A})$ and $P^{\uparrow}_{\mathrm{d}}(\mathcal{A})$ also depend on the structure of the Petri net $\mathcal{N}$. Define

$$\bar{P}^{\mathrm{d}}(\mathcal{A}) := P^{\mathrm{d}}(\mathcal{A}) \cup P^{\uparrow}_{\mathrm{d}}(\mathcal{A}). \tag{4.11}$$

For each of the sets $P^{\downarrow}(\mathcal{A})$, $P^{\uparrow}_{\mathrm{u}}(\mathcal{A})$ and $\bar{P}^{\mathrm{d}}(\mathcal{A})$ we now define a set of places and a set of uncontrollable transitions.

Definition 4.5. *Consider a Petri net $\mathcal{N}$ and a legal set $\mathcal{A} \subseteq \mathbb{N}^n$. Then we define:*

$$I^{\downarrow}(\mathcal{A}) := \bigcup_{p \in P^{\downarrow}(\mathcal{A})} \alpha(p) \subseteq P,$$

$$U^{\downarrow}(\mathcal{A}) := \bigcup_{p \in P^{\downarrow}(\mathcal{A})} \beta(p) \subseteq T^{\mathrm{uc}},$$

$$I^{\uparrow}(\mathcal{A}) := \bigcup_{p \in P^{\uparrow}_{\mathrm{u}}(\mathcal{A})} \left[\left(\bigcup_{p' \in \sigma'(p)} \alpha(p')\right) \cup \{p\}\right] \subseteq P,$$

$$U^{\uparrow}(\mathcal{A}) := \bigcup_{p \in P^{\uparrow}_{\mathrm{u}}(\mathcal{A})} \left[\left(\bigcup_{p' \in \sigma'(p)} \beta(p')\right) \cup (p^{\bullet} \cap T^{\mathrm{uc}})\right] \subseteq T^{\mathrm{uc}},$$

$$I^{\mathrm{d}}(\mathcal{A}) := \bigcup_{p \in \bar{P}^{\mathrm{d}}(\mathcal{A})} \bigcup_{p' \in \sigma(p)} \alpha(p') \subseteq P,$$

$$U^{\mathrm{d}}(\mathcal{A}) := \bigcup_{p \in \bar{P}^{\mathrm{d}}(\mathcal{A})} \left[\left(\bigcup_{p' \in \sigma(p)} \beta(p')\right) \cup (p^{\bullet} \cap T^{\mathrm{uc}})\right] \subseteq T^{\mathrm{uc}}.$$

The sets $I^{\cdot}(\mathcal{A})$ are sets of places. The sets $U^{\cdot}(\mathcal{A})$ are sets of uncontrollable transitions. Although the definition of these six sets may look complicated, they can be easily visualised. This is done in Figure 4.4, where abstraction is made of a Petri net in the following way.

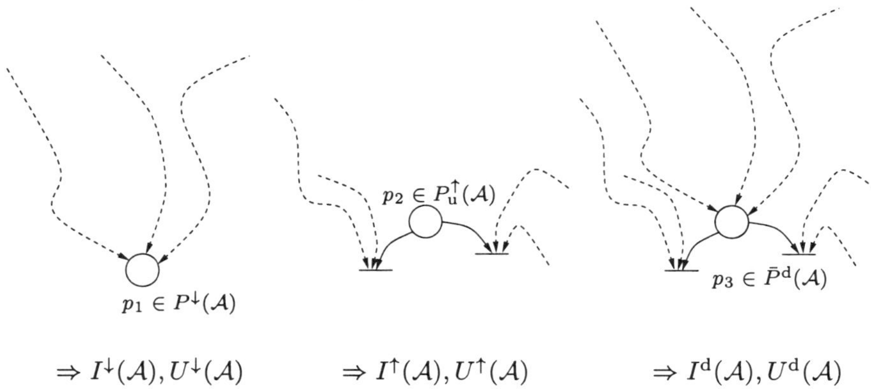

Figure 4.4. Illustration of the sets of Definition 4.5.

The dashed arcs represent uncontrollable paths in $\mathcal{N}$ with an orientation as indicated. Together with the description of the sets $\alpha(\cdot)$, $\beta(\cdot)$, $\sigma(\cdot)$ and $\sigma'(\cdot)$ that follows Definition 4.3, the sets in Definition 4.5 become intuitively clear. In the left most scheme of Figure 4.4, places and uncontrollable transitions in the union of paths over all $p_1 \in P^{\downarrow}(\mathcal{A})$ form $I^{\downarrow}(\mathcal{A})$ and $U^{\downarrow}(\mathcal{A})$, respectively. In the middle, the indicated places and uncontrollable transitions over all $p_2 \in P_{\mathrm{u}}^{\uparrow}(\mathcal{A})$ form $I^{\uparrow}(\mathcal{A})$ and $U^{\uparrow}(\mathcal{A})$, respectively. At the right, the indicated places and uncontrollable transitions over all $p_3 \in \bar{P}^{\mathrm{d}}(\mathcal{A})$ form $I^{\mathrm{d}}(\mathcal{A})$ and $U^{\mathrm{d}}(\mathcal{A})$, respectively. As suggested by the schemes one easily verifies that (see Definition 4.5):

$$\alpha(p) \cup \left[\left(\bigcup_{p' \in \sigma'(p)} \alpha(p') \right) \cup \{p\} \right] = \bigcup_{p' \in \sigma(p)} \alpha(p')$$

and that

$$\beta(p) \cup \left[\left(\bigcup_{p' \in \sigma'(p)} \beta(p') \right) \cup (p^{\bullet} \cap T^{\mathrm{uc}}) \right] = \left(\bigcup_{p' \in \sigma(p)} \beta(p') \right) \cup (p^{\bullet} \cap T^{\mathrm{uc}}).$$

The uncontrollable sub-Petri net $\mathcal{N}_{\mathcal{A}}^{\mathrm{uc}}$, which is found in Section 4.2 below as a consequence of the main reduction theorem, consists of the sets $I^{\downarrow}(\mathcal{A})$ and $U^{\downarrow}(\mathcal{A})$ (corresponding to places in $P^{\downarrow}(\mathcal{A})$), $I^{\uparrow}(\mathcal{A})$ and $U^{\uparrow}(\mathcal{A})$ (corresponding to places in $P_{\mathrm{u}}^{\uparrow}(\mathcal{A})$), and $I^{\mathrm{d}}(\mathcal{A})$ and $U^{\mathrm{d}}(\mathcal{A})$ (corresponding to places in $\bar{P}^{\mathrm{d}}(\mathcal{A})$). Proposition 4.6 follows from Definitions 4.3 and 4.5 and Proposition 4.5.

Proposition 4.6. *For each $\mathcal{A} \subseteq \mathbb{N}^n$ it is true that*

(i) ${}^\bullet U^{\downarrow}(\mathcal{A}) \cup P^{\downarrow}(\mathcal{A}) = I^{\downarrow}(\mathcal{A})$,
(ii) ${}^\bullet I^{\downarrow}(\mathcal{A}) \cap T^{\mathrm{uc}} = U^{\downarrow}(\mathcal{A})$,
(iii) ${}^\bullet U^{\uparrow}(\mathcal{A}) \cup P^{\uparrow}_{\mathrm{u}}(\mathcal{A}) = I^{\uparrow}(\mathcal{A})$,
(iv) ${}^\bullet U^{\mathrm{d}}(\mathcal{A}) \cup \bar{P}^{\mathrm{d}}(\mathcal{A}) = I^{\mathrm{d}}(\mathcal{A})$,
(v) ${}^\bullet I^{\mathrm{d}}(\mathcal{A}) \cap T^{\mathrm{uc}} \subseteq U^{\mathrm{d}}(\mathcal{A})$.

Proof.

(i)
$$\begin{aligned} & {}^\bullet U^{\downarrow}(\mathcal{A}) \cup P^{\downarrow}(\mathcal{A}) & \\ &= \textstyle\bigcup_{p\in P^{\downarrow}(\mathcal{A})} ({}^\bullet\beta(p) \cup \{p\}) & \text{Def. 4.5,} \\ &= \textstyle\bigcup_{p\in P^{\downarrow}(\mathcal{A})} \alpha(p) & \text{Prop. 4.5,} \\ &= I^{\downarrow}(\mathcal{A}) & \text{Def. 4.5.} \end{aligned}$$

(ii)
$$\begin{aligned} & {}^\bullet I^{\downarrow}(\mathcal{A}) \cap T^{\mathrm{uc}} & \\ &= \textstyle\bigcup_{p\in P^{\downarrow}(\mathcal{A})} ({}^\bullet\alpha(p) \cap T^{\mathrm{uc}}) & \text{Def. 4.5,} \\ &= \textstyle\bigcup_{p\in P^{\downarrow}(\mathcal{A})} \beta(p) & \text{Def. 4.3,} \\ &= U^{\downarrow}(\mathcal{A}) & \text{Def. 4.5.} \end{aligned}$$

(iii)
$$\begin{aligned} & {}^\bullet U^{\uparrow}(\mathcal{A}) \cup P^{\uparrow}_{\mathrm{u}}(\mathcal{A}) & \\ &= \textstyle\bigcup_{p\in P^{\uparrow}_{\mathrm{u}}(\mathcal{A})} [(\bigcup_{p'\in\sigma'(p)} {}^\bullet\beta(p')) \cup {}^\bullet(p^\bullet \cap T^{\mathrm{uc}}) \cup \{p\}] & \text{Def. 4.5,} \\ &= \textstyle\bigcup_{p\in P^{\uparrow}_{\mathrm{u}}(\mathcal{A})} [(\bigcup_{p'\in\sigma'(p)} {}^\bullet\beta(p')) \cup (\bigcup_{p'\in\sigma'(p)} \{p'\}) \cup \{p\}] & \text{Def. 4.3,} \\ &= \textstyle\bigcup_{p\in P^{\uparrow}_{\mathrm{u}}(\mathcal{A})} [(\bigcup_{p'\in\sigma'(p)} \alpha(p')) \cup \{p\}] & \text{Prop. 4.5,} \\ &= I^{\uparrow}(\mathcal{A}) & \text{Def. 4.5.} \end{aligned}$$

(iv)
$$\begin{aligned} & {}^\bullet U^{\mathrm{d}}(\mathcal{A}) \cup \bar{P}^{\mathrm{d}}(\mathcal{A}) & \\ &= \textstyle\bigcup_{p\in \bar{P}^{\mathrm{d}}(\mathcal{A})} [(\bigcup_{p'\in\sigma(p)} {}^\bullet\beta(p')) \cup {}^\bullet(p^\bullet \cap T^{\mathrm{uc}}) \cup \{p\}] & \text{Def. 4.5,} \\ &= \textstyle\bigcup_{p\in \bar{P}^{\mathrm{d}}(\mathcal{A})} [\bigcup_{p'\in\sigma(p)} ({}^\bullet\beta(p') \cup \{p'\})] & \text{Def. 4.3,} \\ &= \textstyle\bigcup_{p\in \bar{P}^{\mathrm{d}}(\mathcal{A})} \bigcup_{p'\in\sigma(p)} \alpha(p') & \text{Prop. 4.5,} \\ &= I^{\mathrm{d}}(\mathcal{A}) & \text{Def. 4.5.} \end{aligned}$$

(v)
$$\begin{aligned} & {}^\bullet I^{\mathrm{d}}(\mathcal{A}) \cap T^{\mathrm{uc}} & \\ &= \textstyle\bigcup_{p\in \bar{P}^{\mathrm{d}}(\mathcal{A})} \bigcup_{p'\in\sigma(p)} ({}^\bullet\alpha(p') \cap T^{\mathrm{uc}}) & \text{Def. 4.5,} \\ &= \textstyle\bigcup_{p\in \bar{P}^{\mathrm{d}}(\mathcal{A})} \bigcup_{p'\in\sigma(p)} \beta(p') & \text{Def. 4.3,} \\ &\subseteq U^{\mathrm{d}}(\mathcal{A}) & \text{Def. 4.5.} \end{aligned}$$

□

There is no result for ${}^\bullet I^\uparrow(\mathcal{A})$. Because of $p \notin \sigma'(p)$ it is not necessarily true that all uncontrollable input transitions of $p \in P_\mathrm{u}^\uparrow(\mathcal{A})$ belong to $U^\uparrow(\mathcal{A})$ (see also Figure 4.4).

Define the set of places

$$\bar{I}(\mathcal{A}) := I^\downarrow(\mathcal{A}) \cup I^\uparrow(\mathcal{A}) \cup I^\mathrm{d}(\mathcal{A}), \tag{4.12}$$

and the set of uncontrollable transitions

$$\bar{U}(\mathcal{A}) := U^\downarrow(\mathcal{A}) \cup U^\uparrow(\mathcal{A}) \cup U^\mathrm{d}(\mathcal{A}). \tag{4.13}$$

Proposition 4.7. *For each $\mathcal{A} \subseteq \mathbb{N}^n$ it holds that*

$${}^\bullet\bar{U}(\mathcal{A}) \cup P^\downarrow(\mathcal{A}) \cup P^\uparrow(\mathcal{A}) \cup P^\mathrm{d}(\mathcal{A}) = \bar{I}(\mathcal{A}).$$

Proof. This follows from Proposition 4.6, the definitions of $\bar{I}(\mathcal{A})$ and $\bar{U}(\mathcal{A})$ in (4.12) and (4.13), and $P_\mathrm{u}^\uparrow(\mathcal{A}) \cup \bar{P}^\mathrm{d}(\mathcal{A}) = P^\uparrow(\mathcal{A}) \cup P^\mathrm{d}(\mathcal{A})$:

$$\begin{aligned} &{}^\bullet U^\downarrow(\mathcal{A}) \cup {}^\bullet U^\uparrow(\mathcal{A}) \cup {}^\bullet U^\mathrm{d}(\mathcal{A}) \cup P^\downarrow(\mathcal{A}) \cup P^\uparrow(\mathcal{A}) \cup P^\mathrm{d}(\mathcal{A}) \\ &\qquad\qquad = I^\downarrow(\mathcal{A}) \cup I^\uparrow(\mathcal{A}) \cup I^\mathrm{d}(\mathcal{A}). \end{aligned}$$

□

Example 4.3. Consider the Petri net in Figure 4.3 on page 76 and three legal sets $\mathcal{A}_1$, $\mathcal{A}_2$ and $\mathcal{A}_3$. If

$$\begin{aligned} P^\downarrow(\mathcal{A}_1) &= \{p_1\}, & P_\mathrm{u}^\uparrow(\mathcal{A}_1) &= \emptyset, & \bar{P}^\mathrm{d}(\mathcal{A}_1) &= \emptyset, \\ P^\downarrow(\mathcal{A}_2) &= \emptyset, & P_\mathrm{u}^\uparrow(\mathcal{A}_2) &= \{p_1\}, & \bar{P}^\mathrm{d}(\mathcal{A}_2) &= \emptyset, \\ P^\downarrow(\mathcal{A}_3) &= \emptyset, & P_\mathrm{u}^\uparrow(\mathcal{A}_3) &= \emptyset, & \bar{P}^\mathrm{d}(\mathcal{A}_3) &= \{p_1\}, \end{aligned}$$

then we have that

$$\begin{aligned} \bar{I}(\mathcal{A}_1) &= I^\downarrow(\mathcal{A}_1) = \alpha(p_1) = \{p_1, p_2, p_6\}, \\ \bar{U}(\mathcal{A}_1) &= U^\downarrow(\mathcal{A}_1) = \beta(p_1) = \{t_7, t_{10}\}, \\ \bar{I}(\mathcal{A}_2) &= I^\uparrow(\mathcal{A}_2) = \alpha(p_{11}) \cup \{p_1\} = \{p_1, p_6, p_8, p_9, p_{11}\}, \\ \bar{U}(\mathcal{A}_2) &= U^\uparrow(\mathcal{A}_2) = \beta(p_{11}) \cup \{t_4\} = \{t_4, t_5, t_{10}\}, \\ \bar{I}(\mathcal{A}_3) &= I^\mathrm{d}(\mathcal{A}_3) = \alpha(p_1) \cup \alpha(p_{11}) = \{p_1, p_2, p_6, p_8, p_9, p_{11}\}, \\ \bar{U}(\mathcal{A}_3) &= U^\mathrm{d}(\mathcal{A}_3) = \beta(p_1) \cup \beta(p_{11}) \cup \{t_4\} = \{t_4, t_5, t_7, t_{10}\}. \end{aligned}$$

△

We end this section with two technical properties which are used in Section 4. Define the sets of places

$$V_1(\mathcal{A}) := I^{\downarrow}(\mathcal{A}) \cup I^{\mathrm{d}}(\mathcal{A}) \cup \left(\bigcup_{p \in P_{\mathrm{u}}^{\uparrow}(\mathcal{A})} \bigcup_{p' \in \sigma'(p)} \alpha(p') \right), \tag{4.14}$$

$$V_2(\mathcal{A}) := P_{\mathrm{u}}^{\uparrow}(\mathcal{A}) \setminus V_1(\mathcal{A}). \tag{4.15}$$

These sets are disjoint. Their union is equal to $\bar{I}(\mathcal{A})$ because

$$\left(\cup_{p \in P_{\mathrm{u}}^{\uparrow}(\mathcal{A})} \cup_{p' \in \sigma'(p)} \alpha(p') \right) \cup P_{\mathrm{u}}^{\uparrow}(\mathcal{A}) = I^{\uparrow}(\mathcal{A}).$$

With Definition 4.5, $V_1(\mathcal{A})$ is the set of places which belong to an uncontrollable path that ends in one of the places in

$$P^{\downarrow}(\mathcal{A}) \cup \left(\cup_{p \in \bar{P}^{\mathrm{d}}(\mathcal{A})} \sigma(p) \right) \cup \left(\cup_{p \in P_{\mathrm{u}}^{\uparrow}(\mathcal{A})} \sigma'(p) \right).$$

Define the set of uncontrollable transitions

$$W(\mathcal{A}) := V_2(\mathcal{A})^{\bullet} \cap T^{\mathrm{uc}}. \tag{4.16}$$

Proposition 4.8. *Let $p_1, p_2 \in V_2(\mathcal{A})$. If $p_1 \neq p_2$, then $p_1^{\bullet} \cap p_2^{\bullet} \cap T^{\mathrm{uc}} = \emptyset$.*

Proof. Suppose that $p_1^{\bullet} \cap p_2^{\bullet} \cap T^{\mathrm{uc}} \neq \emptyset$. As $p_1 \neq p_2$, we have that $p_1 \in \sigma'(p_2)$ and $p_2 \in \sigma'(p_1)$. Because of $p_1, p_2 \in P_{\mathrm{u}}^{\uparrow}(\mathcal{A})$ and Proposition 4.5(i), this implies that p_1 and p_2 belong to $\cup_{p \in P_{\mathrm{u}}^{\uparrow}(\mathcal{A})} \cup_{p' \in \sigma'(p)} \alpha(p') \subseteq V_1(\mathcal{A})$. This contradicts $p_1, p_2 \in V_2(\mathcal{A})$. □

In words, two different places in $V_2(\mathcal{A})$ do not have any uncontrollable output transitions in common. With the notation

$$V_2(\mathcal{A}) = \{q_1, \ldots, q_z\}, \tag{4.17}$$

Proposition 4.8 allows to introduce for each $q_j \in V_2(\mathcal{A})$ the notation

$$q_j^{\bullet} \cap T^{\mathrm{uc}} = \{t_{j,1}, \ldots, t_{j,\nu(j)}\}. \tag{4.18}$$

From (4.16) it follows then that

$$W(\mathcal{A}) = \{t_{1,1}, \ldots, t_{1,\nu(1)}, \ldots, t_{z,1}, \ldots, t_{z,\nu(z)}\}.$$

Proposition 4.9. *It is true that*

(i) $\bar{P}^{\mathrm{d}}(\mathcal{A})^{\bullet} \cap W(\mathcal{A}) = \emptyset$,

(ii) ${}^{\bullet}V_1(\mathcal{A}) \cap T^{\mathrm{uc}} \subseteq \bar{U}(\mathcal{A}) \setminus W(\mathcal{A})$,

(iii) $(P_{\mathrm{u}}^{\uparrow}(\mathcal{A}) \cap V_1(\mathcal{A}))^{\bullet} \cap T^{\mathrm{uc}} \subseteq \bar{U}(\mathcal{A}) \setminus W(\mathcal{A})$.

Proof. (i) By contradiction. Suppose that there exists a transition $t \in \bar{P}^{\mathrm{d}}(\mathcal{A})^{\bullet} \cap W(\mathcal{A})$. As t belongs to $W(\mathcal{A})$, there exists a place $p \in V_2(\mathcal{A})$ such that $p \in {}^{\bullet}t$. Because of $t \in \bar{P}^{\mathrm{d}}(\mathcal{A})^{\bullet}$ there exists a place $p' \in \bar{P}^{\mathrm{d}}(\mathcal{A})$ such that $p' \in {}^{\bullet}t$. As t is uncontrollable, it holds that $p \in \sigma(p')$ and that $p \in I^{\mathrm{d}}(\mathcal{A})$. This contradicts $p \in V_2(\mathcal{A})$.

(ii) By contradiction. From Definition 4.5 and Proposition 4.5 and 4.6 it follows that ${}^{\bullet}V_1(\mathcal{A}) \cap T^{\mathrm{uc}} \subseteq \bar{U}(\mathcal{A})$. Suppose that there exists a place $p \in V_1(\mathcal{A})$ and a transition $t \in W(\mathcal{A})$ such that $t \in {}^{\bullet}p \cap T^{\mathrm{uc}}$. Consequently, there exists a place $p' \in V_2(\mathcal{A})$ with $p' \in {}^{\bullet}t$. Therefore, p' belongs to $\alpha(p)$. As $\alpha(\bar{p}) \subseteq V_1(\mathcal{A})$ for each $\bar{p} \in V_1(\mathcal{A})$, it is true that $p' \in V_1(\mathcal{A})$. This leads to a contradiction.

(iii) By contradiction. From Definition 4.5 it follows that $P_{\mathrm{u}}^{\uparrow}(\mathcal{A})^{\bullet} \cap T^{\mathrm{uc}} \subseteq \bar{U}(\mathcal{A})$. Suppose that there exists a transition t in $(P_{\mathrm{u}}^{\uparrow}(\mathcal{A}) \cap V_1(\mathcal{A}))^{\bullet} \cap T^{\mathrm{uc}}$ which also belongs to $W(\mathcal{A}) = V_2(\mathcal{A})^{\bullet} \cap T^{\mathrm{uc}}$. Then there exists a place $p \in V_2(\mathcal{A}) \cap {}^{\bullet}t$ and a place $p' \in P_{\mathrm{u}}^{\uparrow}(\mathcal{A}) \cap V_1(\mathcal{A}) \in {}^{\bullet}t$. As $p' \in P_{\mathrm{u}}^{\uparrow}(\mathcal{A})$, we obtain that $p \in V_1(\mathcal{A})$. This contradicts $p \in V_2(\mathcal{A})$. □

4. Reduction result for $\mathcal{A}^*$

4.1. Reduction theorem

For ease of notation, from now on the argument '$(\mathcal{A})$' of the sets of the previous sections is suppressed when no confusion can arise. Theorem 4.1 is the main result of this chapter.

Theorem 4.1 (Reduction theorem). *For each $\mathcal{A} \subseteq \mathbb{N}^n$ and each $\mathbf{m} \in \mathbb{N}^n$ it holds that*

$$\mathcal{R}_{\infty}^{T^{\mathrm{uc}}}(\mathbf{m}) \subseteq \mathcal{A} \quad \Leftrightarrow \quad \mathcal{R}_{\infty}^{\bar{U}}(\mathbf{m}) \subseteq \mathcal{A}.$$

In words, each state in $\mathcal{R}_{\infty}^{T^{\mathrm{uc}}}(\mathbf{m})$ belongs to $\mathcal{A}$ if and only if each state which is reachable via transitions in $\bar{U}$ (see (4.13)) belongs to $\mathcal{A}$. If $\bar{U}$ is a proper subset of T^{uc}, i.e. if $\bar{U} \subset T^{\mathrm{uc}}$, $\mathcal{A}^*$ can be constructed more efficiently with this theorem than by directly applying its definition (Definition 3.4). Because of Proposition 1.19 (p. 25), it is of no importance which concurrency assumption is used in the proof of this theorem. Unless explicitly stated, the transition bag assumption is used, i.e. $\Omega = \Delta$. Before proving Theorem 4.1 we formulate and prove auxiliary results in Lemma 4.1 and Proposition 4.10.

Lemma 4.1. *Let $\mathbf{m}_1, \mathbf{m}_1' \in \mathbb{N}^n$, $\boldsymbol{\delta} \in \Delta$ and $\mathcal{A} \subseteq \mathbb{N}^n$ with*

$$\boldsymbol{\delta}|_{T^{\mathrm{c}}} = \mathbf{0}, \tag{4.19}$$

$$\mathbf{m}_1 \xrightarrow{\boldsymbol{\delta}} \mathbf{m}_2, \tag{4.20}$$

$$\mathbf{m}'_1|_{V_1} \geq \mathbf{m}_1|_{V_1}, \tag{4.21}$$

$$\mathbf{m}'_1|_{P_\mathrm{u}^\uparrow} \leq \mathbf{m}_1|_{P_\mathrm{u}^\uparrow}. \tag{4.22}$$

Further, consider the vector $\boldsymbol{\delta}' \in \mathbb{Z}^m$ for which (with the notation that precedes Proposition 4.9):

$$\boldsymbol{\delta}'|_{\bar{U}\setminus W} = \boldsymbol{\delta}|_{\bar{U}\setminus W}, \tag{4.23}$$

$$\boldsymbol{\delta}'|_{T\setminus\bar{U}} = \mathbf{0}, \tag{4.24}$$

$$\boldsymbol{\delta}'(t_{j,l}) = \min\left[\boldsymbol{\delta}(t_{j,l}), \mathbf{m}'_1(q_j) - \sum_{v=1}^{l-1} \boldsymbol{\delta}'(t_{j,v})\right] \quad (t_{j,l} \in W; j = 1, \ldots, z; l = 1, \ldots, \nu(j)). \tag{4.25}$$

Then it holds that

$$\mathbf{m}'_1 \xrightarrow{\boldsymbol{\delta}'} \mathbf{m}'_2, \tag{4.26}$$

$$\mathbf{m}'_2|_{V_1} \geq \mathbf{m}_2|_{V_1}, \tag{4.27}$$

$$\mathbf{m}'_2|_{P_\mathrm{u}^\uparrow} \leq \mathbf{m}_2|_{P_\mathrm{u}^\uparrow}. \tag{4.28}$$

Recall that, although $\mathbf{0} \notin \Delta$, we allow the notation $\mathbf{m} \xrightarrow{\mathbf{0}} \mathbf{m}$.

Proof. The proof consists of four parts. In part 1 we show that $\boldsymbol{\delta}' \geq \mathbf{0}$, in part 2 that $\mathbf{m}'_1 \geq \mathbf{F}^-\boldsymbol{\delta}'$ and in parts 3 and 4 that (4.27) and (4.28) hold, respectively.

1. Because of (4.23) and (4.24) it suffices to prove that $\boldsymbol{\delta}'(t_{j,l}) \geq 0$ for all $t_{j,l} \in W$. For $j = 1, \ldots, z$ this is trivial for $l = 1$. Further, it follows from (4.25) for $l = 1, \ldots, \nu(j) - 1$ that

$$\boldsymbol{\delta}'(t_{j,l}) \leq \mathbf{m}'_1(q_j) - \sum_{v=1}^{l-1} \boldsymbol{\delta}'(t_{j,v}),$$

or that

$$0 \leq \mathbf{m}'_1(q_j) - \sum_{v=1}^{l} \boldsymbol{\delta}'(t_{j,v}).$$

From (4.25) we obtain then that $\boldsymbol{\delta}'(t_{j,l+1}) \geq 0$ for $l = 1, \ldots, \nu(j) - 1$. Thus, $\boldsymbol{\delta}' \geq \mathbf{0}$. Note that $\boldsymbol{\delta}' \leq \boldsymbol{\delta}$.

2. The transition vector $\boldsymbol{\delta}'$ is enabled under $\mathbf{m}'_1$ if and only if

$$\mathbf{m}'_1(p) \geq \sum_{t \in p^\bullet} f^-(p,t)\boldsymbol{\delta}'(t) \quad \text{for } p \in P. \tag{4.29}$$

Consider places p in V_1 first. From (4.20) we have that

$$\mathbf{m}_1(p) \geq \sum_{t \in p^\bullet} f^-(p,t)\boldsymbol{\delta}(t) \quad \text{for } p \in P,$$

and, with (4.21), that

$$\mathbf{m}'_1(p) \geq \sum_{t \in p^\bullet} f^-(p,t)\boldsymbol{\delta}(t) \quad \text{for } p \in V_1.$$

Together with $\boldsymbol{\delta}' \leq \boldsymbol{\delta}$ this implies that

$$\mathbf{m}'_1(p) \geq \sum_{t \in p^\bullet} f^-(p,t)\boldsymbol{\delta}'(t) \quad \text{for } p \in V_1.$$

Next, consider places in V_2. As $\boldsymbol{\delta}'|_{T^c} = \mathbf{0}$ (see (4.24)), it suffices to show, with the notation in (4.17) and (4.18), that

$$\mathbf{m}'_1(q_j) \geq \sum_{v=1}^{\nu(j)} f^-(q_j, t_{j,v})\boldsymbol{\delta}'(t_{j,v}) \quad \text{for } q_j \in V_2.$$

Because of $V_2 \subseteq P_{\mathrm{u}}^{\uparrow}$ (see (4.15)) and Definition 4.4 this is equivalent to

$$\mathbf{m}'_1(q_j) \geq \sum_{v=1}^{\nu(j)} \boldsymbol{\delta}'(t_{j,v}) \quad \text{for } q_j \in V_2.$$

This follows from (4.25) for $l = \nu(j)$.

Finally, consider places p in $P \setminus (V_1 \cup V_2) = P \setminus \bar{I}$. Because of ${}^\bullet\bar{U} \subseteq \bar{I}$ (Proposition 4.7) it is true that $p^\bullet \subseteq T \setminus \bar{U}$. From this we obtain, together with (4.24), that $\boldsymbol{\delta}'|_{p^\bullet} = \mathbf{0}$ and thus that

$$\mathbf{m}'_1(p) \geq \sum_{t \in p^\bullet} f^-(p,t)\boldsymbol{\delta}'(t) = 0 \quad \text{for } p \in P \setminus (V_1 \cup V_2).$$

3. From (4.26) it follows that $\mathbf{m}'_2 = \mathbf{m}'_1 + \mathbf{F}\boldsymbol{\delta}'$. Together with Proposition 4.9(ii), (4.21), (4.23) and $\boldsymbol{\delta}|_{T^c} = \boldsymbol{\delta}'|_{T^c} = \mathbf{0}$, this implies that

$$\mathbf{m}'_2(p) \geq \mathbf{m}_1(p) + \sum_{t_1 \in {}^\bullet p} f^-(p,t_1)\boldsymbol{\delta}(t_1) - \sum_{t_2 \in p^\bullet} f^+(t_2,p)\boldsymbol{\delta}'(t_2) \qquad (p \in V_1) \tag{4.30}$$

and, with $\boldsymbol{\delta}' \leq \boldsymbol{\delta}$, that

$$\mathbf{m}_2'(p) \geq \mathbf{m}_1(p) + \sum_{t_1 \in {}^\bullet p} f^+(t_1, p)\boldsymbol{\delta}(t_1) - \sum_{t_2 \in p^\bullet} f^-(p, t_2)\boldsymbol{\delta}(t_2) \quad \text{for } p \in V_1.$$

The right hand side of this inequality is equal to $\mathbf{m}_2(p)$ (see (4.20)). From this (4.27) is obtained.

4. As $\boldsymbol{\delta}' \leq \boldsymbol{\delta}$, we have

$$\mathbf{m}_2 - \mathbf{m}_2' = \mathbf{m}_1 - \mathbf{m}_1' + \mathbf{F}(\boldsymbol{\delta} - \boldsymbol{\delta}') \geq \mathbf{m}_1 - \mathbf{m}_1' - \mathbf{F}^-(\boldsymbol{\delta} - \boldsymbol{\delta}').$$

With Definition 4.4 and $P_{\mathrm{u}}^{\uparrow} \supseteq V_2$ we find that

$$\mathbf{m}_2(q_j) - \mathbf{m}_2'(q_j) \geq \mathbf{m}_1(q_j) - \mathbf{m}_1'(q_j) - \sum_{l=1}^{\nu(j)} [\boldsymbol{\delta}(t_{j,l}) - \boldsymbol{\delta}'(t_{j,l})] \qquad (q_j \in V_2). \quad (4.31)$$

If $\mathbf{m}_1'(q_j) \geq \sum_{v=1}^{\nu(j)} \boldsymbol{\delta}(t_{j,v})$, then it follows by induction from (4.25) that $\boldsymbol{\delta}'(t_{j,l}) = \boldsymbol{\delta}(t_{j,l})$ $(l = 1, \ldots, \nu(j))$. From (4.22) and (4.31) we then find that $\mathbf{m}_2(q_j) \geq \mathbf{m}_2'(q_j)$ for all $q_j \in V_2$.

If

$$\sum_{v=1}^{w} \boldsymbol{\delta}(t_{j,v}) \leq \mathbf{m}_1'(q_j) < \sum_{v=1}^{w+1} \boldsymbol{\delta}(t_{j,v})$$

with $w \in \{0, \ldots, \nu(j) - 1\}$, then it follows by induction from (4.25) that

$$\begin{aligned} \boldsymbol{\delta}'(t_{j,l}) &= \boldsymbol{\delta}(t_{j,l}) && (l = 1, \ldots, w), \\ \boldsymbol{\delta}'(t_{j,w+1}) &= \mathbf{m}_1'(q_j) - \sum_{v=1}^{w} \boldsymbol{\delta}(t_{j,v}), \\ \boldsymbol{\delta}'(t_{j,l}) &= 0 && (l = w + 2, \ldots, \nu(j)). \end{aligned}$$

After substituting this in (4.31), we obtain that $\mathbf{m}_2(q_j) - \mathbf{m}_2'(q_j) \geq \mathbf{m}_1(q_j) - \sum_{l=1}^{\nu(j)} \boldsymbol{\delta}(t_{j,v})$. The right hand side of this inequality is nonnegative because of $\mathbf{m}_1 \geq \mathbf{F}^-\boldsymbol{\delta}$. Hence,

$$\mathbf{m}_2'|_{V_2} \leq \mathbf{m}_2|_{V_2}. \quad (4.32)$$

From Proposition 4.9(iii), (4.23) and (4.30) we obtain that $\mathbf{m}_2'|_{P_{\mathrm{u}}^{\uparrow} \cap V_1} = \mathbf{m}_2|_{P_{\mathrm{u}}^{\uparrow} \cap V_1}$. Together with $P_{\mathrm{u}}^{\uparrow} \supseteq V_2$ and (4.32) this implies (4.28). □

Note that in (4.21) and (4.22) the sets V_1 and $P_{\mathrm{u}}^{\uparrow}$ are not necessarily disjoint.

Proposition 4.10. *For each* $\bar{\mathbf{m}} \in \mathcal{R}_{\infty}^{T^{\mathrm{uc}}}(\mathbf{m})$ *there exists a state* $\bar{\mathbf{m}}'$ *in* $\mathcal{R}_{\infty}^{\bar{U}}(\mathbf{m})$ *such that* $\bar{\mathbf{m}}|_{P^{\downarrow}} \leq \bar{\mathbf{m}}'|_{P^{\downarrow}}$, $\bar{\mathbf{m}}|_{P_{\mathrm{u}}^{\uparrow}} \geq \bar{\mathbf{m}}'|_{P_{\mathrm{u}}^{\uparrow}}$ *and* $\bar{\mathbf{m}}|_{\bar{P}^{\mathrm{d}}} = \bar{\mathbf{m}}'|_{\bar{P}^{\mathrm{d}}}$.

Proof. The proof is constructive. By definition, for all $\bar{\mathbf{m}}$ in $\mathcal{R}_{\infty}^{T^{\mathrm{uc}}}(\mathbf{m})$ there exist transition vectors $\boldsymbol{\delta}_1, \ldots, \boldsymbol{\delta}_k \in \Delta$ with $\boldsymbol{\delta}_i|_{T^{\mathrm{c}}} = \mathbf{0}$ $(i = 1, \ldots, k)$ such that

$$\mathbf{m} \xrightarrow{\boldsymbol{\delta}_1} \mathbf{m}_1 \xrightarrow{\boldsymbol{\delta}_2} \cdots \xrightarrow{\boldsymbol{\delta}_k} \bar{\mathbf{m}} \equiv \mathbf{m}_k. \tag{4.33}$$

We prove that the sequence

$$\mathbf{m} \xrightarrow{\boldsymbol{\delta}_1'} \mathbf{m}_1' \xrightarrow{\boldsymbol{\delta}_2'} \cdots \xrightarrow{\boldsymbol{\delta}_k'} \bar{\mathbf{m}}' \equiv \mathbf{m}_k' \tag{4.34}$$

is valid, with for $i = 1, \ldots, k$ (cfr. (4.23)–(4.25) in Lemma 4.1):

$$\boldsymbol{\delta}_i'|_{\bar{U}\setminus W} = \boldsymbol{\delta}_i|_{\bar{U}\setminus W}, \tag{4.35}$$

$$\boldsymbol{\delta}_i'|_{T\setminus\bar{U}} = \mathbf{0}, \tag{4.36}$$

$$\boldsymbol{\delta}_i'(t_{j,l}) = \min\left[\boldsymbol{\delta}_i(t_{j,l}), \mathbf{m}_{i-1}'(p_j) - \sum_{v=1}^{l-1} \boldsymbol{\delta}_i'(t_{j,v})\right] \quad (t_{j,l} \in W; j = 1, \ldots, z; l = 1, \ldots, \nu(j)). \tag{4.37}$$

The transition vectors $\boldsymbol{\delta}_1$ and $\boldsymbol{\delta}_1'$, and the state $\mathbf{m}$ satisfy the conditions of Lemma 4.1 (see (4.33), (4.34) and (4.35)–(4.37) for $i = 1$). The result of Lemma 4.1, and (4.33) and (4.35)–(4.37) for $i = 2$ make that we can reuse this same lemma for $\boldsymbol{\delta}_2$, $\boldsymbol{\delta}_2'$, $\mathbf{m}_1$ and $\mathbf{m}_1'$. In this way, Lemma 4.1 can be applied k times in succession. The final result is that (4.34) holds and that

$$\mathbf{m}_i'|_{V_1} \geq \mathbf{m}_i|_{V_1}, \tag{4.38}$$

$$\mathbf{m}_i'|_{P_{\mathrm{u}}^{\uparrow}} \leq \mathbf{m}_i|_{P_{\mathrm{u}}^{\uparrow}} \tag{4.39}$$

for $i = 1, \ldots, k$. Equation (4.38) for $i = k$ and $P^{\downarrow} \subseteq I^{\downarrow} \subseteq V_1$ imply that $\bar{\mathbf{m}}|_{P^{\downarrow}} \leq \bar{\mathbf{m}}'|_{P^{\downarrow}}$ and (4.39) for $i = k$ implies that $\bar{\mathbf{m}}|_{P_{\mathrm{u}}^{\uparrow}} \geq \bar{\mathbf{m}}'|_{P_{\mathrm{u}}^{\uparrow}}$. From Definition 4.5 we obtain that $(\bar{P}^{\mathrm{d}})^{\bullet} \cap T^{\mathrm{uc}} \subseteq U^{\mathrm{d}} \subseteq \bar{U}$. Together with Proposition 4.9(i) it then follows that $(\bar{P}^{\mathrm{d}})^{\bullet} \cap T^{\mathrm{uc}} \subseteq \bar{U} \setminus W$. On the other hand, it follows from Proposition 4.9(ii) that ${}^{\bullet}\bar{P}^{\mathrm{d}} \cap T^{\mathrm{uc}} \subseteq {}^{\bullet}I^{\mathrm{d}} \cap T^{\mathrm{uc}} \subseteq {}^{\bullet}V_1 \cap T^{\mathrm{uc}} \subseteq \bar{U} \setminus W$. Consequently, we obtain with (4.35) that $\bar{\mathbf{m}}|_{\bar{P}^{\mathrm{d}}} = \bar{\mathbf{m}}'|_{\bar{P}^{\mathrm{d}}}$. □

Now we are ready to prove Theorem 4.1 (p. 82).

Proof of Theorem 4.1. ($\Rightarrow$) Immediate from Proposition 1.10 (p. 19).

($\Leftarrow$) From Proposition 4.10 and the definition of $P_{\mathrm{u}}^{\uparrow}$ and $\bar{P}^{\mathrm{d}}$ in Definition 4.4 and (4.11) it follows that for each state $\bar{\mathbf{m}}$ in $\mathcal{R}_{\infty}^{T^{\mathrm{uc}}}(\mathbf{m})$ there exists a state $\bar{\mathbf{m}}'$ in $\mathcal{R}_{\infty}^{\bar{U}}(\mathbf{m})$ such that $\bar{\mathbf{m}}|_{P\downarrow} \leq \bar{\mathbf{m}}'|_{P\downarrow}$, $\bar{\mathbf{m}}|_{P\uparrow} \geq \bar{\mathbf{m}}'|_{P\uparrow}$ and $\bar{\mathbf{m}}|_{P^{\mathrm{d}}} = \bar{\mathbf{m}}'|_{P^{\mathrm{d}}}$. As $\bar{\mathbf{m}}'$ belongs to $\mathcal{A}$, we conclude from Proposition 4.4 (p. 74) that $\bar{\mathbf{m}} \in \mathcal{A}$. □

The result of Theorem 4.1 is valid for all Petri nets. They do not need to satisfy any structural conditions.

Corollary 4.1. *Let $\mathcal{A} \subseteq \mathbb{N}^n, \mathbf{m} \in \mathbb{N}^n$ and $T' \subseteq T^{\mathrm{uc}}$. If $T' \supseteq \bar{U}$, then*

$$\mathcal{R}_{\infty}^{T^{\mathrm{uc}}}(\mathbf{m}) \subseteq \mathcal{A} \quad \Leftrightarrow \quad \mathcal{R}_{\infty}^{T'}(\mathbf{m}) \subseteq \mathcal{A}.$$

Proof. Immediate from Proposition 1.10 and Theorem 4.1. □

Corollary 4.2. *Let $\mathcal{A} \subseteq \mathbb{N}^n, \mathbf{m} \in \mathbb{N}^n$ and $T' \subseteq T^{\mathrm{uc}}$. It holds that:*

$$\mathcal{R}_{\infty}^{\bar{U}}(\mathbf{m}) \subseteq \mathcal{A} \quad \Rightarrow \quad \mathcal{R}_{\infty}^{T'}(\mathbf{m}) \subseteq \mathcal{A}.$$

Proof. If $\bar{U} \subseteq T'$ then this result follows from Corollary 4.1 and Theorem 4.1. If $T' \subseteq \bar{U}$, then with Proposition 1.10 it is true that $\mathcal{R}_{\infty}^{T'}(\mathbf{m}) \subseteq \mathcal{R}_{\infty}^{\bar{U}}(\mathbf{m}) \subseteq \mathcal{A}$. □

4.2. Consequences and discussion

4.2.1 Influencing net of the legal set

From the definition of $\mathcal{A}^*$ in Definition 3.4 and Theorem 4.1 we obtain that

$$\mathcal{A}^* = \{\mathbf{m} \in \mathbb{N}^n \mid \mathcal{R}_{\infty}^{\bar{U}}(\mathbf{m}) \subseteq \mathcal{A}\}. \tag{4.40}$$

Therefore, for the construction of $\mathcal{A}^*$ it suffices to determine $\mathcal{R}_{\infty}^{\bar{U}}(\mathbf{m})$ instead of $\mathcal{R}_{\infty}^{T^{\mathrm{uc}}}(\mathbf{m})$. As $\bar{U} \subseteq T^{\mathrm{uc}}$ it is true for all $\mathbf{m} \in \mathbb{N}^n$ that $\mathcal{R}_{\infty}^{\bar{U}}(\mathbf{m}) \subseteq \mathcal{R}_{\infty}^{T^{\mathrm{uc}}}(\mathbf{m})$. To construct the reachable set $\mathcal{R}_{\infty}^{\bar{U}}(\mathbf{m})$ it suffices to consider the sub-Petri net

$$(P, \bar{U}, f^+|_{\bar{U}\times P}, f^-|_{P\times\bar{U}}) \tag{4.41}$$

of the uncontrollable Petri net $\mathcal{N}^{\mathrm{uc}} = (P, T^{\mathrm{uc}}, f^+|_{T^{\mathrm{uc}}\times P}, f^-|_{P\times T^{\mathrm{uc}}})$ in (3.24).

In the Petri net (4.41) we do not need all places. To check whether a transition vector $\boldsymbol{\delta}$ with $\boldsymbol{\delta}|_{T\setminus\bar{U}} = \mathbf{0}$ is enabled, knowledge of the state of places in ${}^\bullet\bar{U}$ suffices. To verify that a state $\mathbf{m}' \in \mathcal{R}_\infty^{\bar{U}}(\mathbf{m})$ belongs to $\mathcal{A}$, it suffices to know, because of Proposition 4.4, the state of places in

$$\tilde{P}(\mathcal{A}) := P^{\downarrow}(\mathcal{A}) \cup P^{\uparrow}(\mathcal{A}) \cup P^{\mathrm{d}}(\mathcal{A}). \tag{4.42}$$

Therefore, the set of places $P \setminus ({}^\bullet\bar{U} \cup \tilde{P}(\mathcal{A}))$, together with the corresponding arcs, can be removed from the Petri net in (4.41).

From Proposition 4.7 we have (again suppressing the argument '$(\mathcal{A})$') that

$$\bar{I} = {}^\bullet\bar{U} \cup \tilde{P}.$$

The conclusion is that, to construct $\mathcal{A}^*$, it suffices to study the (uncontrollable) sub-Petri net

$$\mathcal{N}_{\mathcal{A}}^{\mathrm{uc}} := (\bar{I}, \bar{U}, f^+|_{\bar{U}\times\bar{I}}, f^-|_{\bar{I}\times\bar{U}}) \tag{4.43}$$

of $\mathcal{N}^{\mathrm{uc}}$. We refer to $\mathcal{N}_{\mathcal{A}}^{\mathrm{uc}}$ as the *influencing net* of $\mathcal{A}$. Consequently, if $\mathcal{N}_{\mathcal{A}}^{\mathrm{uc}}$ is a proper sub-Petri net of $\mathcal{N}^{\mathrm{uc}}$, we have simplified the construction of $\mathcal{A}^*$.

The structure of $\mathcal{N}_{\mathcal{A}}^{\mathrm{uc}}$ is of great importance in the remainder of the book. Define $\mathbf{F}_{\mathcal{A}}^+$ and $\mathbf{F}_{\mathcal{A}}^-$ as the incidence matrices of $\mathcal{N}_{\mathcal{A}}^{\mathrm{uc}}$. Their difference is denoted $\mathbf{F}_{\mathcal{A}} := \mathbf{F}_{\mathcal{A}}^+ - \mathbf{F}_{\mathcal{A}}^-$. These matrices have dimension $\sharp\bar{I} \times \sharp\bar{U}$.

The above discussion allows us to formulate an invariance property of the supremal controllable subset of a given legal set.

Proposition 4.11. *Let $\mathcal{A} \subseteq \mathbb{N}^n$. It is true that:*

$$P \setminus \bar{I}(\mathcal{A}) \subseteq P^{\mathrm{i}}(\mathcal{A}^*).$$

Example 4.4. For the Petri net of Figure 4.3 on page 76 and the legal sets $\mathcal{A}_1$ and $\mathcal{A}_2$ of Example 4.2, the influencing nets $\mathcal{N}_{\mathcal{A}_1}^{\mathrm{uc}}$ and $\mathcal{N}_{\mathcal{A}_2}^{\mathrm{uc}}$ are given in Figure 4.5. △

Example 4.5. We revisit the three examples of Section 1. The invariance properties of the legal sets in (4.1), (4.4) and (4.6) were given in Example 4.1 (p. 73). In Figure 4.1(a) (p. 70) one has $\bar{I}(\mathcal{A}_1) = I^{\downarrow}(\mathcal{A}_1) = \alpha(p_1) = \{p_1, p_2\}$, $\bar{U}(\mathcal{A}_1) = U^{\downarrow}(\mathcal{A}_1) = \beta(p_1) = \{t_1\}$ and $V_2(\mathcal{A}_1) = W(\mathcal{A}_1) = \emptyset$. In Figure 4.1(b) one has $\bar{I}(\mathcal{A}_2) = I^{\uparrow}(\mathcal{A}_2) = \alpha(p_3) \cup \{p_1\} = \{p_1, p_3\}$, $\bar{U}(\mathcal{A}_2) = U^{\uparrow}(\mathcal{A}_2) = \beta(p_3) \cup \{t_2\} = \{t_2\}$, $V_2(\mathcal{A}_2) = \{p_1\}$ and $W(\mathcal{A}_2) = \{t_2\}$. In Figure 4.1(c) one has $\bar{I}(\mathcal{A}_3) = I^{\mathrm{d}}(\mathcal{A}_3) = \alpha(p_1) \cup \alpha(p_6) = \{p_1, p_2, p_6\}$, $\bar{U}(\mathcal{A}_3) = U^{\mathrm{d}}(\mathcal{A}_3) = \beta(p_1) \cup \beta(p_6) \cup \{t_3\} = \{t_2, t_3\}$ and $V_2(\mathcal{A}_3) = W(\mathcal{A}_3) = \emptyset$. These sets of places and transitions form the influencing nets indicated in Figure 4.1. △

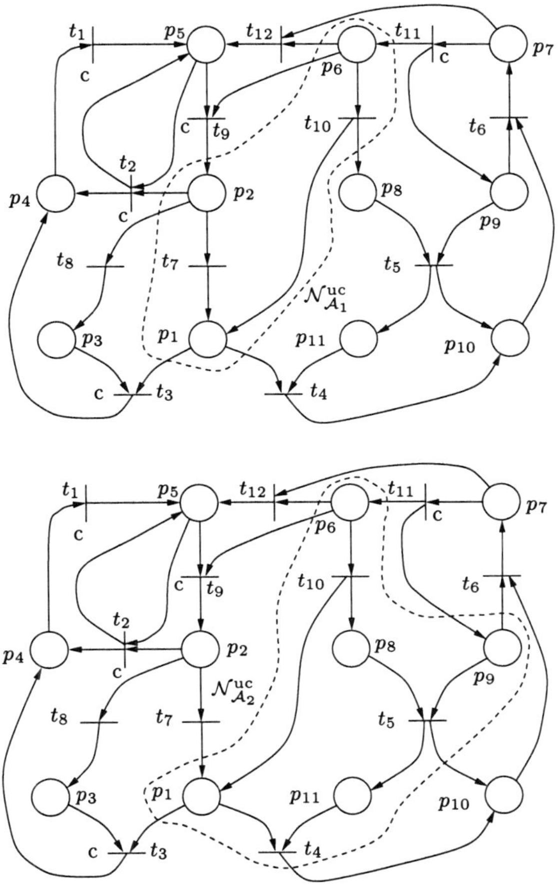

Figure 4.5. The Petri net in Figure 4.3 with the influencing nets $\mathcal{N}^{uc}_{\mathcal{A}_1}$ and $\mathcal{N}^{uc}_{\mathcal{A}_2}$ of $\mathcal{A}_1$ and $\mathcal{A}_2$ in Example 4.2 (p. 76).

It is possible that the invariance properties of the legal set are not known completely, i.e. one only finds an uncontrollable supernet $\hat{\mathcal{N}}^{uc}$ of the influencing net. If this Petri net is a proper sub-Petri net of $\mathcal{N}^{uc}$, then the problem of constructing the supremal controllable subset is made easier with respect to the original problem. Due to the incomplete knowledge of the invariance properties, the profit is not as large as when one would use $\mathcal{N}^{uc}_{\mathcal{A}}$ to construct $\mathcal{A}^*$.

4.2.2 Construction of influencing nets

To construct the influencing net $\mathcal{N}_{\mathcal{A}}^{\mathrm{uc}}$ of a given legal set $\mathcal{A}$, one needs in first instance the set of places $\bar{I}$ and the set of uncontrollable transitions $\bar{U}$. These sets consist of $I^{\downarrow}$, $I^{\uparrow}$ and I^{d}, and $U^{\downarrow}$, $U^{\uparrow}$ and U^{d}, respectively. To construct these six sets, we need, following Definition 4.5 on page 77, the sets

$$\begin{aligned} &\alpha(p) \text{ and } \beta(p) && \text{for } p \in P^{\downarrow} \cup \bigcup_{p' \in P_{\mathrm{u}}^{\uparrow}} \sigma'(p') \cup \bigcup_{p' \in \bar{P}^{\mathrm{d}}} \sigma(p'), && (4.44)\\ &\sigma'(p) && \text{for } p \in P_{\mathrm{u}}^{\uparrow}, \\ &\sigma(p) && \text{for } p \in \bar{P}^{\mathrm{d}}. \end{aligned}$$

Once one has the sets $\alpha(\cdot)$, the sets $\beta(\cdot)$ are obtained immediately (see Definition 4.3). The sets $\sigma(\cdot)$ and $\sigma'(\cdot)$ can be easily constructed as well. The computational effort to construct these sets is negligible. Consequently, in essence, the construction of the influencing net $\mathcal{N}_{\mathcal{A}}^{\mathrm{uc}}$ can be reduced to the construction of the sets $P^{\downarrow}$, $P_{\mathrm{u}}^{\uparrow}$ and $\bar{P}^{\mathrm{d}}$ or, with Definition 4.4 and (4.11), the sets $P^{\downarrow}$, $P^{\uparrow}$ and P^{d}, and the sets $\alpha(\cdot)$ in (4.44). In general, the construction of $P^{\downarrow}$, $P^{\uparrow}$ and P^{d} (see Definition 4.2) for a given legal set, i.e. to determine its invariance properties, is the most difficult part. Section 6 is dedicated to this problem.

Recall from Definition 4.3 that for $p \in P$, $\alpha(p)$ is the set of places p' for which there exists an uncontrollable path that starts in p' and ends in p. Thus, it suffices to consider the uncontrollable sub-Petri net $\mathcal{N}^{\mathrm{uc}}$ of $\mathcal{N}$. Then, p' belongs to $\alpha(p)$ if and only if in $\mathcal{N}^{\mathrm{uc}}$ there exists a path from p' to p. This is a standard problem in graph theory (see for example [29, Chapter 9]). Its complexity is linear.

4.3. Further extensions

In the proof of Lemma 4.1 we used the property that the set $P_{\mathrm{u}}^{\uparrow}(\mathcal{A})$ (see Definition 4.4) only contains places $P^{\uparrow}(\mathcal{A})$ of which the arc weights to uncontrollable output transitions equal one. This condition is essential as is illustrated next.

Consider the uncontrollable Petri net in Figure 4.6. The state $\mathbf{m}_0$ is as indicated. The legal set is $\mathcal{A} = \{\mathbf{m} \in \mathbb{N}^3 \mid \mathbf{m}(p_1) \geq 1\}$. Thus one has $P^{\uparrow} = \{p_1\}$ and $P^{\mathrm{i}} = \{p_2, p_3\}$. We show that attention cannot be limited to states reachable via t_2 only. We have that

$$\min_{\mathbf{m}' \in \mathcal{R}_{\infty}^{t_2}(\mathbf{m}_0)} \mathbf{m}'(p_1) = 1, \tag{4.45}$$

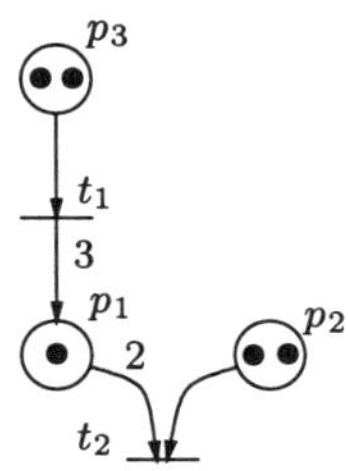

Figure 4.6. An uncontrollable Petri net with initial state $\mathbf{m}_0 = (1, 2, 2)$ to illustrate the definition of $P_{\mathrm{u}}^{\uparrow}(\mathcal{A})$.

because $\mathcal{R}_{\infty}^{t_2}(\mathbf{m}_0) = \{\mathbf{m}_0\}$. On the other hand, via t_1 and t_2, we find that

$$\min_{\mathbf{m}' \in \mathcal{R}_{\infty}^{\{t_1,t_2\}}(\mathbf{m}_0)} \mathbf{m}'(p_1) = 0. \tag{4.46}$$

This minimum, which is smaller than the one in (4.45), is obtained for the state $(0, 0, 1)$, which is reached from $\mathbf{m}_0$ by executing t_1 once and t_2 twice.

This example shows that places p in $P^{\uparrow}$ which have an uncontrollable output transition t with $f^{-}(p, t) > 1$ should be treated like places in P^{d}. This is exactly what we have done in the reduction theorem by defining $\bar{P}^{\mathrm{d}} = P^{\mathrm{d}} \cup P_{\mathrm{d}}^{\uparrow}$ in (4.11).

If in Figure 4.6 the arc weight between t_1 and p_1 were equal to four instead of to three, then the minimum in (4.46) equals one, and thus the result in (4.45). That is why the result of Theorem 4.1 can be made stronger in some cases. Namely, if for $p \in P_{\mathrm{d}}^{\uparrow}$ it holds that (recall from Section 2 in Chapter 1 the notation $a \mid b$ for 'a divides b')

$$\forall t \in {}^{\bullet}p \cap T^{\mathrm{uc}}, \forall t' \in p^{\bullet} \cap T^{\mathrm{uc}} : f^{-}(p, t') \mid f^{+}(t, p),$$

then p can be considered as belonging to $P_{\mathrm{u}}^{\uparrow}$, instead of to $P_{\mathrm{d}}^{\uparrow}$. As this case is so exotic, we do not give a formal proof here and do not consider it later in this book.

5. Reduction of the control design

In this section we use invariance properties of the legal set to facilitate the design of (maximally) permissive control laws. The present approach is similar to the one of the previous section. As a matter of fact, the result for $\mathcal{A}^*$ in (4.40) and the corresponding influencing net $\mathcal{N}_{\mathcal{A}}^{\mathrm{uc}}$ form a first step towards reduction of the control design. We show that for

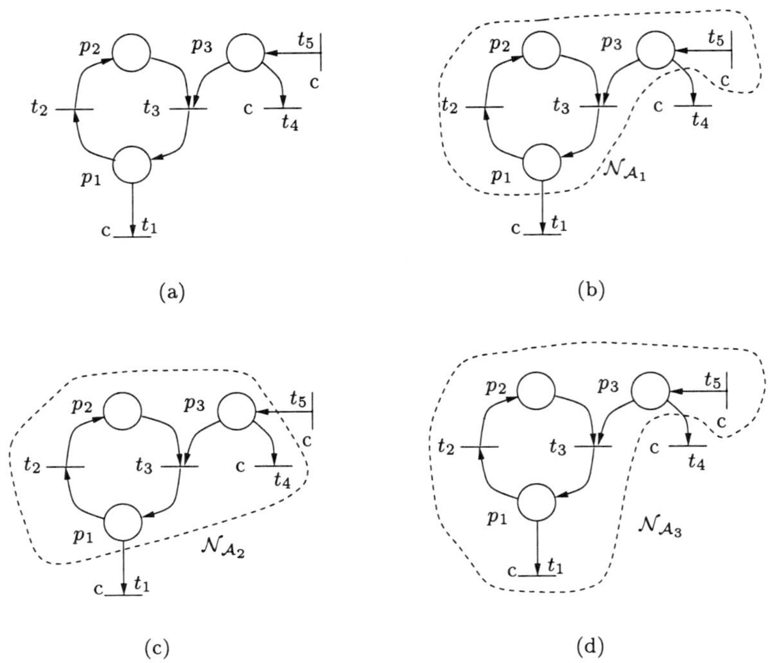

Figure 4.7. An example that illustrates how control design can be simplified by using invariance properties of the legal set.

control design for a legal set $\mathcal{A}$ it suffices to study a sub-Petri net $\mathcal{N}_{\mathcal{A}}$ of $\mathcal{N}$.

5.1. An intuitive example

Consider the Petri net $\mathcal{N}$ in Figure 4.7(a). Controllable transitions are indicated with a label 'c'. Transitions t_1, t_4 and t_5 are controllable. Transitions t_2 and t_3 are uncontrollable. We consider three cases:

1. If $P^{\downarrow}(\mathcal{A}_1) = \{p_1\}$ and $P^{\mathrm{i}}(\mathcal{A}_1) = \{p_2, p_3\}$, then a maximally permissive control law does not influence t_1 and t_4, because in the worst case (with respect to $\mathcal{A}_1$) they are not executed. Transition t_1 only decreases the state of p_1. Transition t_4 decreases the state of p_3 and thus the maximal number of times that t_3 can be executed.
2. If $P^{\uparrow}(\mathcal{A}_2) = \{p_3\}$ and $P^{\mathrm{i}}(\mathcal{A}_2) = \{p_1, p_2\}$, then a maximally permissive control law does not influence t_1 and t_5, because in the worst case (with respect to $\mathcal{A}_2$) they are not executed. Transition

t_5 only increases the state of p_3. Transition t_1 decreases the state of p_1 and thus the maximal number of times that t_2 and t_3 can be executed.

3. If $P^{\mathrm{d}}(\mathcal{A}_3) = \{p_1\}$ and $P^{\mathrm{i}}(\mathcal{A}_3) = \{p_2, p_3\}$, then one finds by combining the reasonings in 1 and 2 that a maximally permissive control law does not influence t_4.

Therefore, to design control laws, it suffices to study the respective sub-Petri nets $\mathcal{N}_{\mathcal{A}_1}$, $\mathcal{N}_{\mathcal{A}_2}$ and $\mathcal{N}_{\mathcal{A}_3}$ of $\mathcal{N}$ (see Figures 4.7(b) to (d)).

5.2. Sets of controllable transitions

Define for each place $p \in P$ the set of controllable transitions

$$\gamma(p) := {}^\bullet\alpha(p) \cap T^{\mathrm{c}}.$$

This set contains all controllable input transitions of places in $\alpha(p)$. It is the controllable counterpart of $\beta(\cdot)$ (cfr. Definition 4.3). The next property immediately follows from the definitions of $\beta(p)$ and $\gamma(p)$.

Proposition 4.12. *For each $p \in P$ it is true that ${}^\bullet\alpha(p) = \beta(p) \cup \gamma(p)$.*

Definition 4.6 introduces for each of the sets $P^{\downarrow}(\mathcal{A})$, $P^{\uparrow}_{\mathrm{u}}(\mathcal{A})$ and $\bar{P}^{\mathrm{d}}(\mathcal{A})$ a set of controllable transitions.

Definition 4.6. *Define for $\mathcal{A} \subseteq \mathbb{N}^n$ the sets of controllable transitions*

$$\begin{aligned}
N^{\downarrow}(\mathcal{A}) &:= \cup_{p \in P^{\downarrow}(\mathcal{A})}\gamma(p),\\
N^{\uparrow}(\mathcal{A}) &:= \cup_{p \in P^{\uparrow}_{\mathrm{u}}(\mathcal{A})}[\cup_{p' \in \sigma'(p)}\gamma(p') \cup (p^\bullet \cap T^{\mathrm{c}})],\\
N^{\mathrm{d}}(\mathcal{A}) &:= \cup_{p \in \bar{P}^{\mathrm{d}}(\mathcal{A})}[\cup_{p' \in \sigma(p)}\gamma(p') \cup (p^\bullet \cap T^{\mathrm{c}})].
\end{aligned}$$

These sets are, with the aid of Figure 4.4, easy to visualise. Analogous to the proof of Proposition 4.6, with Proposition 4.12 we can prove the following.

Proposition 4.13. *For all $\mathcal{A} \subseteq \mathbb{N}^n$ it is true that*

(i) ${}^\bullet I^{\downarrow}(\mathcal{A}) = U^{\downarrow}(\mathcal{A}) \cup N^{\downarrow}(\mathcal{A})$,

(ii) ${}^\bullet I^{\mathrm{d}}(\mathcal{A}) \subseteq U^{\mathrm{d}}(\mathcal{A}) \cup N^{\mathrm{d}}(\mathcal{A})$.

Define the set of controllable transitions

$$\bar{N}(\mathcal{A}) := N^{\downarrow}(\mathcal{A}) \cup N^{\uparrow}(\mathcal{A}) \cup N^{\mathrm{d}}(\mathcal{A}). \tag{4.47}$$

As already mentioned in Section 4.1, we suppress the argument '$(\mathcal{A})$' when there is no danger for confusion.

Example 4.6. For the Petri net in Figure 4.3 (p. 76) and the legal sets $\mathcal{A}_1, \mathcal{A}_2$ and $\mathcal{A}_3$ of Example 4.3 (p. 80) it holds that

$$\begin{aligned}
\gamma(p_1) &= \{t_9, t_{11}\},\\
\gamma(p_{11}) &= \{t_{11}\},\\
\bar{N}(\mathcal{A}_1) &= N^{\downarrow}(\mathcal{A}_1) = \gamma(p_1) = \{t_9, t_{11}\},\\
\bar{N}(\mathcal{A}_2) &= N^{\uparrow}(\mathcal{A}_2) = \gamma(p_{11}) \cup \{t_3\} = \{t_3, t_{11}\},\\
\bar{N}(\mathcal{A}_3) &= N^{\mathrm{d}}(\mathcal{A}_3) = \gamma(p_1) \cup \gamma(p_{11}) \cup \{t_3\} = \{t_3, t_9, t_{11}\}.
\end{aligned}$$

$\triangle$

5.3. Reduction theorem for control design

Recall from (3.22) that for a Petri net with T^{uc} as set of uncontrollable transitions, concurrency assumption Ω, control set $\mathcal{U}$ and legal set $\mathcal{A}$, the set of permissive control values is given by:

$$\mathcal{U}_{\mathbf{m}}^{\mathcal{A}} = \{\zeta \in \mathcal{U} \mid \mathcal{R}_{\infty}^{T^{\mathrm{uc}}}(\mathcal{R}_{1,\Omega}^{T^{\mathrm{c}}}(\zeta, \mathbf{m})) \subseteq \mathcal{A}\}. \tag{4.48}$$

By using Theorem 4.2 the design of (maximally) permissive supervisors can be simplified.

Theorem 4.2. *Consider a Petri net $\mathcal{N} = (P, T, f^+, f^-, T^{\mathrm{uc}})$ with concurrency assumption Ω and control set $\mathcal{U}$, and a legal set $\mathcal{A}$. For all $\mathbf{m} \in \mathbb{N}^n, \mathcal{A} \subseteq \mathbb{N}^n$ and $\zeta \in \mathcal{U}$ it is true that*

$$\mathcal{R}_{\infty}^{T^{\mathrm{uc}}}(\mathcal{R}_{1,\Omega}^{T^{\mathrm{c}}}(\zeta, \mathbf{m})) \subseteq \mathcal{A} \quad \Leftrightarrow \quad \mathcal{R}_{\infty}^{\bar{U}}(\mathcal{R}_{1,\Omega}^{\bar{N}}(\zeta, \mathbf{m})) \subseteq \mathcal{A}.$$

Proof. For the ∞ part of the reachable sets we use, as in Theorem 4.1, the transition bag assumption.

($\Rightarrow$) This follows from Proposition 1.10.

($\Leftarrow$) By definition, for each state $\bar{\mathbf{m}}$ in $\mathcal{R}_{\infty}^{T^{\mathrm{uc}}}(\mathcal{R}_{1,\Omega}^{T^{\mathrm{c}}}(\zeta, \mathbf{m})))$, there exist transition vectors $\boldsymbol{\delta}_0 \in \Omega$ and $\boldsymbol{\delta}_1, \ldots, \boldsymbol{\delta}_k \in \Delta$ with $\boldsymbol{\delta}_0|_{T^{\mathrm{uc}}} = \mathbf{0}$, and $\boldsymbol{\delta}_i|_{T^{\mathrm{c}}} = \mathbf{0}$ $(i = 1, \ldots, k)$ such that

$$\mathbf{m} \xrightarrow{\boldsymbol{\delta}_0, \zeta} \mathbf{m}_0 \xrightarrow{(\boldsymbol{\delta}_1, \ldots, \boldsymbol{\delta}_k)} \bar{\mathbf{m}}. \tag{4.49}$$

We prove that

$$\mathbf{m} \xrightarrow{\boldsymbol{\delta}_0', \zeta} \mathbf{m}_0' \xrightarrow{(\boldsymbol{\delta}_1', \ldots, \boldsymbol{\delta}_k')} \bar{\mathbf{m}}' \tag{4.50}$$

holds, with

$$\boldsymbol{\delta}_0'|_{\bar{N}} = \boldsymbol{\delta}_0|_{\bar{N}} \quad \text{and} \quad \boldsymbol{\delta}_0'|_{T \setminus \bar{N}} = \mathbf{0}, \tag{4.51}$$

and for $i = 1, \dots, k$ (see also (4.35)–(4.37) in the proof of Proposition 4.10):

$$\boldsymbol{\delta}'_i|_{\bar{U}\setminus W} = \boldsymbol{\delta}_i|_{\bar{U}\setminus W}, \tag{4.52}$$

$$\boldsymbol{\delta}'_i|_{T\setminus\bar{U}} = \mathbf{0}, \tag{4.53}$$

$$\boldsymbol{\delta}'_i(t_{j,l}) = \min\left[\boldsymbol{\delta}_i(t_{j,l}), \mathbf{m}'_{i-1}(p_j) - \sum_{v=1}^{l-1} \boldsymbol{\delta}'_i(t_{j,v})\right]$$

$$(t_{j,l} \in W; j = 1, \dots, z; l = 1, \dots, \nu(j)). \tag{4.54}$$

From $\boldsymbol{\delta}_0 \in \Omega_{\mathbf{m}} \cap \zeta$, $\boldsymbol{\delta}'_0 \le \boldsymbol{\delta}_0$ and the fact that both $\Omega_{\mathbf{m}}$ and ζ are down-sets in $(\Omega, \le)$, we have that $\boldsymbol{\delta}'_0 \in \Omega_{\mathbf{m}} \cap \zeta$. Next, we look at the relation between $\mathbf{m}_0(p)$ and $\mathbf{m}'_0(p)$ for places p in $\bar{I}$. By definition, we have from (4.50) that $\mathbf{m}'_0 = \mathbf{m} + \mathbf{F}\boldsymbol{\delta}'_0$ or that

$$\mathbf{m}'_0(p) = \mathbf{m}(p) + \sum_{t \in {}^\bullet p} f^+(t,p)\boldsymbol{\delta}'_0(t) - \sum_{t \in p^\bullet} f^-(p,t)\boldsymbol{\delta}'_0(t) \qquad (p \in P). \tag{4.55}$$

From the definition of V_1 in (4.14), Definition 4.5 and Propositions 4.12 and 4.13 it follows that

$$^\bullet V_1 \cap T^c \subseteq N^\downarrow \cup N^d \cup \left[\cup_{p \in P^\uparrow} \cup_{p' \in \sigma'(p)} \gamma(p')\right] \subseteq \bar{N}.$$

With (4.51) and (4.55) we then find that

$$\mathbf{m}'_0(p) = \mathbf{m}(p) + \sum_{t \in {}^\bullet p} f^+(t,p)\boldsymbol{\delta}_0(t) - \sum_{t \in p^\bullet} f^-(p,t)\boldsymbol{\delta}'_0(t) \qquad (p \in V_1).$$

Together with $\boldsymbol{\delta}'_0 \le \boldsymbol{\delta}_0$ and $\mathbf{m}_0 = \mathbf{m} + \mathbf{F}\boldsymbol{\delta}_0$ we obtain that

$$\mathbf{m}'_0|_{V_1} \ge \mathbf{m}_0|_{V_1}. \tag{4.56}$$

From the definition of $N^\uparrow$ in Definition 4.6 it follows that $(P^\uparrow_{\mathrm{u}})^\bullet \cap T^c \subseteq \bar{N}$. By applying this to (4.55) we find:

$$\mathbf{m}'_0(p) = \mathbf{m}(p) + \sum_{t \in {}^\bullet p} f^+(t,p)\boldsymbol{\delta}'_0(t) - \sum_{t \in p^\bullet} f^-(p,t)\boldsymbol{\delta}_0(t) \qquad (p \in P^\uparrow_{\mathrm{u}}),$$

and, with $\boldsymbol{\delta}'_0 \le \boldsymbol{\delta}_0$ and $\mathbf{m}_0 = \mathbf{m} + \mathbf{F}\boldsymbol{\delta}_0$, that

$$\mathbf{m}'_0|_{P^\uparrow_{\mathrm{u}}} \le \mathbf{m}_0|_{P^\uparrow_{\mathrm{u}}}. \tag{4.57}$$

Because of (4.56) and (4.57), the rest of the proof is similar to that of Lemma 4.1, Proposition 4.10 and Theorem 4.1. □

Corollary 4.3. *Let $\mathcal{A} \subseteq \mathbb{N}^n$, $\mathbf{m} \in \mathbb{N}^n$, $\zeta \in \mathcal{U}$, $T' \subseteq T^{\mathrm{uc}}$ and $T'' \subseteq T^{\mathrm{c}}$. It is true that:*

$$\mathcal{R}_{\infty}^{\bar{U}}(\mathcal{R}_{1,\Omega}^{\bar{N}}(\zeta, \mathbf{m})) \subseteq \mathcal{A} \quad \Rightarrow \quad \mathcal{R}_{\infty}^{T'}(\mathcal{R}_{1,\Omega}^{T''}(\zeta, \mathbf{m})) \subseteq \mathcal{A}.$$

From (3.22) we know that when designing (maximally) permissive control laws the condition $\mathcal{R}_{\infty}^{T^{\mathrm{uc}}}(\mathcal{R}_{1,\Omega}^{T^{\mathrm{c}}}(\zeta, \mathbf{m})) \subseteq \mathcal{A}$ should be checked. By using Theorem 4.2 this condition can be simplified, we have

$$\mathcal{U}_{\mathbf{m}}^{\mathcal{A}} = \{\zeta \in \mathcal{U} \mid \mathcal{R}_{\infty}^{\bar{U}}(\mathcal{R}_{1,\Omega}^{\bar{N}}(\zeta, \mathbf{m})) \subseteq \mathcal{A}\}. \tag{4.58}$$

for the set of permissive control values for $\mathbf{m}$. As it suffices to consider the reachable set via (controllable) transitions in $\bar{N}$, in case of fine and on/off control each maximally permissive control law is equivalent to one which does not impose any constraints on transitions in $T^{\mathrm{c}} \setminus \bar{N}$.

Proposition 4.14. *Consider the set of control values $\mathcal{U}$ in (3.12), i.e. fine control is used. It holds for all $\mathcal{A} \subseteq \mathbb{N}^n$ and for all $\mathbf{m} \in \mathcal{A}^*$ that*

$$\forall \xi \in \max(\mathcal{E}_{\mathbf{m}}^{\mathcal{A}}, \leq_{\mathbf{m}}) : \exists \zeta \in \xi : \zeta \in \cap_{t \in T^{\mathrm{c}} \setminus \bar{N}} \mathcal{O}^{\uparrow}(\Delta, \leq_t).$$

Proposition 4.15. *Consider the set of control values $\mathcal{U}$ in (3.13), i.e. on/off control is used. It holds for all $\mathcal{A} \subseteq \mathbb{N}^n$ and for all $\mathbf{m} \in \mathcal{A}^*$ that*

$$\forall \xi \in \max(\mathcal{E}_{\mathbf{m}}^{\mathcal{A}}, \leq_{\mathbf{m}}) : \exists \zeta_{T'} \in \xi : T' \subseteq \bar{N}.$$

Another consequence of Theorem 4.2 is that to check condition (4.58) not the whole Petri net $\mathcal{N}$ needs to be looked at. To know whether transitions in $\bar{N}$ are enabled, it suffices to know the state of places in $\bar{O}(\mathcal{A}) := {}^\bullet\bar{N}(\mathcal{A})$. If we add the set of transitions $\bar{N}$ and the set of places $\bar{O}$ to the influencing net $\mathcal{N}_{\mathcal{A}}^{\mathrm{uc}} = (\bar{I}, \bar{U}, f^+|_{\bar{U} \times \bar{I}, \bar{U}}, f^-|_{\bar{I} \times \bar{U}})$, then it follows from Theorem 4.2 that for control design it suffices to study the sub-net

$$\mathcal{N}_{\mathcal{A}} := (\bar{P}, \bar{T}, f^+|_{\bar{T} \times \bar{P}}, f^-|_{\bar{T} \times \bar{I}}, \bar{U}) \tag{4.59}$$

of $\mathcal{N}$. Herein is

$$\bar{P}(\mathcal{A}) := \bar{I}(\mathcal{A}) \cup \bar{O}(\mathcal{A}) \quad \text{and} \quad \bar{T}(\mathcal{A}) := \bar{U}(\mathcal{A}) \cup \bar{N}(\mathcal{A}). \tag{4.60}$$

Note that arcs from transitions in $\bar{T}$ to places in $\bar{O} \setminus \bar{I}$ do not belong to this Petri net. This is because, to determine whether a given state $\mathbf{m}$ belongs to $\mathcal{A}^*$, the value of $\mathbf{m}|_{\bar{I}}$ contains enough information. We refer to $\mathcal{N}_{\mathcal{A}}$ as the *controlled influencing net* of $\mathcal{A}$.

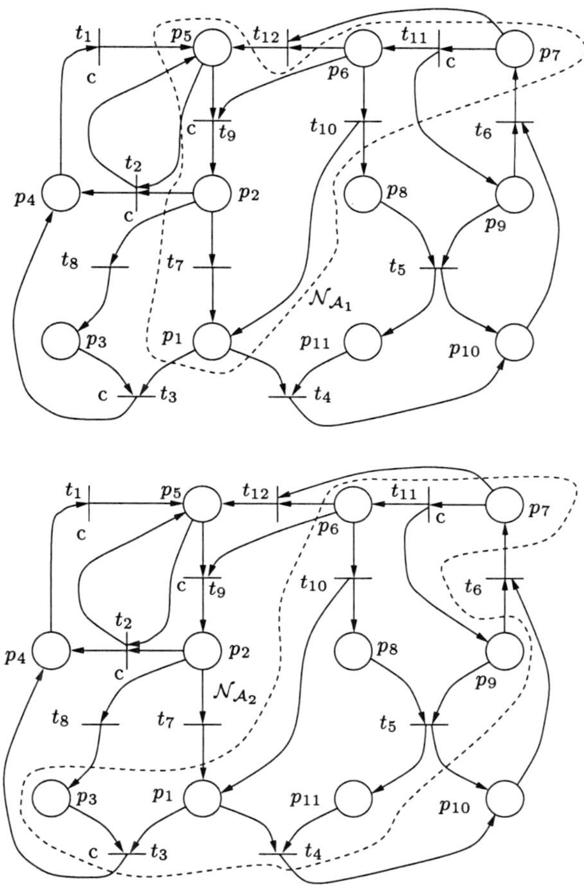

Figure 4.8. The Petri net of Figure 4.3 with the controlled influencing nets $\mathcal{N}_{\mathcal{A}_1}$ and $\mathcal{N}_{\mathcal{A}_2}$ for the legal sets $\mathcal{A}_1$ and $\mathcal{A}_2$ of Example 4.3.

Example 4.7. For the Petri net in Figure 4.3 (p. 76) and the legal sets of Example 4.3 (p. 80) one has

$$\begin{aligned}\bar{O}(\mathcal{A}_1) &= {}^\bullet N^{\downarrow}(\mathcal{A}_1) = \{p_5, p_6, p_7\},\\ \bar{O}(\mathcal{A}_2) &= {}^\bullet N^{\uparrow}(\mathcal{A}_2) = \{p_1, p_3, p_7\},\\ \bar{O}(\mathcal{A}_3) &= {}^\bullet N^{\mathrm{d}}(\mathcal{A}_3) = \{p_1, p_3, p_5, p_6, p_7\}.\end{aligned}$$

The controlled influencing nets $\mathcal{N}_{\mathcal{A}_1}$ and $\mathcal{N}_{\mathcal{A}_2}$ are given in Figure 4.8. △

As this example illustrates, it is possible that $\bar{I} \cap \bar{O} \neq \emptyset$.

Because of the fact that the control problem is reduced to $\mathcal{N}_{\mathcal{A}}$, the control value only needs to be changed/recalculated when the state of $\bar{P}$ changes. That is why transitions which do not belong to ${}^\bullet\bar{P} \cup \bar{P}^\bullet$, do not need to be observed by the supervisory controller.

6. Structural and invariance properties of the legal set

In this section we first study the structure of up- and down-sets in $(\mathbb{N}^n, \leq)$ (Section 6.1). Further, we look at the problem of determining the sets $P^{\mathrm{i}}, P^{\downarrow}, P^{\uparrow}$ and P^{d} of Definition 4.2 when $\mathcal{A}$ is defined by one linear inequality (Section 6.2) or by disjunctions of linear inequalities (Section 6.3).

6.1. Up- and down-sets in $(\mathbb{N}^n, \leq)$

Here, we study the structure of up- and down-sets in $(\mathbb{N}^n, \leq)$ in order to improve the understanding of their invariance properties.

Proposition 4.16. *Let $V \subseteq \mathbb{N}^n$. If $(V, \leq)$ is an antichain, then V is a finite set.*

Proof. By contradiction. We prove that if there exists an infinite antichain in $(\mathbb{N}^n, \leq)$, then there exists an infinite antichain in $(\mathbb{N}, \leq)$. And this is not true, because $(\mathbb{N}, \leq)$ is a chain.

This is proven by induction. We show that if there exists an infinite antichain in $(\mathbb{N}^n, \leq)$, then there exists an infinite antichain in $(\mathbb{N}^{n-1}, \leq)$.

Suppose that $(V, \leq)$ is an infinite antichain in $(\mathbb{N}^n, \leq)$. Pick an element $\mathbf{y} \in V$. By the definition of an antichain, each other element $\mathbf{x} \in V$ is not comparable to $\mathbf{y}$. If we use the notation $\mathbf{x} = (x_1, \ldots, x_n)$ and $\mathbf{y} = (y_1, \ldots, y_n)$, then this is equivalent to

$$\begin{aligned}
\neg(\mathbf{x} \geq \mathbf{y} \vee \mathbf{x} \leq \mathbf{y}) \quad &\Leftrightarrow \quad \neg\left(\bigwedge_{i=1}^{n} x_i \geq y_i \vee \bigwedge_{j=1}^{n} x_j \leq y_j\right), \\
&\Leftrightarrow \quad \neg\left(\bigwedge_{i=1}^{n} x_i \geq y_i\right) \wedge \neg\left(\bigwedge_{j=1}^{n} x_j \leq y_j\right), \\
&\Leftrightarrow \quad \left(\bigvee_{i=1}^{n} x_i < y_i\right) \wedge \left(\bigvee_{j=1}^{n} x_j > y_j\right), \\
&\Leftrightarrow \quad \bigvee_{\substack{i=1,\ldots,n, \\ j=1,\ldots,n; j\neq i}} (x_i < y_i \wedge x_j > y_j). \qquad (4.61)
\end{aligned}$$

Define for each $i \in \{1, \ldots, n\}$, for each $j \in \{1, \ldots, n\}$ with $j \neq i$ and for each $a \in \mathbb{N}$ with $a < y_i$, the set

$$X_{i,j}^a = \{\mathbf{x} \in \mathbb{N}^n \mid x_i = a \wedge x_j > y_j\}. \tag{4.62}$$

Consequently, we obtain from (4.61) that

$$V \setminus \{\mathbf{y}\} \subseteq \bigcup_{\substack{i=1,\ldots,n,\\ j=1,\ldots,n; j\neq i}} \bigcup_{a\in\mathbb{N}, a<y_i} X_{i,j}^a. \tag{4.63}$$

As $V \setminus \{\mathbf{y}\}$ is an infinite set and as the right hand side of (4.63) is a finite union, there exist i', j' and a' such that $X_{i',j'}^{a'} \cap V$ is an infinite set. As $X_{i',j'}^{a'}$ is isomorphic to $\mathbb{N}^{n-1}$, we can conclude without much effort that $X_{i',j'}^{a'}$ is an infinite antichain in $(\mathbb{N}^{n-1}, \leq)$. □

Define for $\mathbf{x} \in \mathbb{N}^n$ the sets $\uparrow\mathbf{x} := \{\mathbf{x}' \in \mathbb{N}^n \mid \mathbf{x} \leq \mathbf{x}'\}$ and $\downarrow\mathbf{x} := \{\mathbf{x}' \in \mathbb{N}^n \mid \mathbf{x} \geq \mathbf{x}'\}$.

Proposition 4.17. *If $X \subseteq \mathbb{N}^n$ and $X \neq \emptyset$, then for each $\mathbf{x} \in X$ there exists an $\bar{\mathbf{x}} \in \min(X, \leq)$ such that $\bar{\mathbf{x}} \leq \mathbf{x}$.*

Proof. By contradiction. Suppose that

$$\exists \mathbf{x} \in X : \forall \mathbf{y} \in \min(X, \leq) : \mathbf{y} \not\leq \mathbf{x}.$$

From this we obtain that:

$$\downarrow\mathbf{x} \cap \min(X, \leq) = \emptyset. \tag{4.64}$$

We consider two cases:

1. $\downarrow\mathbf{x} \cap X = \{\mathbf{x}\}$. Then there does not exist any $\mathbf{x}'$ in X such that $\mathbf{x}' \leq \mathbf{x}$ and $\mathbf{x}' \neq \mathbf{x}$. Consequently, $\mathbf{x} \in \min(X, \leq)$, which contradicts (4.64).

2. $\downarrow\mathbf{x} \cap X \supseteq \{\mathbf{x}, \mathbf{x}_1\}$ with $\mathbf{x}_1 \neq \mathbf{x}$. It is clear that $\mathbf{x}_1 \leq \mathbf{x}$. Thus we have that $\downarrow\mathbf{x}_1 \subseteq \downarrow\mathbf{x}$ and with (4.64) that

$$\downarrow\mathbf{x}_1 \cap \min(X, \leq) = \emptyset. \tag{4.65}$$

Note that (4.65) has exactly the same form as (4.64). That is why for (4.65) we can again consider two cases, etc. Because $\downarrow\mathbf{x}$ is a finite set and because $\mathbf{x}_1 \leq \mathbf{x}$ and $\mathbf{x}_1 \neq \mathbf{x}$, this reasoning leads to a contradiction in a finite number of steps. □

Theorem 4.3. *Each up- and each down-set in $(\mathbb{N}^n, \leq)$ can be described by a finite number of disjunctions and conjunctions of linear inequality constraints with integer coefficients.*

Proof. From Proposition 4.17 it follows that it holds for $X \in \mathcal{O}^{\uparrow}(\mathbb{N}^n, \leq)$ that

$$X \subseteq \bigcup_{\bar{\mathbf{x}} \in \min(X, \leq)} \uparrow \bar{\mathbf{x}}. \tag{4.66}$$

If $\mathbf{x} \in \uparrow \bar{\mathbf{x}}$ with $\bar{\mathbf{x}} \in \min(X, \leq) \subseteq X$, then, because of the definition of up-set, $\uparrow \bar{\mathbf{x}} \subseteq X$ and thus $\mathbf{x} \in X$. Hence,

$$X = \bigcup_{\bar{\mathbf{x}} \in \min(X, \leq)} \uparrow \bar{\mathbf{x}}.$$

The set $\min(X, \leq)$ is an antichain in $(\mathbb{N}^n, \leq)$ and therefore finite (Proposition 4.16). Together with Proposition 1.5 (p. 7) this finishes the proof. □

6.2. One linear inequality

Consider a legal set $\mathcal{A}$ which is described by one linear inequality:

$$\mathcal{A} = \{\mathbf{m} \in \mathbb{N}^n \mid \mathbf{A}^T \mathbf{m} \leq b\} \quad \text{with } \mathbf{A} \in \mathbb{Z}^n, b \in \mathbb{Z}. \tag{4.67}$$

In this section we determine for legal sets of this form the sets $P^{\mathrm{i}}, P^{\downarrow}, P^{\uparrow}$ and P^{d} of Definition 4.2.

Before giving necessary and sufficient conditions for $\mathcal{A} = \emptyset$ and for $\mathcal{A} = \mathbb{N}^n$, we formulate necessary and sufficient conditions for this problem in $\mathbb{Q}^n$. Recall that $\mathbb{Q}^n_+ = \{\mathbf{x} \in \mathbb{Q}^n \mid \mathbf{0} \leq \mathbf{x}\}$ is the closed first orthant of $\mathbb{Q}^n$. Define

$$\mathcal{H} := \{\mathbf{x} \in \mathbb{Q}^n_+ \mid \mathbf{A}^T \mathbf{x} \leq b\}. \tag{4.68}$$

The set $\mathcal{H}$ is described by the same inequality as $\mathcal{A}$ in (4.67). The two following properties are proven in Chapter 8 (Proposition 8.1 and 8.2 on p. 180), because they are corollaries of Theorem 8.2.

Proposition 4.18. *For $\mathcal{H}$ in (4.68) it is true that $\mathcal{H} = \emptyset$ if and only if $\mathbf{0} \leq \mathbf{A}$ and $b < 0$.*

Proposition 4.19. *For $\mathcal{H}$ in (4.68) it is true that $\mathcal{H} = \mathbb{Q}^n_+$ if and only if $\mathbf{A} \leq \mathbf{0}$ and $0 \leq b$.*

Next, we show that both conditions are also necessary and sufficient in $\mathbb{N}^n$.

Proposition 4.20. *For $\mathcal{A}$ in (4.67) it is true that $\mathcal{A} = \emptyset$ if and only if $\mathbf{0} \leq \mathbf{A}$ and $b < 0$.*

Proof. ($\Leftarrow$) This follows from Proposition 4.18.

($\Rightarrow$) Because of Proposition 4.18 it suffices to prove that if there exists an $\mathbf{x}$ in $\mathbb{Q}_+^n$ with $\mathbf{A}^T\mathbf{x} \leq b$, then there exists an $\mathbf{m}$ in $\mathbb{N}^n$ with $\mathbf{A}^T\mathbf{m} \leq b$. Define $\mathbf{m}$ such that for each $p \in P$ (recall from Section 2 in Chapter 1 that $\lceil\cdot\rceil$ is the ceiling operator):

$$\mathbf{m}(p) = \begin{cases} \lceil\mathbf{x}(p)\rceil & \text{if } \mathbf{A}(p) \leq 0, \\ 0 & \text{if } \mathbf{A}(p) > 0. \end{cases}$$

It holds that $\mathbf{m} \in \mathbb{N}^n$ and that $\mathbf{A}^T\mathbf{m} \leq \mathbf{A}^T\mathbf{x} \leq b$. Thus, $\mathbf{m}$ belongs to $\mathcal{A}$. □

Proposition 4.21. *For $\mathcal{A}$ in* (4.67) *it is true that $\mathcal{A} = \mathbb{N}^n$ if and only if $\mathbf{A} \leq \mathbf{0}$ and $0 \leq b$.*

Proof. ($\Leftarrow$) This follows from Proposition 4.19.

($\Rightarrow$) Because of Proposition 4.19 it suffices to prove that if there exists an $\mathbf{x}$ in $\mathbb{Q}_+^n$ with $\mathbf{A}^T\mathbf{x} > b$, then there exists an $\mathbf{m}$ in $\mathbb{N}^n$ with $\mathbf{A}^T\mathbf{m} > b$.

Without loss of generality, assume that all components of $\mathbf{x}$ are fractional. Define $\bar{\mathbf{m}}_k$ ($k = 1, \ldots, 2^n$) as the lattice points of the cell to which $\mathbf{x}$ belongs. Then, $\mathbf{x}$ can be written as a convex sum of these points:

$$\mathbf{x} = \sum_{k=1}^{2^n} \lambda_k \bar{\mathbf{m}}_k \quad \text{with } 0 \leq \lambda_k \quad (k = 1, \ldots, 2^n) \text{ and } \sum_{k=1}^{2^n} \lambda_k = 1. \qquad (4.69)$$

As the following reasoning leads to a contradiction, it follows from (4.69) that there exists a $k \in \{1, \ldots, 2^n\}$ such that $\mathbf{A}^T\bar{\mathbf{m}}_k > b$. Suppose that

$$\mathbf{A}^T\bar{\mathbf{m}}_i \leq b \quad \text{for all } k = 1, \ldots, 2^n.$$

When we add all 2^n inequalities after multiplying each of them with $\lambda_k \geq 0$, we obtain with (4.69) that

$$\sum_{k=1}^{2^n} \lambda_k \mathbf{A}^T\bar{\mathbf{m}}_k = \mathbf{A}^T \sum_{k=1}^{2^n} \lambda_k \bar{\mathbf{m}}_k = \mathbf{A}^T\mathbf{x} \leq b.$$

This contradicts $\mathbf{A}^T\mathbf{x} > b$. □

Next we determine the invariance properties of $\mathcal{A}$.

Proposition 4.22. *Consider a set $\mathcal{A}$ as in* (4.67). *Let $\mathcal{A} \neq \mathbb{N}^n$ and $p \in P$. Then, $\mathcal{A} \in \mathcal{O}^{\downarrow}(\mathbb{N}^n, \leq_p)$ if and only if $\mathbf{A}(p) \geq 0$.*

Proof. ($\Leftarrow$) Immediate.

($\Rightarrow$) By contradiction. Suppose that $\mathbf{A}(p) < 0$. As $\mathcal{A} \neq \mathbb{N}^n$ there exists an $\mathbf{m}_1 \in \mathbb{N}^n \setminus \mathcal{A}$. Because of $\mathbf{A}(p) < 0$ there exists a state $\mathbf{m}_2 \geq_p \mathbf{m}_1$ which belongs to $\mathcal{A}$. Together with $\mathcal{A} \in \mathcal{O}^{\downarrow}(\mathbb{N}^n, \leq_p)$ this implies that $\mathbf{m}_1 \in \mathcal{A}$. This contradicts $\mathbf{m}_1 \in \mathbb{N}^n \setminus \mathcal{A}$. □

Proposition 4.23. *Consider a set $\mathcal{A}$ as in* (4.67)*. Let $\mathcal{A} \neq \emptyset$ and $p \in P$. Then, $\mathcal{A} \in \mathcal{O}^{\uparrow}(\mathbb{N}^n, \leq_p)$ if and only if $\mathbf{A}(p) \leq 0$.*

Proof. ($\Leftarrow$) Immediate.

($\Rightarrow$) By contradiction. Suppose that $\mathbf{A}(p) > 0$. As $\mathcal{A} \neq \emptyset$, there exists a state $\mathbf{m}_1 \in \mathcal{A}$. Because of $\mathbf{A}(p) > 0$, there exists a state $\mathbf{m}_2 \geq_p \mathbf{m}_1$ which does not belong to $\mathcal{A}$. This contradicts $\mathcal{A} \in \mathcal{O}^{\uparrow}(\mathbb{N}^n, \leq_p)$. □

The next proposition follows from Definition 4.2 and Propositions 4.22 and 4.23.

Proposition 4.24. *Consider a set $\mathcal{A}$ as in* (4.67)*. If $\mathcal{A} \neq \emptyset$ and $\mathcal{A} \neq \mathbb{N}^n$, then*

$$\begin{aligned} P^{\mathrm{i}} &= \{p \in P \mid \mathbf{A}(p) = 0\}, \\ P^{\downarrow} &= \{p \in P \mid \mathbf{A}(p) > 0\}, \\ P^{\uparrow} &= \{p \in P \mid \mathbf{A}(p) < 0\}, \\ P^{\mathrm{d}} &= \emptyset. \end{aligned}$$

As already mentioned in Section 2, $P^{\mathrm{i}}(\emptyset) = P^{\mathrm{i}}(\mathbb{N}^n) = P$.

6.3. Unions of linear inequalities

Because of the discrete nature of the state space for general legal sets and of Theorem 4.3 for for up- and down-sets, legal sets which are described by a finite number of disjunctions and conjunctions of linear inequalities with integer coefficients form an important class for supervision of Petri nets. Each legal set of this type can be written as (see [33, §1.38]):

$$\mathcal{A} = \bigcap_{i=1}^{\beta} \bigcup_{j=1}^{\nu_i} \{\mathbf{m} \in \mathbb{N}^n \mid \mathbf{A}_{i,j}^T \mathbf{m} \leq b_{i,j}\},$$

with $\mathbf{A}_{i,j} \in \mathbb{Z}^n$ and $b_{i,j} \in \mathbb{Z}$ for $i = 1, \ldots, \beta$ and $j = 1, \ldots, \nu_i$. Thanks to the decomposition result of Proposition 3.13 (p. 66) for intersections of legal sets, we can concentrate on the construction of supremal controllable subsets of legal sets of the form:

$$\mathcal{A} = \bigcup_{i=1}^{\nu} \mathcal{A}_i, \tag{4.70}$$

with for $i = 1, \ldots, \nu$:

$$\mathcal{A}_i = \{\mathbf{m} \in \mathbb{N}^n \mid \mathbf{A}_i^T \mathbf{m} \leq b_i\} \quad \text{and } \mathbf{A}_i \in \mathbb{Z}^n, b_i \in \mathbb{Z}. \tag{4.71}$$

The next proposition is valid for general legal sets and is used further in this section.

Proposition 4.25. *If $\mathcal{A} = \cup_{i=1}^{\nu} \mathcal{A}_i$ with $\mathcal{A}_i \subseteq \mathbb{N}^n$ for $i = 1, \ldots, \nu$, then it holds for each p in P that*

(i) $\mathcal{A} \notin \mathcal{O}^{\downarrow}(\mathbb{N}^n, \leq_p) \Rightarrow \exists i \in \{1, \ldots, \nu\} : \mathcal{A}_i \notin \mathcal{O}^{\downarrow}(\mathbb{N}^n, \leq_p)$,

(ii) $\mathcal{A} \notin \mathcal{O}^{\uparrow}(\mathbb{N}^n, \leq_p) \Rightarrow \exists i \in \{1, \ldots, \nu\} : \mathcal{A}_i \notin \mathcal{O}^{\uparrow}(\mathbb{N}^n, \leq_p)$.

Proof. This follows from Propositions 1.3 and 1.4. □

First we study unions of down-sets and unions of up-sets $\mathcal{A}_i$ in $(\mathbb{N}^n, \leq)$ of the type (4.71). The general case is studied in Section 6.3.3.

6.3.1 Unions of down-sets

Consider the legal set

$$\mathcal{A} = \bigcup_{i=1}^{\nu} \mathcal{A}_i, \tag{4.72}$$

with $\nu \geq 2$ and for $i = 1, \ldots, \nu$:

$$\mathcal{A}_i = \{\mathbf{m} \in \mathbb{N}^n \mid \mathbf{A}_i^T \mathbf{m} \leq b_i\} \quad \text{and } \mathbf{A}_i \in \mathbb{N}^n, b_i \in \mathbb{Z}. \tag{4.73}$$

Because of $\mathbf{A}_i \in \mathbb{N}^n$, $\mathcal{A}_i$ belongs to $\mathcal{O}^{\downarrow}(\mathbb{N}^n, \leq)$ for $i = 1, \ldots, \nu$. The following proposition determines the invariance properties of this kind of legal sets.

Proposition 4.26. *Consider $\mathcal{A}$ in (4.72)-(4.73). If*

$$\mathcal{A}_i \nsubseteq \bigcup_{\substack{j \in \{1, \ldots, \nu\}, \\ j \neq i}} \mathcal{A}_j \quad \text{for } i = 1, \ldots, \nu, \tag{4.74}$$

then

$$P^{\mathrm{i}}(\mathcal{A}) = \bigcap_{i=1}^{\nu} P^{\mathrm{i}}(\mathcal{A}_i), \qquad P^{\downarrow}(\mathcal{A}) = \bigcup_{i=1}^{\nu} P^{\downarrow}(\mathcal{A}_i),$$
$$P^{\uparrow}(\mathcal{A}) = \emptyset, \qquad P^{\mathrm{d}}(\mathcal{A}) = \emptyset.$$

Condition (4.74) expresses nonredundancy of each of the sets $\mathcal{A}_i$ in $\mathcal{A}$. Each set $\mathcal{A}_i$ contains at least one state which does not belong to any of the sets $\mathcal{A}_j$ ($j = 1, \ldots, \nu; j \neq i$). A necessary condition for (4.74) is that

$$\mathcal{A}_i \neq \emptyset \quad \text{and} \quad \mathcal{A}_i \neq \mathbb{N}^n \quad \text{for } i = 1, \ldots, \nu, \tag{4.75}$$

or, with (4.73) and Propositions 4.20 and 4.21, that $\mathbf{A}_i \neq \mathbf{0}$ and $b_i \geq 0$ for $i = 1, \ldots, \nu$. The sets $P^{\mathrm{i}}(\mathcal{A}_i)$ and $P^{\downarrow}(\mathcal{A}_i)$ are given by Proposition 4.24.

Proof. For each $p \in P$ it follows from (4.72)–(4.73) that $\mathcal{A} \in \mathcal{O}^{\downarrow}(\mathbb{N}^n, \leq_p)$. Consequently, it follows from Definition 4.2 that $P^{\uparrow}(\mathcal{A}) = \emptyset$ and that $P^{\mathrm{d}}(\mathcal{A}) = \emptyset$. Because of $P^{\uparrow}(\mathcal{A}_i) = P^{\mathrm{d}}(\mathcal{A}_i) = \emptyset$ for $i = 1, \ldots, \nu$ and because the union of the sets in Definition 4.2 is equal to P, it suffices to show that $P^{\downarrow}(\mathcal{A}) = \cup_{i=1}^{\nu} P^{\downarrow}(\mathcal{A}_i)$.

First, we prove that $\cup_{i=1}^{\nu} P^{\downarrow}(\mathcal{A}_i) \subseteq P^{\downarrow}(\mathcal{A})$. Because of $\mathcal{A}_i \neq \emptyset$ and $\mathcal{A}_i \neq \mathbb{N}^n$ (see (4.75)) it holds that (see Proposition 4.24):

$$\begin{aligned} p \in P^{\downarrow}(\mathcal{A}_i) \quad &\Leftrightarrow \quad \mathbf{A}_i(p) > 0, \\ &\Rightarrow \quad \forall \mathbf{m} \in \mathcal{A}_i : \exists \mathbf{m}' \in \mathbb{N}^n, \mathbf{m}' \geq_p \mathbf{m} : \mathbf{m}' \notin \mathcal{A}_i, \\ &\Rightarrow \quad \forall \mathbf{m} \in \mathcal{A}_i \setminus \cup_{\substack{j \in \{1,\ldots,k\} \\ j \neq i}} \mathcal{A}_j : \\ &\qquad\qquad \exists \mathbf{m}' \in \mathbb{N}^n, \mathbf{m}' \geq_p \mathbf{m} : \mathbf{m}' \notin \mathcal{A}_i. \end{aligned}$$

As $\mathcal{A}_i \in \mathcal{O}^{\downarrow}(\mathbb{N}^n, \leq)$ for $i = 1, \ldots, \nu$ and $\mathbf{m}' \geq \mathbf{m}$, it holds that $\mathbf{m}' \notin \mathcal{A}$. Consequently, we have from (4.74) that

$$\exists \mathbf{m} \in \mathcal{A}, \exists \mathbf{m}' \in \mathbb{N}^n, \mathbf{m}' \geq_p \mathbf{m} : \mathbf{m}' \notin \mathcal{A},$$

or that $\mathcal{A} \notin \mathcal{O}^{\uparrow}(\mathbb{N}^n, \leq_p)$. As $\mathcal{A}$ is a down-set in $(\mathbb{N}^n, \leq_p)$, we can conclude that $p \in P^{\downarrow}(\mathcal{A})$.

Next, we prove that $P^{\downarrow}(\mathcal{A}) \subseteq \cup_{i=1}^{\nu} P^{\downarrow}(\mathcal{A}_i)$. As $\mathcal{A} \in \mathcal{O}^{\downarrow}(\mathbb{N}^n, \leq_p)$ and $\mathcal{A}_i \in \mathcal{O}^{\downarrow}(\mathbb{N}^n, \leq_p)$ for $i = 1, \ldots, \nu$, it suffices to prove that

$$\mathcal{A} \notin \mathcal{O}^{\uparrow}(\mathbb{N}^n, \leq_p) \quad \Rightarrow \quad \exists i \in \{1, \ldots, \nu\} : \mathcal{A}_i \notin \mathcal{O}^{\uparrow}(\mathbb{N}^n, \leq_p).$$

This follows from Proposition 4.25(ii). □

Example 4.8. Figure 4.9 illustrates the necessity of condition (4.74) in Proposition 4.26. One has that $n = 2$, $\nu = 2$ and $\mathcal{A}_2 \subseteq \mathcal{A}_1$. From $P^{\mathrm{i}}(\mathcal{A}_1) = \{p_2\}$, $P^{\downarrow}(\mathcal{A}_1) = \{p_1\}$ and $P^{\downarrow}(\mathcal{A}_2) = \{p_1, p_2\}$ we find that $P^{\mathrm{i}}(\mathcal{A}_1 \cup \mathcal{A}_2) = \{p_2\} \neq P^{\mathrm{i}}(\mathcal{A}_1) \cap P^{\mathrm{i}}(\mathcal{A}_2) = \emptyset$. △

6.3.2 Unions of up-sets

Consider the legal set

$$\mathcal{A} = \bigcup_{i=1}^{\nu} \mathcal{A}_i, \tag{4.76}$$

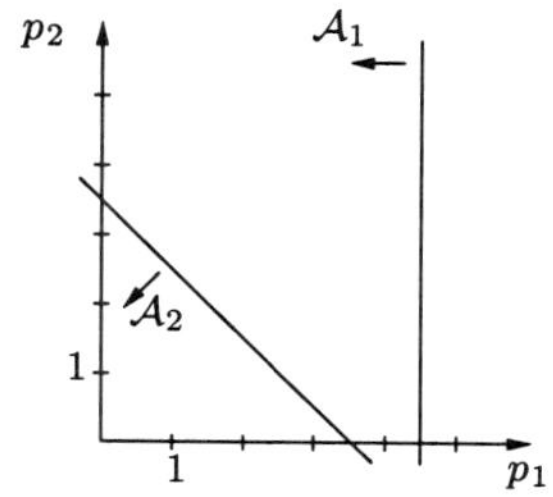

Figure 4.9. Two legal sets $\mathcal{A}_1$ and $\mathcal{A}_2$ which illustrate the necessity of condition (4.74) in Proposition 4.26.

with $\nu \geq 2$ and for $i = 1, \ldots, \nu$:

$$\mathcal{A}_i = \{\mathbf{m} \in \mathbb{N}^n \mid \mathbf{A}_i^T \mathbf{m} \leq b_i\} \quad \text{with } -\mathbf{A}_i \in \mathbb{N}^n, b_i \in \mathbb{Z}. \tag{4.77}$$

Because of $-\mathbf{A}_i \in \mathbb{N}^n$, it is true that $\mathcal{A}_i \in \mathcal{O}^\uparrow(\mathbb{N}^n, \leq)$ for $i = 1, \ldots, \nu$.

Proposition 4.27. *Consider $\mathcal{A}$ as in* (4.76)–(4.77). *If $\mathcal{A}$ is not equal to $\mathbb{N}^n$, then it holds that*

$$P^{\mathrm{i}}(\mathcal{A}) = \bigcap_{i=1}^{\nu} P^{\mathrm{i}}(\mathcal{A}_i), \qquad P^\uparrow(\mathcal{A}) = \bigcup_{i=1}^{\nu} P^\uparrow(\mathcal{A}_i),$$
$$P^\downarrow(\mathcal{A}) = \emptyset, \qquad P^{\mathrm{d}}(\mathcal{A}) = \emptyset.$$

A necessary condition for $\mathcal{A} \neq \mathbb{N}^n$ is that

$$\mathcal{A}_i \neq \mathbb{N}^n \quad \text{for } i = 1, \ldots, \nu, \tag{4.78}$$

or, with (4.77) and Proposition 4.21, that $b_i < 0$ for $i = 1, \ldots, \nu$. The sets $P^{\mathrm{i}}(\mathcal{A}_i)$ and $P^\uparrow(\mathcal{A}_i)$ are given by Proposition 4.24.

Proof. From (4.77) and Proposition 4.23 it follows that

$$\mathcal{A}_i \in \mathcal{O}^\uparrow(\mathbb{N}^n, \leq_p) \quad \text{for } i = 1, \ldots, \nu \text{ and } p \in P, \tag{4.79}$$

and thus, with (4.76) and Proposition 1.4, that

$$\mathcal{A} \in \mathcal{O}^\uparrow(\mathbb{N}^n, \leq_p) \quad \text{for } p \in P. \tag{4.80}$$

Consequently, we have with Definition 4.2 that $P^\downarrow(\mathcal{A}) = P^{\mathrm{d}}(\mathcal{A}) = \emptyset$. Now, it suffices to show that $P^\uparrow(\mathcal{A}) = \cup_{i=1}^{\nu} P^\uparrow(\mathcal{A}_i)$.

First, we prove that $P^\uparrow(\mathcal{A}) \supseteq \cup_{i=1}^{\nu} P^\uparrow(\mathcal{A}_i)$. It suffices to prove for $i = 1, \ldots, \nu$ that

$$p \in P^\uparrow(\mathcal{A}_i) \quad \Rightarrow \quad p \in P^\uparrow(\mathcal{A}),$$

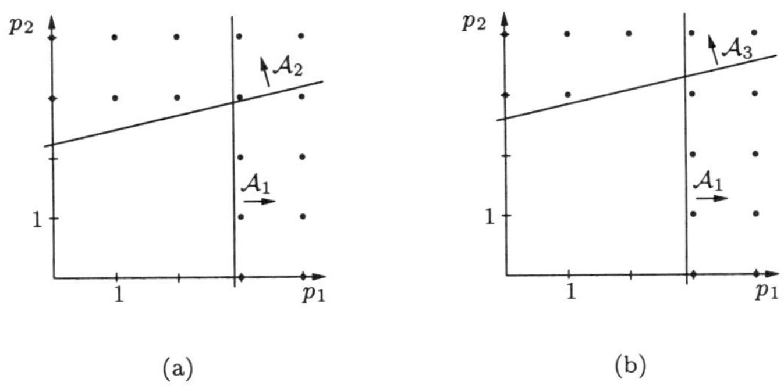

Figure 4.10. Illustration of the fact that for the general case the invariance properties of $\mathcal{A}$ in (4.70)–(4.71) are hard to determine.

or, because of (4.79) and (4.80), that

$$\mathcal{A}_i \notin \mathcal{O}^{\downarrow}(\mathbb{N}^n, \leq_p) \quad \Rightarrow \quad \mathcal{A} \notin \mathcal{O}^{\downarrow}(\mathbb{N}^n, \leq_p).$$

From (4.78), (4.79) and $\mathcal{A}_i \notin \mathcal{O}^{\downarrow}(\mathbb{N}^n, \leq_p)$ it follows with Proposition 4.22 that $\mathbf{A}_i(p) < 0$. As $\mathcal{A}$ is not equal to $\mathbb{N}^n$, there exists a state $\mathbf{m}^* \in \mathbb{N}^n \setminus \mathcal{A}$. As $\mathbf{A}_i(p) < 0$, there exists a state $\bar{\mathbf{m}} \geq_p \mathbf{m}^*$ with $\bar{\mathbf{m}} \in \mathcal{A}_i \subseteq \mathcal{A}$. Consequently, $\mathcal{A}$ is not a down-set in $(\mathbb{N}^n, \leq_p)$.

Finally, we prove that $P^{\uparrow}(\mathcal{A}) \subseteq \cup_{i=1}^{\nu} P^{\uparrow}(\mathcal{A}_i)$. Because of (4.79) and (4.80), it suffices to show that $\mathcal{A} \notin \mathcal{O}^{\downarrow}(\mathbb{N}^n, \leq_p)$ implies that there exists an $i \in \{1, \ldots, \nu\}$ such that $\mathcal{A}_i \notin \mathcal{O}^{\downarrow}(\mathbb{N}^n, \leq_p)$. This follows from Proposition 4.25(i). □

6.3.3 The general case

The case in which $\mathcal{A}$ is described by a finite number of disjunctions of general linear inequalities as in (4.70)–(4.71), is much more difficult to handle. We illustrate this with an example.

Example 4.9. In Figure 4.10(a) one has $P^{\uparrow}(\mathcal{A}_1 \cup \mathcal{A}_2) = \{p_1, p_2\}$ and $P^{\mathrm{i}}(\mathcal{A}_1 \cup \mathcal{A}_2) = P^{\downarrow}(\mathcal{A}_1 \cup \mathcal{A}_2) = P^{\mathrm{d}}(\mathcal{A}_1 \cup \mathcal{A}_2) = \emptyset$. In Figure 4.10(b) one has $P^{\uparrow}(\mathcal{A}_1 \cup \mathcal{A}_3) = \{p_2\}$, $P^{\mathrm{d}}(\mathcal{A}_1 \cup \mathcal{A}_3) = \{p_1\}$ and $P^{\mathrm{i}}(\mathcal{A}_1 \cup \mathcal{A}_3) = P^{\downarrow}(\mathcal{A}_1 \cup \mathcal{A}_3) = \emptyset$, because $(2,3) \notin \mathcal{A}_1 \cup \mathcal{A}_2$. The difference in invariance properties is due to a small shift of the supporting line of $\mathcal{A}_3$. △

Before stating the main result of this subsection, we state a proposition that is used in its proof.

Proposition 4.28. *Let $\mathcal{A}$ be as in* (4.70)–(4.71) *and $p \in P$. If $\mathcal{A}_i \neq \emptyset$ and $\mathbf{A}_i(p) \leq 0$ for $i = 1, \ldots, \nu$, then p belongs to $P^{\mathrm{i}}(\mathcal{A}) \cup P^{\uparrow}(\mathcal{A})$.*

Proof. This follows from Propositions 1.4 and 4.23. □

Often, the sets $\mathcal{A}_i$ in (4.70)–(4.71) depend on different places. In this case it is possible to determine the invariance properties of $\mathcal{A}$.

Proposition 4.29. *Consider $\mathcal{A}$ as in (4.70)–(4.71). If $\mathcal{A} \neq \mathbb{N}^n$ and if*

$$P^{\mathrm{i}}(\mathcal{A}_i) \cup P^{\mathrm{i}}(\mathcal{A}_j) = P \qquad (i, j = 1, \ldots, \nu; i \neq j), \tag{4.81}$$

then

$$P^{\mathrm{i}}(\mathcal{A}) = \bigcap_{i=1}^{\nu} P^{\mathrm{i}}(\mathcal{A}_i), \tag{4.82}$$

$$P^{\downarrow}(\mathcal{A}) = \bigcup_{i=1}^{\nu} P^{\downarrow}(\mathcal{A}_i), \tag{4.83}$$

$$P^{\uparrow}(\mathcal{A}) = \bigcup_{i=1}^{\nu} P^{\uparrow}(\mathcal{A}_i), \tag{4.84}$$

$$P^{\mathrm{d}}(\mathcal{A}) = \emptyset. \tag{4.85}$$

Proof. The proof consists of three parts. We prove (4.83) in part 1, (4.84) in part 2 and (4.82) and (4.85) in part 3.

1. First we show that $P^{\downarrow}(\mathcal{A}) \subseteq \cup_{i=1}^{\nu} P^{\downarrow}(\mathcal{A}_i)$. By definition, p belongs to $P^{\downarrow}(\mathcal{A})$ if and only if

$$\begin{aligned} & \mathcal{A} \in \mathcal{O}^{\downarrow}(\mathbb{N}^n, \leq_p) \wedge \mathcal{A} \notin \mathcal{O}^{\uparrow}(\mathbb{N}^n, \leq_p), \\ \Rightarrow \quad & \mathcal{A} \in \mathcal{O}^{\downarrow}(\mathbb{N}^n, \leq_p) \wedge \bigvee_{i=1}^{\nu} \mathcal{A}_i \notin \mathcal{O}^{\uparrow}(\mathbb{N}^n, \leq_p) \qquad \text{Prop. 4.25(i)}, \\ \Leftrightarrow \quad & \bigvee_{i=1}^{\nu} \left(\mathcal{A} \in \mathcal{O}^{\downarrow}(\mathbb{N}^n, \leq_p) \wedge \mathcal{A}_i \notin \mathcal{O}^{\uparrow}(\mathbb{N}^n, \leq_p) \right). \end{aligned} \tag{4.86}$$

Next we show by contraposition that

$$\mathcal{A} \in \mathcal{O}^{\downarrow}(\mathbb{N}^n, \leq_p) \quad \Rightarrow \quad \bigwedge_{i=1}^{\nu} \mathcal{A}_i \in \mathcal{O}^{\downarrow}(\mathbb{N}^n, \leq_p). \tag{4.87}$$

Suppose $\mathcal{A}_j \notin \mathcal{O}^{\downarrow}(\mathbb{N}^n, \leq_p)$ for some $j \in \{1, \ldots, \nu\}$. We prove that this implies $\mathcal{A} \notin \mathcal{O}^{\downarrow}(\mathbb{N}^n, \leq_p)$. By definition, $\mathcal{A}_j$ is not a down-set in $(\mathbb{N}^n, \leq_p)$ if and only if

$$\exists \mathbf{m} \in \mathcal{A}_j, \exists \mathbf{m}' \in \mathbb{N}^n, \mathbf{m}' \leq_p \mathbf{m} : \mathbf{m}' \notin \mathcal{A}_j. \tag{4.88}$$

As $\mathcal{A} \neq \mathbb{N}^n$, there exists a state

$$\mathbf{m}_j \notin \bigcup_{\substack{i\in\{1,\dots,\nu\},\\ i\neq j}} \mathcal{A}_i. \tag{4.89}$$

Define the state $\bar{\mathbf{m}}$ such that

$$\bar{\mathbf{m}}|_{P\setminus P^{\mathrm{i}}(\mathcal{A}_j)} = \mathbf{m}'|_{P\setminus P^{\mathrm{i}}(\mathcal{A}_j)},$$
$$\bar{\mathbf{m}}|_{P^{\mathrm{i}}(\mathcal{A}_j)} = \mathbf{m}_j|_{P^{\mathrm{i}}(\mathcal{A}_j)}.$$

From Proposition 4.4 and (4.81), (4.88) and (4.89) it follows that $\bar{\mathbf{m}} \notin \mathcal{A}$. Thus (4.87) is proven. After substituting (4.87) in (4.86), we find that $p \in \cup_{i=1}^{\nu} P^{\downarrow}(\mathcal{A}_i)$.

Next, we show that $P^{\downarrow}(\mathcal{A}) \supseteq \cup_{i=1}^{\nu} P^{\downarrow}(\mathcal{A}_i)$. It suffices to prove that for $i = 1, \dots, \nu$:

$$p \in P^{\downarrow}(\mathcal{A}_i) \quad \Rightarrow \quad p \in P^{\downarrow}(\mathcal{A}).$$

By definition, p belongs to $P^{\downarrow}(\mathcal{A})$ if and only if

$$\mathcal{A} \in \mathcal{O}^{\downarrow}(\mathbb{N}^n, \leq_p) \tag{4.90}$$

and

$$\mathcal{A} \notin \mathcal{O}^{\uparrow}(\mathbb{N}^n, \leq_p). \tag{4.91}$$

First, we prove (4.90). As p belongs to $P^{\downarrow}(\mathcal{A}_i)$, we have that $\mathcal{A}_i$ is a down-set in $(\mathbb{N}^n, \leq_p)$. From (4.81) we get that p belongs to $P^{\mathrm{i}}(\mathcal{A}_j)$ and thus that $\mathcal{A}_j \in \mathcal{O}^{\downarrow}(\mathbb{N}^n, \leq_p)$ for $j = 1, \dots, \nu$ with $j \neq i$. Then, Proposition 1.3 implies (4.90). The proof of (4.91) is similar to that of (4.87).

2. Equation (4.84) is proven in a similar manner as (4.83) in part 1.

3. From (4.83) and (4.84) we have that

$$P^{\mathrm{i}}(\mathcal{A}) \cup P^{\mathrm{d}}(\mathcal{A}) = \bigcap_{i=1}^{\nu} P^{\mathrm{i}}(\mathcal{A}_i).$$

For each p in $\cap_{i=1}^{\nu} P^{\mathrm{i}}(\mathcal{A}_i)$ it holds with Propositions 1.3 and 1.4 that

$$\mathcal{A} \in \mathcal{O}^{\downarrow}(\mathbb{N}^n, \leq_p) \cap \mathcal{O}^{\uparrow}(\mathbb{N}^n, \leq_p)$$

and thus that $p \in P^{\mathrm{i}}(\mathcal{A})$. From which (4.82) and (4.85) follow. □

7. Notes and references

So far, the idea of reduction has only been looked at for legal down-sets, e.g. [5], [7] and [18]. Part of the results in this chapter are published in [37].

Chapter 5

ACYCLIC PETRI NETS

By themselves, acyclic Petri nets (Definition 1.8, p. 17) form a limited class of Petri nets. So far, they received little attention in the literature (*unfolding* of Petri nets is an exception) because it was generally thought that their practical importance would be limited. Our motivation to study acyclic Petri nets is that Petri nets $\mathcal{N}$ and legal sets $\mathcal{A}$ with an acyclic uncontrollable part $\mathcal{N}^{\text{uc}}$ or $\mathcal{N}_{\mathcal{A}}^{\text{uc}}$ form an important class of systems, because the number of uncontrollable transitions in a Petri net is often rather limited. In the course of the book it is made clear that a lot of important results can be deduced for supervisory control of such Petri nets.

In Sections 1 and 2 we partition the set of places and the set of transitions of an acyclic Petri net. We show, by using these partitions, that by properly ordering the rows and columns of the incidence matrices $\mathbf{F}^{-}$ and $\mathbf{F}^{+}$, these matrices obtain a block triangular structure. A standard result for acyclic Petri nets says that the necessary condition for reachability in Proposition 1.9 (p. 13) is also sufficient. The present approach allows us to give a new proof for this property (Section 3). Further, we formulate reachability properties which are important for the construction of supervisory controllers discussed in Chapters 6 and 7.

In Section 4 we give, as an alternative for solving a set of Diophantine equalities and inequalities, a new algorithm to check whether, in an acyclic Petri net, a given state is reachable from another given state. Acyclic Petri nets which do not contain any choice places are used in Chapters 6 and 7, and they are therefore studied in Section 5. Finally, in Section 6 the construction of the supremal controllable subset $\mathcal{A}^{*}$ for Petri nets with an acyclic uncontrollable part $\mathcal{N}_{\mathcal{A}}^{\text{uc}}$, is formulated

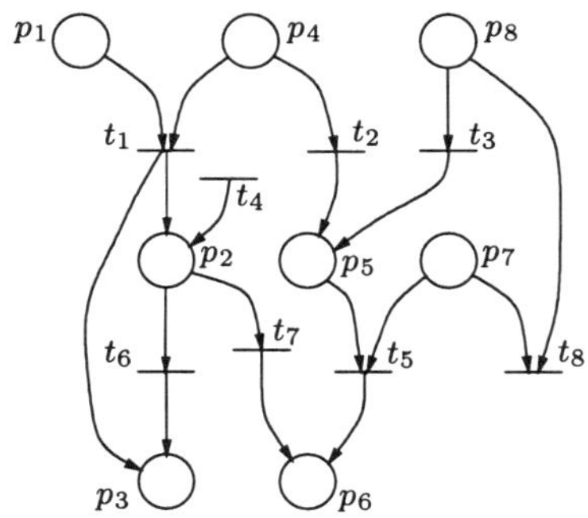

Figure 5.1. An acyclic Petri net with all arc weights equal to one.

as a discrete optimisation problem instead of a Petri net reachability problem.

1. Partitioning of the sets of places and transitions

An acyclic Petri net is a Petri that does not contain any cycles (Definition 1.8, p. 17). An example of an acyclic Petri net is given in Figure 5.1. In this section we partition the set of places and the set of transitions of such a Petri net.

Definition 5.1. *Define for an acyclic Petri net* $\mathcal{N} = (P, T, f^+, f^-)$ *and for each* $k \in \mathbb{N} \setminus \{0\}$ *a set of transitions* $T_k \subseteq T$ *and a set of places* $P_k \subseteq P$ *as*

$$\begin{aligned} T_k &:= \{t \in T \setminus \cup_{i=1}^{k-1} T_i \mid {}^\bullet t \subseteq \cup_{i=1}^{k-1} P_i\}, \\ P_k &:= \{p \in P \setminus \cup_{i=1}^{k-1} P_i \mid {}^\bullet p \subseteq \cup_{i=1}^{k} T_i\}. \end{aligned} \tag{5.1}$$

It is clear that all sets T_k and P_k with $k \in \mathbb{N} \setminus \{0\}$ are mutually disjoint.

Proposition 5.1. *Let* $i, j \in \mathbb{N} \setminus \{0\}$. *If* $i \neq j$, *then* $P_i \cap P_j = \emptyset$ *and* $T_i \cap T_j = \emptyset$.

The set T_1 is the set of all source transitions in $\mathcal{N}$, i.e. it is the set of transitions which do not contain any input places (Definition 1.9, p. 17). The set P_1 is the set of places in $\mathcal{N}$ of which all input transitions belong to T_1. In particular, all places without input transitions belong to P_1.

Example 5.1. Consider the acyclic Petri net in Figure 5.1. All arcs have weight one. By applying the partitioning of Definition 5.1, we obtain that

$$\begin{aligned} T_1 &= \{t_4\}, & P_1 &= \{p_1, p_4, p_7, p_8\}, \\ T_2 &= \{t_1, t_2, t_3, t_8\}, & P_2 &= \{p_2, p_5\}, \\ T_3 &= \{t_5, t_6, t_7\}, & P_3 &= \{p_3, p_6\}. \end{aligned}$$

All other sets T_i and P_i for $i \geq 4$ are empty. △

Proposition 5.2. *For an acyclic Petri net and the sets in Definition 5.1 it holds that:*

(i) $\forall k \in \mathbb{N} \setminus \{0,1\} : \cup_{i=1}^{k-1} T_i \neq T \Rightarrow T_k \neq \emptyset$,

(ii) $\forall k \in \mathbb{N} \setminus \{0\} : \cup_{i=1}^{k-1} P_i \neq P \Rightarrow P_k \neq \emptyset$.

Proof. (i) By contradiction. Suppose that $T_k = \emptyset$ with $k \in \mathbb{N} \setminus \{0,1\}$. From this we obtain with Definition 5.1 that

$$\forall t \in T \setminus \cup_{i=1}^{k-1} T_i : {}^\bullet t \not\subseteq \cup_{i=1}^{k-1} P_i, \tag{5.2}$$

or that

$$\forall t \in T \setminus \cup_{i=1}^{k-1} T_i, \exists p_0 \in P \setminus \cup_{i=1}^{k-1} P_i : p_0 \in {}^\bullet t.$$

As p_0 belongs to $P \setminus \cup_{i=1}^{k-1} P_i$ and as $k \in \mathbb{N} \setminus \{0,1\}$, it holds that ${}^\bullet p_0 \not\subseteq \cup_{i=1}^{k-1} T_i$ (Definition 5.1) or that

$$\exists t_1 \in T \setminus \cup_{i=1}^{k-1} T_i : t_1 \in {}^\bullet p_0.$$

This reasoning can be repeated for t_1 (see (5.2)). As $\cup_{i=1}^{k-1} T_i \neq T$, in this way a path $\pi = (t_l, \ldots, t_1, p_0, t)$ with $l+1 > m = \sharp T$ can be constructed. Consequently, π contains at least $m+1$ transitions of which at least two are equal. Hence, $\mathcal{N}$ contains a cycle.

(ii) The proof is analogous to that of (i). □

For $k = 2$ the result of Proposition 5.2(i) is that $T_2 \neq \emptyset$ if $T_1 \neq T$. For the sequence of sets of transitions $(T_1, T_2, \ldots, T_\lambda, T_{\lambda+1}, \ldots)$, a natural number $\lambda \geq 1$ can be defined such that $T_k \neq \emptyset$ if $1 < k \leq \lambda$ and $T_k = \emptyset$ if $k > \lambda$. Note that T_1 can be empty although T_2 is not. If $T_1 \neq \emptyset$, then T is partitioned into λ sets. If $T_1 = \emptyset$, then T is partitioned into $\lambda - 1$ sets.

For $k = 1$ the result of Proposition 5.2(ii) is that $P_1 \neq \emptyset$, because P is not empty. For the sequence of sets of places $(P_1, P_2, \ldots, P_{\lambda'}, P_{\lambda'+1}, \ldots)$, a natural number $\lambda' \geq 1$ can be defined such that $P_k \neq \emptyset$ if $1 \leq k \leq \lambda'$ and $P_k = \emptyset$ if $k > \lambda'$. Consequently, P is partitioned into λ' sets.

Proposition 5.3. *Wit λ and λ' as defined above, it holds that $\lambda = \lambda'$ or $\lambda = \lambda' + 1$.*

Proof. From Definition 5.1 it follows for each $k \in \mathbb{N} \setminus \{0\}$ that

$$\begin{aligned} \cup_{i=1}^{k} T_i = T \quad &\Rightarrow \quad \cup_{i=1}^{k} P_i = P, \\ \cup_{i=1}^{k} P_i = P \quad &\Rightarrow \quad \cup_{i=1}^{k+1} T_i = T. \end{aligned}$$

Together with the definition of λ and λ', this implies that $\lambda = \lambda'$ or that $\lambda = \lambda' + 1$. □

Corollary 5.1. *For all $k = 1, \ldots, \lambda$ it is true that*

$$^{\bullet}T_k \subseteq \cup_{i=1}^{k-1} P_i, \qquad T_k^{\bullet} \subseteq \cup_{i=k}^{\lambda'} P_i,$$

and for all $k' = 1, \ldots, \lambda'$ that

$$^{\bullet}P_{k'} \subseteq \cup_{i=1}^{k'} T_i, \qquad P_{k'}^{\bullet} \subseteq \cup_{i=k'+1}^{\lambda} T_i.$$

Corollary 5.2. *For each path*

$$(p_k, t_k, p_{k-1}, t_{k-1}, \ldots, p_1, t_1, p_0)$$

in an acyclic Petri net, with $p_j \in P_{i_j}$ for $j = 1, \ldots, k$ and $t_j \in T_{i'_j}$ for $j' = 1, \ldots, k$, it holds that

$$1 \leq i_k < i'_k \leq i_{k-1} < i'_{k-1} \leq \cdots \leq i_1 < i'_1 \leq i_0 \leq \lambda'.$$

Similar properties for paths which start or end in a transition follow immediately. Note that for each pair of places or transitions which belong to the same set P_i or T_i, there does not exist a path which links these places or transitions.

All the above properties can be verified in the acyclic Petri net of Figure 5.1 and Example 5.1. In particular, $\lambda = \lambda' = 3$.

2. Structure of the incidence matrices

Now we apply the partitions of P and T of the previous section to structure the incidence matrices $\mathbf{F}^+$ and $\mathbf{F}^-$ of an acyclic Petri net. To limit notational complexity, we assume in first instance that $T_1 \neq \emptyset$ and that $\lambda = \lambda'$. Order rows in $\mathbf{F}^+$ and $\mathbf{F}^-$ such that places in P_1 come first, then places in P_2, and so forth. Order columns in $\mathbf{F}^+$ and $\mathbf{F}^-$ such that transitions in T_1 come first, then transitions in T_2, and so forth. Corollary 5.1 implies that $\mathbf{F}^+$ and $\mathbf{F}^-$ now have the following block triangular structure:

$$\mathbf{F}^+ = \begin{bmatrix} \mathbf{F}^+_{1,1} & \mathbf{O}_{1,2} & \mathbf{O}_{1,3} & \cdots & \mathbf{O}_{1,\lambda} \\ \mathbf{F}^+_{2,1} & \mathbf{F}^+_{2,2} & \mathbf{O}_{2,3} & \cdots & \mathbf{O}_{2,\lambda} \\ \vdots & \vdots & \ddots & \ddots & \vdots \\ \mathbf{F}^+_{\lambda-1,1} & \cdots & \cdots & \ddots & \mathbf{O}_{\lambda-1,\lambda} \\ \mathbf{F}^+_{\lambda,1} & \cdots & \cdots & \cdots & \mathbf{F}^+_{\lambda,\lambda} \end{bmatrix},$$

$$\mathbf{F}^- = \begin{bmatrix} \mathbf{O}_{1,1} & \mathbf{F}^-_{1,2} & \mathbf{F}^-_{1,3} & \cdots & \mathbf{F}^-_{1,\lambda} \\ \mathbf{O}_{2,1} & \mathbf{O}_{2,2} & \mathbf{F}^-_{2,3} & \cdots & \mathbf{F}^-_{2,\lambda} \\ \vdots & \vdots & \ddots & \ddots & \vdots \\ \mathbf{O}_{\lambda-1,1} & \cdots & \cdots & \ddots & \mathbf{F}^-_{\lambda-1,\lambda} \\ \mathbf{O}_{\lambda,1} & \cdots & \cdots & \cdots & \mathbf{O}_{\lambda,\lambda} \end{bmatrix}.$$

Note that the order of places and transitions within one particular set P_i or T_i is of no importance whatsoever to the general structure of these matrices.

The matrices $\mathbf{F}^+_{i,j}$ and $\mathbf{F}^-_{i,j}$ belong to $\mathbb{N}^{\sharp P_i \times \sharp T_j}$ and the matrices $\mathbf{O}_{i,j}$ are zero matrices with dimension $\sharp P_i \times \sharp T_j$. For $i = 1, \ldots, \lambda$ the ith row of matrices in both $\mathbf{F}^+$ and $\mathbf{F}^-$ corresponds to places in P_i. For $i = 1, \ldots, \lambda$ the ith column of matrices in both $\mathbf{F}^+$ and $\mathbf{F}^-$ corresponds to transitions in T_i.

From the definitions of the sets P_i and T_i if follows that each row in $\mathbf{F}^+_{i,i}$ with $i > 1$ contains at least one nonzero element. If not, this would imply that the corresponding place belongs to a set $P_{i'}$ with $i' < i$. Analogously, each column in $\mathbf{F}^-_{i-1,i}$ with $i > 1$ contains at least one nonzero element. If not, the corresponding transition would belong to a set $T_{i'}$ with $i' < i$.

If $\lambda = \lambda' + 1$, then the last row of matrices in $\mathbf{F}^+$ and $\mathbf{F}^-$ should be left out. If $T_1 = \emptyset$, then the first column of matrices in $\mathbf{F}^+$ and $\mathbf{F}^-$ should be left out.

Example 5.2. Consider again the Petri net in Figure 5.1. By using the results of Example 5.1, order rows and columns in $\mathbf{F}^+$ and $\mathbf{F}^-$, respectively, as:

$$\underbrace{p_1, p_4, p_7, p_8}_{P_1}, \underbrace{p_2, p_5}_{P_2}, \underbrace{p_3, p_6}_{P_3} \quad \text{and} \quad \underbrace{t_4}_{T_1}, \underbrace{t_1, t_2, t_3, t_8}_{T_2}, \underbrace{t_5, t_6, t_7}_{T_3}.$$

We obtain for the incidence matrices that:

$$\mathbf{F}^+ = \left(\begin{array}{c|cccc|ccc} 0 & 0 & 0 & 0 & 0 & 0 & 0 & 0 \\ 0 & 0 & 0 & 0 & 0 & 0 & 0 & 0 \\ 0 & 0 & 0 & 0 & 0 & 0 & 0 & 0 \\ 0 & 0 & 0 & 0 & 0 & 0 & 0 & 0 \\ \hline 1 & 1 & 0 & 0 & 0 & 0 & 0 & 0 \\ 0 & 0 & 1 & 1 & 0 & 0 & 0 & 0 \\ \hline 0 & 1 & 0 & 0 & 0 & 0 & 1 & 0 \\ 0 & 0 & 0 & 0 & 0 & 1 & 0 & 1 \end{array}\right),$$

$$\mathbf{F}^- = \left(\begin{array}{c|cccc|ccc} 0 & 1 & 0 & 0 & 0 & 0 & 0 & 0 \\ 0 & 1 & 1 & 0 & 0 & 0 & 0 & 0 \\ 0 & 0 & 0 & 0 & 1 & 1 & 0 & 0 \\ 0 & 0 & 0 & 1 & 1 & 0 & 0 & 0 \\ \hline 0 & 0 & 0 & 0 & 0 & 0 & 1 & 1 \\ 0 & 0 & 0 & 0 & 0 & 1 & 0 & 0 \\ \hline 0 & 0 & 0 & 0 & 0 & 0 & 0 & 0 \\ 0 & 0 & 0 & 0 & 0 & 0 & 0 & 0 \end{array}\right).$$

The submatrices are indicated with horizontal and vertical lines. △

3. Reachability in acyclic Petri nets

3.1. Transition bag assumption

In this section, as well as in the two following sections, reachability properties of acyclic Petri nets are studied. In Theorem 5.1 it is shown that the necessary condition for reachability of Proposition 1.9 (p. 13) is also sufficient for acyclic Petri nets. This result is not new, but our proof is. The present proof is constructive, uses the structural results of the two previous sections and allows to formulate important reachability properties which are used in Chapters 6 and 7 for the construction of supremal controllable subsets.

At first, the transition bag assumption is used, i.e. $\Omega = \Delta$. The present version of the theorem is more general than those found in the literature, because we consider acyclic Petri nets as a part of a larger Petri net. This formulation is useful for control design.

Theorem 5.1. *Let $\mathcal{N} = (P, T, f^+, f^-)$ be a Petri net, $T' \subseteq T$ and $\mathbf{m}_0, \mathbf{m}_\mathrm{f} \in \mathbb{N}^n$. If the sub-Petri net*

$$\mathcal{N}' = ({}^\bullet T', T, f^+|_{T' \times {}^\bullet T'}, f^-|_{{}^\bullet T' \times T'})$$

of $\mathcal{N}$ is acyclic, then $\mathbf{m}_\mathrm{f} \in \mathcal{R}_\infty^{T'}(\mathbf{m}_0)$ if and only if there exists a vector $\mathbf{y} \in \mathbb{N}^m$ with $\mathbf{y}|_{T\setminus T'} = \mathbf{0}$ such that

$$\mathbf{m}_\mathrm{f} = \mathbf{m}_0 + \mathbf{F}\mathbf{y}. \tag{5.3}$$

Proof. ($\Rightarrow$) This follows from Proposition 1.9.

($\Leftarrow$) Partition $P' := {}^\bullet T'$ and T' as in Definition 5.1. Define for $i = 1, \ldots, \lambda$ vectors $\boldsymbol{\delta}_i \in \mathbb{N}^m$ such that

$$\boldsymbol{\delta}_i|_{T_i} = \mathbf{y}|_{T_i}, \tag{5.4}$$

$$\boldsymbol{\delta}_i|_{T\setminus T_i} = \mathbf{0}. \tag{5.5}$$

As $\mathbf{m} \xrightarrow{\mathbf{0}} \mathbf{m}$ holds for all $\mathbf{m}$, we assume that $\boldsymbol{\delta}_i \neq \mathbf{0}$ ($i = 1, \ldots, \lambda$). We show that the sequence $(\boldsymbol{\delta}_1, \ldots, \boldsymbol{\delta}_\lambda)$ in (5.4)–(5.5) is enabled under $\mathbf{m}_0$. This is equivalent to

$$\mathbf{m}_0 \xrightarrow{\boldsymbol{\delta}_1} \mathbf{m}_1 \xrightarrow{\boldsymbol{\delta}_2} \cdots \xrightarrow{\boldsymbol{\delta}_{\lambda-1}} \mathbf{m}_{\lambda-1} \xrightarrow{\boldsymbol{\delta}_\lambda} \mathbf{m}_\lambda, \tag{5.6}$$

or to

$$\boldsymbol{\delta}_i \in \Delta_{\mathbf{m}_{i-1}} \quad \text{for } i = 1, \ldots, \lambda. \tag{5.7}$$

From (5.3) it follows that $\mathbf{m}_0 + \mathbf{F}\mathbf{y} \geq \mathbf{0}$ or that

$$\mathbf{m}_0(p) + \sum_{t \in {}^\bullet p} f^+(t,p)\mathbf{y}(t) \geq \sum_{t \in p^\bullet} f^-(p,t)\mathbf{y}(t) \quad \text{for } p \in P. \tag{5.8}$$

From Definition 5.1 we have that ${}^\bullet T_1 \cap P' = \emptyset$. This implies that ${}^\bullet T_1 = \emptyset$. This is because $T_1 \neq \emptyset$ and $T_1 \subseteq T'$ imply that ${}^\bullet T' \neq \emptyset$ and together with ${}^\bullet T_1 \subseteq {}^\bullet T'$ that ${}^\bullet T_1 \cap {}^\bullet T' = {}^\bullet T_1 \cap P' = \emptyset$. Consequently, it follows from $\mathbf{y}|_{T\setminus T_1} = \mathbf{0}$ that

$$\mathbf{m}_0(p) \geq \sum_{t \in p^\bullet} f^-(p,t)\boldsymbol{\delta}_1(t) = 0 \quad \text{for } p \in P,$$

and that $\boldsymbol{\delta}_1 \in \Delta_{\mathbf{m}_0}$. The rest of the proof is done by induction. The induction hypothesis is that

$$\mathbf{m}_0 \xrightarrow{\boldsymbol{\delta}_1} \mathbf{m}_1 \xrightarrow{\boldsymbol{\delta}_2} \cdots \xrightarrow{\boldsymbol{\delta}_{i-1}} \mathbf{m}_{i-1}. \tag{5.9}$$

To prove that $\boldsymbol{\delta}_i \in \Delta_{\mathbf{m}_{i-1}}$, it suffices to show (see (5.4) and (5.5)) that

$$\mathbf{m}_{i-1}(p) \geq \sum_{t \in p^\bullet} f^-(p,t)\boldsymbol{\delta}_i(t) \quad \text{for } p \in {}^\bullet T_i. \tag{5.10}$$

From (5.9) we obtain that (see (5.4) and (5.5) for the second inequality)

$$\begin{aligned}
\mathbf{m}_{i-1}(p) &= \mathbf{m}_0(p) + \sum_{t \in {}^\bullet p} \left[f^+(t,p) \sum_{j=1}^{i-1} \boldsymbol{\delta}_j(t) \right] \\
&\qquad - \sum_{t \in p^\bullet} \left[f^-(p,t) \sum_{j=1}^{i-1} \boldsymbol{\delta}_j(t) \right] \qquad (p \in {}^\bullet T_i), \\
&= \mathbf{m}_0(p) + \sum_{t \in {}^\bullet p \cap (\cup_{j=1}^{i-1} T_j)} f^+(t,p)\mathbf{y}(t) \\
&\qquad - \sum_{t \in p^\bullet \cap (\cup_{j=1}^{i-1} T_j)} f^-(p,t)\mathbf{y}(t) \qquad (p \in {}^\bullet T_i).
\end{aligned} \tag{5.11}$$

From Corollary 5.1 we know that

$$^\bullet({}^\bullet T_i) \subseteq \cup_{j=1}^{i-1} {}^\bullet P_j \subseteq T \setminus \left(\cup_{j=1}^{i-1} T_j\right) = (T \setminus T') \dot{\cup} \left(\cup_{j=1}^{i-1} T_j\right).$$

Together with $\mathbf{y}|_{T \setminus T'} = \mathbf{0}$, it follows from (5.11) that

$$\mathbf{m}_{i-1}(p) = \mathbf{m}_0(p) + \sum_{t \in {}^\bullet p} f^+(t,p)\mathbf{y}(t) - \sum_{t \in p^\bullet \cap (\cup_{j=1}^{i-1} T_j)} f^-(p,t)\mathbf{y}(t) \qquad (p \in {}^\bullet T_i).$$

After substitution in (5.10), it suffices to prove that

$$\mathbf{m}_0(p) + \sum_{t \in {}^\bullet p} f^+(t,p)\mathbf{y}(t) \geq \sum_{t \in p^\bullet \cap (\cup_{j=1}^{i} T_j)} f^-(p,t)\mathbf{y}(t) \qquad (p \in {}^\bullet T_i).$$

This follows from (5.8). As $\sum_{j=1}^{\lambda} \boldsymbol{\delta}_j(t) = \mathbf{y}(t)$ for each $t \in T$ (see (5.4) and (5.5)), it holds that $\mathbf{m}_\lambda = \mathbf{m}_\mathrm{f}$ in (5.6). □

Checking condition (5.3) amounts to solving a set of linear Diophantine equalities and inequalities.

To limit notational complexity, in the rest of this section and in the two following sections we only consider, in contrast to Theorem 5.1, acyclic Petri nets which are not embedded in a larger net. The generalisation of the results to the more general case is straightforward.

Owing to the constructive nature of the proof of Theorem 5.1 two important corollaries can be formulated. Given a solution $\mathbf{y} \in \mathbb{N}^m$ of (5.3) it is a natural question how the final marking $\mathbf{m}_\mathrm{f}$ can be reached from $\mathbf{m}_0$, i.e. how to construct a sequence of transition vectors with sum $\mathbf{y}$ which is enabled under $\mathbf{m}_0$. The following corollary provides the answer to this problem.

Corollary 5.3. *Consider an acyclic Petri net and states $\mathbf{m}_0$ and $\mathbf{m}_\mathrm{f}$. If $\mathbf{m}_\mathrm{f} = \mathbf{m}_0 + \mathbf{F}\mathbf{y}$ with $\mathbf{y} \in \mathbb{N}^m$, then it is true that*

$$\mathbf{m}_0 \xrightarrow{(\boldsymbol{\delta}_1,\ldots,\boldsymbol{\delta}_\lambda)} \mathbf{m}_\mathrm{f},$$

where

$$\boldsymbol{\delta}_i|_{T_i} = \mathbf{y}|_{T_i} \quad \textit{and} \quad \boldsymbol{\delta}_i|_{T\setminus T_i} = \mathbf{0} \qquad (i = 1, \ldots, \lambda).$$

Corollary 5.4. *In an acyclic Petri net it holds for each state $\mathbf{m}_0$ that*

$$\mathcal{R}_\infty(\mathbf{m}_0) = \mathcal{R}_\lambda(\mathbf{m}_0) = \mathcal{R}_1^{T_\lambda}(\cdots(\mathcal{R}_1^{T_2}(\mathcal{R}_1^{T_1}(\mathbf{m}_0)))\cdots).$$

(Recall from Chapter 1 that $\mathcal{R}_\cdot^\cdot(\cdot)$ are reachable sets under the transition bag assumption.)

Starting from an initial state $\mathbf{m}_0$ each reachable state $\mathbf{m}_\mathrm{f}$ can be reached in at most λ steps. First a transition vector with transitions in T_1 is executed, then a transition vector with transitions in T_2, and so forth until T_λ. Note that λ only depends on the structure of the Petri net, thus on $\mathbf{F}^+$ and $\mathbf{F}^-$, and not on the initial and final states $\mathbf{m}_0$ and $\mathbf{m}_\mathrm{f}$.

If ${}^\bullet P_1 = \emptyset$, then the result of Corollary 5.4 can be made stronger; each reachable state is then reachable in at most $\lambda - 1$ steps.

Proposition 5.4. *Consider an acyclic Petri net. If ${}^\bullet P_1 = \emptyset$ and $\lambda > 1$, then*

$$\mathcal{R}_\infty(\mathbf{m}_0) = \mathcal{R}_{\lambda-1}(\mathbf{m}_0) = \mathcal{R}_1^{T_\lambda}(\cdots(\mathcal{R}_1^{T_3}(\mathcal{R}_1^{T_1\cup T_2}(\mathbf{m}_0)))\cdots).$$

Proof. With the notation of the proof of Theorem 5.1 we show that $\boldsymbol{\delta}_1 + \boldsymbol{\delta}_2 \in \Delta_{\mathbf{m}_0}$ if ${}^\bullet P_1 = \emptyset$. Because of ${}^\bullet T_1 = \emptyset$ and (5.4) and (5.5) for $i = 1$ and $i = 2$ it suffices to prove that

$$\mathbf{m}_0(p) \geq \sum_{t\in p^\bullet} f^-(p,t)\boldsymbol{\delta}_2(t) \quad \text{for } p \in {}^\bullet T_2. \tag{5.12}$$

From (5.6) it follows that $\boldsymbol{\delta}_2 \in \Delta_{\mathbf{m}_1}$ and because ${}^\bullet T_1$ is empty, that $\mathbf{m}_1(p) = \mathbf{m}_0(p) + \sum_{t\in {}^\bullet p} f^+(t,p)\boldsymbol{\delta}_1(t)$ for all $p \in P$. This implies that

$$\mathbf{m}_0(p) + \sum_{t\in {}^\bullet p} f^+(t,p)\boldsymbol{\delta}_1(t) \geq \sum_{t\in p^\bullet} f^-(p,t)\boldsymbol{\delta}_2(t) \quad \text{for } p \in {}^\bullet T_2. \tag{5.13}$$

As ${}^\bullet({}^\bullet T_2) \subseteq {}^\bullet P_1 = \emptyset$ (see Corollary 5.1), the second term on the left hand side of (5.13) equals zero. This implies (5.12). □

If there are no source transitions, i.e. if $T_1 = \emptyset$, then it holds that ${}^\bullet P_1 = \emptyset$. For the Petri net in Figure 5.1 one can verify that ${}^\bullet P_1$ is empty: each reachable state is reachable in not more than two steps.

For $i = 1, \ldots, \lambda$ define $\tau_i := \sharp T_i$ and

$$\tau_i' := \sum_{k=1}^{i} \tau_k = \sharp \left(\cup_{k=1}^{i} T_k \right). \tag{5.14}$$

Thus, $m = \sharp T = \tau_\lambda'$. Rename transitions in T such that $T = \{t_1, \ldots, t_m\}$ and that

$$t_{\tau_{i-1}'+1}, \ldots, t_{\tau_i'} \in T_i \quad \text{for } i = 1, \ldots, \lambda. \tag{5.15}$$

Herein is $\tau_0' := 0$. Proposition 5.5 follows from Corollary 5.4 and Proposition 1.7 (p. 10).

Proposition 5.5. *In an acyclic Petri net it holds for each state* $\mathbf{m}_0$ *that*

$$\mathcal{R}_\infty(\mathbf{m}_0) = \mathcal{R}_1^{t_m}(\cdots(\mathcal{R}_1^{t_1}(\mathbf{m}_0))\cdots). \tag{5.16}$$

Each reachable state from $\mathbf{m}_0$ can be reached by first executing t_1 a number of times, then t_2 a number of times, and so forth until t_m. Note that in the notation which precedes Proposition 5.5 there is freedom in the order of the transitions within a set T_i. In total there are

$$\prod_{i=1}^{\lambda} (\sharp T_i)!$$

possibilities. Proposition 5.5 and the freedom in the ordering of transitions is important for the results in Chapter 7.

3.2. Other concurrency assumptions

So far, all presented reachability results in this chapter were formulated under the transition bag assumption. In Chapter 1 also general concurrency assumptions were discussed. From Proposition 1.19 (p. 25) it follows that Theorem 5.1 is also valid for a general concurrency assumption $\Omega \in \mathcal{O}^\downarrow(\Delta, \leq)$ when $\Delta_N \subseteq \Omega$.

With the aid of Proposition 1.7 (p. 10), in the proof of Theorem 5.1 all transition vectors $\boldsymbol{\delta}_i$ $(i = 1, \ldots, \lambda)$ can be decomposed into transition vectors that belong to Ω. Consequently, if $\mathbf{y} \in \mathbb{N}^m$ satisfies (5.3) and the concurrency assumption is used, i.e. $\Omega = \Delta_C$, then $\mathbf{m}_f$ can be reached in

$$\sum_{j=1}^{\lambda} \max_{t \in T_j} (\mathbf{y}(t))$$

steps. Under the no concurrency assumption, i.e. $\Omega = \Delta_N$, $\mathbf{m}_f$ can be reached in

$$\sum_{t \in T} \mathbf{y}(t) \tag{5.17}$$

steps. Note that both values, in contrast to the result of Corollary 5.4, depend on $\mathbf{y}$ and thus on $\mathbf{m}_0$ and $\mathbf{m}_f$.

4. A reachability algorithm

In this section an algorithm to check point-to-point reachability in acyclic Petri nets is presented, as an alternative to solving a set of linear Diophantine equalities and inequalities. Proposition 5.6 provides a necessary and sufficient condition for the finiteness of all reachable sets $\mathcal{R}_\infty(\mathbf{m}_0)$ ($\mathbf{m}_0 \in \mathbb{N}^n$) in an acyclic Petri net.

Proposition 5.6. *Consider an acyclic Petri net $\mathcal{N}$ in which each source transition has at least one output place (to rule out trivial cases). The reachable sets $\mathcal{R}_\infty(\mathbf{m}_0)$ are finite for all $\mathbf{m}_0 \in \mathbb{N}^n$ if and only if there are no source transitions in $\mathcal{N}$.*

Proof. ($\Leftarrow$) As T_1 is empty, it follows from Proposition 5.4 that

$$\mathcal{R}_\infty(\mathbf{m}_0) = \mathcal{R}_1^{T_\lambda}(\cdots(\mathcal{R}_1^{T_2}(\mathbf{m}_0))\cdots)$$

and that the set of transition vectors which are zero on $T \backslash T_2$ and enabled under $\mathbf{m}_0$ is finite. This also holds for the set of transition vectors which are zero on $T \setminus T_{i+1}$ and enabled under $\mathbf{m}_i$ ($i = 2, \ldots, \lambda - 1$). Herein, $\mathbf{m}_i$ belongs to $\mathcal{R}_1^{T_i}(\cdots(\mathcal{R}_1^{T_2}(\mathbf{m}_0))\cdots)$. Together with the finiteness of λ, this finishes the proof.

($\Rightarrow$) By contraposition. Suppose that T_1 is not empty. As each source transition has at least one output place, the state of each of these output places can grow without bound. □

Consequently, if there are no source transitions in an acyclic Petri net, then the reachable set of any initial state is finite and reachability problems can be answered via enumeration. This enumeration process can be structured with the aid of Proposition 5.4.

By applying the necessary and sufficient condition of Theorem 5.1, it can be checked whether in an acyclic Petri net $\mathbf{m}_{\mathrm{f}}$ can be reached from $\mathbf{m}_0$. It suffices to show that the set of linear equalities $\mathbf{m}_{\mathrm{f}} = \mathbf{m}_0 + \mathbf{F}\mathbf{y}$ has a solution $\mathbf{y} \in \mathbb{N}^m$. This is a standard problem in number theory. Propositions 5.7 and 5.8 give an alternative to this solution method in the case $\mathcal{R}_\infty(\mathbf{m}_0)$ is finite. The transition bag assumption is adopted.

Proposition 5.7. *Consider an acyclic Petri net and states* $\mathbf{m}_0$ *and* $\mathbf{m}_{\mathrm{f}}$. *If*

$$\tilde{\mathbf{m}} \in \mathcal{R}_1^{T_i}(\cdots(\mathcal{R}_1^{T_1}(\mathbf{m}_0))\cdots)$$

with $i \in \{1, \ldots, \lambda - 1\}$ *and if there exists a place* p *in* $\cup_{j=1}^{i} P_j$ *such that* $\tilde{\mathbf{m}}(p) < \mathbf{m}_{\mathrm{f}}(p)$, *then* $\mathbf{m}_{\mathrm{f}} \notin \mathcal{R}_1^{T_\lambda}(\cdots(\mathcal{R}_1^{T_{i+1}}(\tilde{\mathbf{m}}))\cdots)$.

Proof. As the input transitions of a place $p \in \cup_{j=1}^{i} P_j$ belong to $\cup_{j=1}^{i} T_j$ (Corollary 5.1), it holds for each state $\bar{\mathbf{m}} \in \mathcal{R}_1^{T_\lambda}(\cdots(\mathcal{R}_1^{T_{i+1}}(\tilde{\mathbf{m}}))\cdots)$ that $\bar{\mathbf{m}}(p) \leq \tilde{\mathbf{m}}(p)$. If there exists a p in $\cup_{j=1}^{i} P_j$ such that $\tilde{\mathbf{m}}(p) < \mathbf{m}_{\mathrm{f}}(p)$, then it holds that $\bar{\mathbf{m}}(p) < \mathbf{m}_{\mathrm{f}}(p)$, from which the result follows. □

Intuitively, this result follows from the fact that when transition vectors are executed as in Corollary 5.3, then the state of a place $p \in P_i$ does not decrease until $\boldsymbol{\delta}_i$ is executed and does not increase afterwards.

Define for each place $p \in P_i$ $(i = 1, \ldots, \lambda')$ the natural number

$$\omega(p) := \begin{cases} \max\{j \mid t \in T_j, t \in p^\bullet\} & \text{if } p^\bullet \neq \emptyset, \\ i & \text{if } p^\bullet = \emptyset. \end{cases}$$

If $p^\bullet \neq \emptyset$, then $\omega(p)$ is the highest index j of a set T_j to which an output transition of p belongs. All input and output transitions of $p \in P_i$ belong to $\cup_{j=1}^{\omega(p)} T_j$.

Proposition 5.8. *Consider an acyclic Petri net. If*

$$\tilde{\mathbf{m}} \in \mathcal{R}_1^{T_i}(\cdots(\mathcal{R}_1^{T_1}(\mathbf{m}_0))\cdots)$$

with $i \in \{1, \ldots, \lambda - 1\}$ *and if there exists a place* p *in* $\cup_{j=1}^{i-1} P_j$ *such that* $\omega(p) \leq i$ *and* $\tilde{\mathbf{m}}(p) \neq \mathbf{m}_{\mathrm{f}}(p)$, *then* $\mathbf{m}_{\mathrm{f}} \notin \mathcal{R}_1^{T_\lambda}(\cdots(\mathcal{R}_1^{T_{i+1}}(\tilde{\mathbf{m}}))\cdots)$.

Proof. If $p \in \cup_{j=1}^{i-1} P_j$ and $\omega(p) \leq i$, then it holds for each

$$\bar{\mathbf{m}} \in \mathcal{R}_1^{T_\lambda}(\cdots(\mathcal{R}_1^{T_{i+1}}(\tilde{\mathbf{m}}))\cdots)$$

that $\bar{\mathbf{m}}(p) = \tilde{\mathbf{m}}(p)$. Together with $\tilde{\mathbf{m}}(p) \neq \mathbf{m}_{\mathrm{f}}(p)$, this finishes the proof. □

When $\mathcal{R}_\infty(\mathbf{m}_0)$ is finite (see Proposition 5.6), Propositions 5.7 and 5.8 can be used in a branch and bound algorithm to check whether $\mathbf{m}_\mathrm{f}$ is reachable from $\mathbf{m}_0$.

First construct $\bar{\mathcal{R}}_1^{T_2}(\mathbf{m}_0) \subseteq \mathcal{R}_1^{T_2}(\mathbf{m}_0)$ as the set of all states which are reachable from $\mathbf{m}_0$ by the execution of one transition vector which only contains transition in T_2, and which are not rejected by Proposition 5.7 or 5.8. In an analogous manner, construct the set $\bar{\mathcal{R}}_1^{T_3}(\bar{\mathcal{R}}_1^{T_2}(\mathbf{m}_0))$. Repeat this procedure until $\mathbf{m}_\mathrm{f}$ is found, until for a $j \in \{2, \ldots, \lambda\}$ the set $\bar{\mathcal{R}}_1^{T_j}(\cdots(\bar{\mathcal{R}}_1^{T_2}(\mathbf{m}_0))\cdots)$ is empty or until $\mathbf{m}_\mathrm{f}$ is not found in the set $\mathcal{R}_1^{T_\lambda}(\bar{\mathcal{R}}_1^{T_{\lambda-1}}(\cdots(\bar{\mathcal{R}}_1^{T_2}(\mathbf{m}_0))\cdots))$.

The usefulness of this algorithm strongly depends on the values of $\omega(p) - i$ (see Proposition 5.8). An example where this algorithm is useful, is when all matrices in $\mathbf{F}^+$ and $\mathbf{F}^-$ are zero except $\mathbf{F}^+_{i,i}$ and $\mathbf{F}^-_{i-1,i}$ with $i > 1$. The *generalised tree structure* in [24] belongs to this class of acyclic Petri nets. This structure is defined with the additional condition that none of the places has more than one output transition. The number of choice places also influences the usefulness of the here presented algorithm.

5. Acyclic Petri nets free of choice places

In this section we study acyclic Petri nets which do not contain any choice places (Definition 1.9, p. 17), i.e.

$$\sharp(p^\bullet) \leq 1 \quad \text{for all } p \in P. \tag{5.18}$$

Each place has at most one outgoing arc. This class of acyclic Petri nets is used in Chapter 7 when $\mathcal{A}^*$ is constructed for a class of Petri nets with acyclic influencing nets. Moreover, because of the following property, it plays an important role in the decomposition results of Chapter 6.

Proposition 5.9. *Consider an acyclic Petri net that satisfies* (5.18), *i.e. it is free of choice places. Let* $\mathbf{m}_0 \in \mathbb{N}^n$ *and* $\mathbf{y}_1, \mathbf{y}_2 \in \mathbb{N}^m$. *If* $\mathbf{m}_0 + \mathbf{F}\mathbf{y}_1 \geq \mathbf{0}$ *and* $\mathbf{m}_0 + \mathbf{F}\mathbf{y}_2 \geq \mathbf{0}$, *then* $\mathbf{y} = \sup(\mathbf{y}_1, \mathbf{y}_2)$ *is such that* $\mathbf{m}_0 + \mathbf{F}\mathbf{y} \geq \mathbf{0}$.

Proof. With Corollary 5.3 we have that

$$\mathbf{m}_0^j \xrightarrow{\boldsymbol{\delta}_1^j} \cdots \xrightarrow{\boldsymbol{\delta}_\lambda^j} \mathbf{m}_\lambda^j \quad \text{for } j = 1, 2, \tag{5.19}$$

with $\mathbf{m}_0 = \mathbf{m}_0^1 = \mathbf{m}_0^2$ and

$$\boldsymbol{\delta}_i^j|_{T_i} = \mathbf{y}_j|_{T_i} \quad \text{and} \quad \boldsymbol{\delta}_i^j|_{T\setminus T_i} = \mathbf{0} \quad \text{for } i = 1, \ldots, \lambda; j = 1, 2. \tag{5.20}$$

We show that

$$\mathbf{m}_0 \xrightarrow{\boldsymbol{\delta}_1} \cdots \xrightarrow{\boldsymbol{\delta}_\lambda} \mathbf{m}_\lambda \tag{5.21}$$

with

$$\boldsymbol{\delta}_i := \sup(\boldsymbol{\delta}_i^1, \boldsymbol{\delta}_i^2) \quad \text{for } i = 1, \ldots, \lambda. \tag{5.22}$$

As $(\Delta \cup \{\mathbf{0}\}, \leq)$ is a lattice, $\boldsymbol{\delta}_i$ is well defined for $i = 1, \ldots, \lambda$. Because of (5.20) it is clear that

$$\sum_{i=1}^{\lambda} \boldsymbol{\delta}_i = \sum_{i=1}^{\lambda} \sup(\boldsymbol{\delta}_i^1, \boldsymbol{\delta}_i^2) = \sup(\mathbf{y}_1, \mathbf{y}_2) = \mathbf{y}.$$

By induction we show that (5.21) holds or that

$$\boldsymbol{\delta}_i \in \Delta_{\mathbf{m}_{i-1}} \quad \text{for } i = 1, \ldots, \lambda. \tag{5.23}$$

As ${}^\bullet T_1$ is empty, this holds for $i = 1$. The induction hypothesis is that $\boldsymbol{\delta}_l \in \Delta_{\mathbf{m}_{l-1}}$ holds for $l = 1, \ldots, i-1$. Next, we show that $\boldsymbol{\delta}_i$ belongs to $\Delta_{\mathbf{m}_{i-1}}$. It follows from (5.20) and (5.22) that $\boldsymbol{\delta}_i|_{T \setminus T_i} = \mathbf{0}$. Consequently, $\boldsymbol{\delta}_i \in \Delta_{\mathbf{m}_{i-1}}$ is equivalent to

$$\mathbf{m}_{i-1}(p) \geq \sum_{t \in p^\bullet \cap T_i} f^-(p,t)\boldsymbol{\delta}_i(t) \qquad (p \in {}^\bullet T_i). \tag{5.24}$$

Because of (5.18), each place p in P has at most one output transition. If it exists, this transition is denoted t_p. Then (5.24) is equivalent to

$$\mathbf{m}_{i-1}(p) \geq f^-(p,t_p)\boldsymbol{\delta}_i(t_p) \qquad (p \in {}^\bullet T_i, p^\bullet \neq \emptyset), \tag{5.25}$$

and, because of (5.22), to

$$\mathbf{m}_{i-1}(p) \geq f^-(p,t_p) \max\left(\boldsymbol{\delta}_i^1(t_p), \boldsymbol{\delta}_i^2(t_p)\right) \qquad (p \in {}^\bullet T_i, p^\bullet \neq \emptyset). \tag{5.26}$$

Analogously, it follows from (5.19) that

$$\mathbf{m}_{i-1}^j(p) \geq f^-(p,t_p)\boldsymbol{\delta}_i^j(t_p) \qquad (j = 1,2; p \in {}^\bullet T_i; p^\bullet \neq \emptyset). \tag{5.27}$$

On the other hand, if follows from the induction hypothesis and (5.19), (5.20) and (5.21) that

$$\begin{aligned}\mathbf{m}_{i-1}(p) - \mathbf{m}_{i-1}^j(p) = &\sum_{t \in {}^\bullet p \cap (\cup_{v=1}^{i-1} T_v)} f^+(t,p) \left[\sum_{v=1}^{i-1} \left(\boldsymbol{\delta}_i(t) - \boldsymbol{\delta}_i^j(t)\right)\right] \\ + &\sum_{t \in p^\bullet \cap (\cup_{v=1}^{i-1} T_v)} f^-(p,t) \left[\sum_{v=1}^{i-1} \left(\boldsymbol{\delta}_i^j(t) - \boldsymbol{\delta}_i(t)\right)\right] \qquad (j = 1,2; p \in {}^\bullet T_i).\end{aligned} \tag{5.28}$$

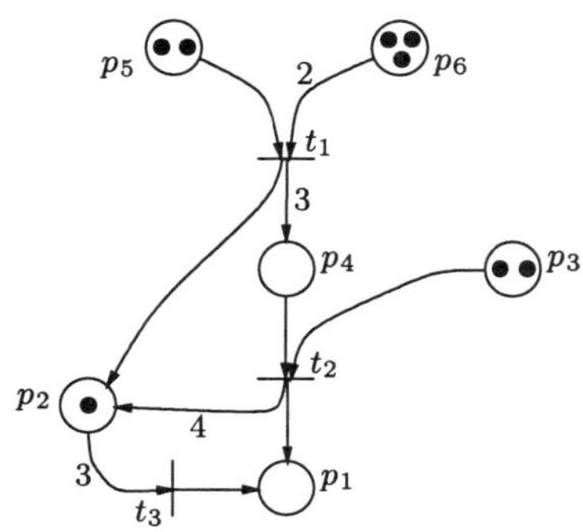

Figure 5.2. An acyclic Petri net which is free of choice places. The indicated state is $\mathbf{m}_0 = (0, 1, 2, 0, 2, 3)$.

From (5.18) we obtain for $p \in {}^\bullet T_i$ that $p^\bullet \subseteq T_i$. Thus, the second term on the right hand side of (5.28) is equal to zero. Consequently, we have that

$$\mathbf{m}_{i-1}(p) \geq \mathbf{m}_{i-1}^j(p) \qquad (j = 1, 2; p \in {}^\bullet T_i).$$

This is equivalent to

$$\mathbf{m}_{i-1}(p) \geq \max(\mathbf{m}_{i-1}^1(p), \mathbf{m}_{i-1}^2(p)) \qquad (p \in {}^\bullet T_i).$$

From this we obtain, together with (5.27), that

$$\mathbf{m}_{i-1}(p) \geq f^-(p, t_p) \max(\boldsymbol{\delta}_i^1(t_p), \boldsymbol{\delta}_i^2(t_p)) \qquad (p \in {}^\bullet T_i, p^\bullet \neq \emptyset).$$

And this is (5.26). □

Example 5.3. The acyclic Petri net in Figure 5.2 is free of choice places. For the indicated state $\mathbf{m}_0$ and the vectors $\mathbf{y}_1 = (1, 1, 1)$ and $\mathbf{y}_2 = (1, 2, 0)$ (the ith element in both $\mathbf{y}_1$ and $\mathbf{y}_2$ corresponds to transition t_i), it holds that $\mathbf{m}_0 + \mathbf{F}\mathbf{y}_i \geq \mathbf{0}$ for $i = 1, 2$. It can easily be verified that $\mathbf{y} = \sup(\mathbf{y}_1, \mathbf{y}_2) = (1, 2, 1)$ satisfies $\mathbf{m}_0 + \mathbf{F}\mathbf{y} \geq \mathbf{0}$. △

From Proposition 5.9 it follows that in an acyclic Petri net free of choice places the supremum of two sequences

$$(\boldsymbol{\delta}_1^j, \ldots, \boldsymbol{\delta}_\lambda^j) \in \bar{\Delta}_{\mathbf{m}_0} \quad \text{with } \boldsymbol{\delta}_i^j|_{T \setminus T_i} = \mathbf{0} \ (j = 1, 2; i = 1, \ldots, \lambda) \qquad (5.29)$$

is also enabled under $\mathbf{m}_0$.

We say that a partially ordered set is *sup-closed* if the supremum of each pair of elements belongs to this set. As $\bar{\Delta}_{\mathbf{m}_0}$ is a finite set if there are no source transitions and as this set is sup-closed under the product order, the set of enabled sequences $(\boldsymbol{\delta}_2,\ldots,\boldsymbol{\delta}_\lambda) \in \bar{\Delta}_{\mathbf{m}_0}$ with $\boldsymbol{\delta}_i|_{T\setminus T_i} = \mathbf{0}$ $(i = 1,\ldots,\lambda)$ has a maximum $(\boldsymbol{\delta}_2^{\mathrm{M}},\ldots,\boldsymbol{\delta}_\lambda^{\mathrm{M}})$ if $T^{\mathrm{so}} = \emptyset$. Without much effort one obtains:

$$\boldsymbol{\delta}_i^{\mathrm{M}}(t) = \min_{p\in{}^\bullet t}\left(\left\lfloor\frac{\mathbf{m}_{i-1}^{\mathrm{M}}(p)}{f^-(p,t)}\right\rfloor\right) \qquad (t \in T_i; i = 2,\ldots,\lambda),$$

where

$$\mathbf{m}_0 = \mathbf{m}_1^{\mathrm{M}} \xrightarrow{(\boldsymbol{\delta}_2^{\mathrm{M}},\ldots,\boldsymbol{\delta}_{i-1}^{\mathrm{M}})} \mathbf{m}_{i-1}^{\mathrm{M}}.$$

This implies that the set $(\{\mathbf{y} \in \mathbb{N}^m \mid \mathbf{m}_0 + \mathbf{F}\mathbf{y} \geq \mathbf{0}\}, \leq)$ has a maximum, namely $\mathbf{y}^{\mathrm{M}} := \sum_{i=2}^{\lambda} \boldsymbol{\delta}_i^{\mathrm{M}}$. Note that $\Delta_{\mathbf{m}_\lambda^{\mathrm{M}}} = \emptyset$, i.e. there are no enabled transition vectors under $\mathbf{m}_\lambda^{\mathrm{M}}$.

Example 5.4. In the case of Figure 5.2, one finds $\mathbf{y}^{\mathrm{M}} = (1,2,3)$ and $\mathbf{m}_\lambda^{\mathrm{M}} = \mathbf{m}_0 + \mathbf{F}\mathbf{y}^{\mathrm{M}} = (5,1,0,1,1,1)$. Verify that $\Delta_{\mathbf{m}_\lambda^{\mathrm{M}}} = \emptyset$. △

6. Construction of the supremal controllable subset

Petri nets $\mathcal{N}$ and legal sets $\mathcal{A}$ for which the influencing net

$$\mathcal{N}_{\mathcal{A}}^{\mathrm{uc}} = (\bar{I}, \bar{U}, f^+|_{\bar{U}\times\bar{I}}, f^-|_{\bar{I}\times\bar{U}})$$

is acyclic, form an important class of systems and they are studied in more detail in Chapter 7. To be able to apply the reachability results of this chapter to the construction of the supremal controllable subset $\mathcal{A}^*$ of $\mathcal{A}$, the follow property is important (cfr. Theorem 5.1).

Proposition 5.10. *Let $\mathcal{N} = (P,T,f^+,f^-,T^{\mathrm{uc}})$ be a Petri net and $\mathcal{A} \subseteq \mathbb{N}^n$. Then $\mathcal{N}_{\mathcal{A}}^{\mathrm{uc}}$ is acyclic if and only if*

$$\mathcal{N}^{\bar{U}} := ({}^\bullet\bar{U}, \bar{U}, f^+|_{\bar{U}\times{}^\bullet\bar{U}}, f^-|_{{}^\bullet\bar{U}\times\bar{U}})$$

is acyclic.

Proof. $(\Rightarrow)$ As ${}^\bullet\bar{U} \subseteq \bar{I}$, this directly follows from the fact that $\mathcal{N}^{\bar{U}}$ is a sub-Petri net of $\mathcal{N}_{\mathcal{A}}^{\mathrm{uc}}$.

$(\Leftarrow)$ Suppose that there exists a cycle $\pi = (p_1,t_1,p_2,\ldots,p_l,t_l)$ in $\mathcal{N}_{\mathcal{A}}^{\mathrm{uc}}$. Then, $p_i \in \bar{I}$ and $t_i \in \bar{U}$ for $i = 1,\ldots,l$. Consequently, p_i belongs to ${}^\bullet\bar{U}$ for $i = 1,\ldots,l$ and π is a cycle in $\mathcal{N}^{\bar{U}}$. □

As already discussed Chapter 4 (Section 6.3), for the design of algorithms which construct the supremal controllable subset of a given legal set, in practice, it suffices to focus on legal sets which are described by a finite disjunction of linear inequalities. In the rest of this section we show that when in addition the influencing net is acyclic, then the problem to decide whether a given state belongs to the supremal controllable subset can be written as an integer linear programming problem.

Consider a legal set given by

$$\mathcal{A} = \bigcup_{i=1}^{\nu} \{\mathbf{m} \in \mathbb{N}^n \mid \mathbf{A}_i^T \mathbf{m} \leq b_i\} \quad \text{with } \mathbf{A}_i \in \mathbb{Z}^n, b_i \in \mathbb{Z} \ \ (i = 1, \ldots, \nu).$$

By definition, a state $\mathbf{m}$ belongs to $\mathcal{A}^*$ if and only if

$$\bigvee_{i=1}^{\nu} \mathbf{A}_i^T \mathbf{m}' \leq b_i \quad \text{for all } \mathbf{m}' \in \mathcal{R}_{\infty}^{T^{\mathrm{uc}}}(\mathbf{m}).$$

With the reduction result of Chapter 3 (Theorem 4.1) this is equivalent to

$$\bigvee_{i=1}^{\nu} \mathbf{A}_i^T|_{\bar{I}} \mathbf{m}'|_{\bar{I}} \leq b_i \quad \text{for all } \mathbf{m}' \in \mathcal{R}_{\infty}^{\bar{U}}(\mathbf{m}). \tag{5.30}$$

Herein, $\mathbf{A}_i^T|_{\bar{I}}$ and $\mathbf{m}|_{\bar{I}}$ are the restrictions of $\mathbf{A}_i^T$ and $\mathbf{m}$ to $\bar{I}$, respectively. Assume $\mathcal{N}_{\mathcal{A}}^{\mathrm{uc}}$ acyclic. From Proposition 5.10 we find that $\mathcal{N}^{\bar{U}} = ({}^{\bullet}\bar{U}, \bar{U}, f^+|_{\bar{U}\times{}^{\bullet}\bar{U}}, f^-|_{{}^{\bullet}\bar{U}\times\bar{U}})$ is also acyclic. Therefore, with Theorem 5.1, condition 5.30 can be rewritten as

$$\begin{aligned} & \bigvee_{i=1}^{\nu} \mathbf{A}_i^T|_{\bar{I}} \, (\mathbf{m}|_{\bar{I}} + \mathbf{F}_{\mathcal{A}}\mathbf{y}) \leq b_i & & (\mathbf{y} \in \mathbb{N}^{\sharp\bar{U}}, \mathbf{m}|_{\bar{I}} + \mathbf{F}_{\mathcal{A}}\mathbf{y} \geq \mathbf{0}), \\ \Leftrightarrow \quad & \bigvee_{i=1}^{\nu} \mathbf{A}_i^T|_{\bar{I}} \, (\mathbf{m}|_{\bar{I}} + \mathbf{F}_{\mathcal{A}}\mathbf{y}) - b_i \leq 0 & & (\mathbf{y} \in \mathbb{N}^{\sharp\bar{U}}, \mathbf{m}|_{\bar{I}} + \mathbf{F}_{\mathcal{A}}\mathbf{y} \geq \mathbf{0}). \end{aligned}$$

Herein, $\mathbf{F}_{\mathcal{A}}$ is the incidence matrix of $\mathcal{N}_{\mathcal{A}}^{\mathrm{uc}}$. We obtain that

$$\mathbf{m} \in \mathcal{A}^* \quad \Leftrightarrow \quad \max_{\substack{\mathbf{y} \in \mathbb{N}^{\sharp\bar{U}}, \\ \mathbf{0} \leq \mathbf{m}|_{\bar{I}} + \mathbf{F}_{\mathcal{A}}\mathbf{y}}} \left[\min_{i=1,\ldots,\nu} \left(\mathbf{A}_i^T|_{\bar{I}} \, (\mathbf{m}|_{\bar{I}} + \mathbf{F}_{\mathcal{A}}\mathbf{y}) - b_i \right) \right] \leq 0. \tag{5.31}$$

If $\mathcal{A}$ is defined by one linear inequality ($\nu = 1$), then we immediately obtain on the right hand side of (5.31) an integer linear programming problem whose outcome should be nonpositive. To check this condition one can either solve the integer linear programming problem or look for a sufficiently low upper bound of this optimisation problem.

If $\nu > 1$, then, by using a standard result from linear optimisation, the left hand side of the inequality in (5.31) can also be written as an integer linear programming problem. This is because each optimisation problem of a piecewise linear convex function over a polyhedron can be written as a linear programming problem. As the (pointwise) minimum of any number of linear functions is a convex function [4, p. 17], this result can be applied here. By introducing one extra variable $z \in \mathbb{Z}$, we obtain in (5.31):

$$\min_{i=1,\ldots,\nu} \left(\mathbf{A}_i^T|_{\bar{I}} \left(\mathbf{m}|_{\bar{I}} + \mathbf{F}_{\mathcal{A}}\mathbf{y}\right) - b_i\right) = \max_{\substack{z \le \mathbf{A}_i^T|_{\bar{I}}(\mathbf{m}|_{\bar{I}} + \mathbf{F}_{\mathcal{A}}\mathbf{y}) - b_i \\ i=1,\ldots,\nu}} z.$$

Thus $\mathbf{m}$ belongs to $\mathcal{A}^*$ if and only if

$$\max z \le 0 \tag{5.32}$$

$$\text{subject to} \begin{cases} z \le \mathbf{A}_i^T|_{\bar{I}} \left(\mathbf{m}|_{\bar{I}} + \mathbf{F}_{\mathcal{A}}\mathbf{y}\right) - b_i \quad (i = 1, \ldots, \nu) \\ \mathbf{y} \in \mathbb{N}^{\sharp \bar{U}} \\ \mathbf{0} \le \mathbf{m}|_{\bar{I}} + \mathbf{F}_{\mathcal{A}}\mathbf{y}. \end{cases}$$

We have cast the Petri net reachability problem, namely to determine whether a given state belongs to the supremal controllable subset of the legal set, into a standard optimisation problem. For this algorithm to be applicable two conditions should be satisfied:

1. The legal set $\mathcal{A}$ is described by a finite number of conjunctions and disjunctions of linear inequalities.
2. The influencing net $\mathcal{N}_{\mathcal{A}}^{\mathrm{uc}}$ is acyclic.

Although the above discussion is applicable to solve the problem for any given state, it is in general not applicable to the construction of $\mathcal{A}^*$ itself. For an infinite legal set, one would obtain an infinite (countable) number of integer linear programming inequalities. However if $\mathcal{A}$ is finite, we have obtained a ready-to-use algorithm for the construction of its supremal controllable subset.

As discussed in Section 6 of Chapter 4, it is not always trivial to determine the invariance properties of a legal set $\mathcal{A}$. Consequently, in some cases the influencing net $\mathcal{N}_{\mathcal{A}}^{\mathrm{uc}}$ is not completely known. Then the above result can be applied to any acyclic uncontrollable Petri net $\mathcal{N}_{\epsilon}^{\mathrm{uc}} = (P_\epsilon, T_\epsilon, f_\epsilon^+, f_\epsilon^-, T_\epsilon)$ which contains $\mathcal{N}_{\mathcal{A}}^{\mathrm{uc}}$. The cost of not knowing $\mathcal{N}_{\mathcal{A}}^{\mathrm{uc}}$ exactly is that the dimension of the optimisation problem becomes larger, namely $\sharp T_\epsilon + 1$, instead of $\sharp \bar{U} + 1$. As $\mathcal{N}_{\mathcal{A}}^{\mathrm{uc}}$ is a sub-Petri net of $\mathcal{N}_{\epsilon}^{\mathrm{uc}}$, it is true that $\sharp \bar{U} \le \sharp T_\epsilon$. This shows that it is beneficial to know the invariance properties of the legal set as precise as possible in order to approximate $\mathcal{N}_{\mathcal{A}}^{\mathrm{uc}}$ as accurately as possible.

7. Notes and references

As acylic Petri nets have not been studied a lot, a reference list is short: [20], [24] and [36]. The necessary and sufficient condition in Theorem 5.1 is proven in a different way in [20] and [24]. The former proof can also be found in [28].

In Section 6, we have improved the result of [24] in three ways. On the one hand the acyclicity condition is not on $\mathcal{N}^{\mathrm{uc}}$ but on the influencing net $\mathcal{N}^{\mathrm{uc}}_{\mathcal{A}}$ (by using the reduction result of Chapter 3). On the other hand $\mathbf{y}$ in (5.32) belongs to $\mathbb{N}^{\sharp\bar{U}}$ and not to $\mathbb{N}^{\sharp T^{\mathrm{uc}}}$, which implies that the maximalisation problem has a smaller dimension. Finally, we also cover the case in which the legal set is given by a finite disjunction of linear inequalities.

In [39] we have published most of the results in Sections 1 to 4.

Chapter 6

DECOMPOSITION OF THE CONTROL DESIGN

In practice, the state of a Petri net must satisfy the conjunction and disjunction of a number of conditions. This corresponds to a legal set which is defined by intersections and unions of sets of states. In this chapter we study the possibility of simplifying the construction of the supremal controllable subset and the design of (maximally) permissive control laws by decomposing them into subproblems. To this purpose the aforementioned structure of the legal set is used. In particular, we present conditions under which the supremal controllable subset of the union of a number of sets is equal to the union of the supremal controllable subsets of these sets. For control design we formulate conditions under which the set of all permissive control values for the union of a number of sets is equal to the union of the sets of all permissive control values for each of these sets. In analogy to the structure of Chapter 4, we first formulate the decomposition result for the supremal controllable subset (Sections 1 to 5) and afterwards that for control design (Section 6).

1. Introduction

Modular design tries to tackle the complexity of a problem by splitting it into several subproblems. In the first part of this chapter we investigate how this principle is applicable to the computation of the supremal controllable subset $\mathcal{A}^*$ of a given legal set $\mathcal{A}$.

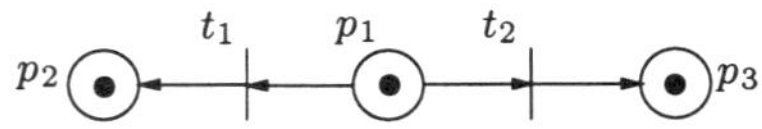

Figure 6.1. An uncontrollable Petri net with initial state $\mathbf{m}_0 = (1,1,1)$.

When $\mathcal{A}$ is equal to the intersection of ν sets $\mathcal{A}_i$, then we know from Proposition 3.13 (p. 66) that

$$\mathcal{A}^* = \left(\bigcap_{i=1}^{\nu} \mathcal{A}_i\right)^* = \bigcap_{i=1}^{\nu} \mathcal{A}_i^*. \tag{6.1}$$

The supremal controllable subset of the intersection equals the intersection of the supremal controllable subsets. The set $\mathcal{A}^*$ can be obtained by calculating the sets $\mathcal{A}_i^*$ $(i = 1, \ldots, \nu)$ first and then taking the intersection of these ν sets. As there is equality in (6.1), this modular design can be done without loss of maximal permissiveness of the corresponding control laws.

For unions of legal sets the situation is different. If $\mathcal{A}$ is equal to the union of ν sets $\mathcal{A}_i$, then we obtain from the definition of $\mathcal{A}^*$ (Definition 3.4, p. 62) that

$$\mathcal{A}^* = \left(\bigcup_{i=1}^{\nu} \mathcal{A}_i\right)^* \supseteq \bigcup_{i=1}^{\nu} \mathcal{A}_i^*. \tag{6.2}$$

It is possible that the subset relation is proper, as it is illustrated next.

Example 6.1. Consider the uncontrollable Petri net in Figure 6.1 and two legal sets $\mathcal{A}_i = \{\mathbf{m} \in \mathbb{N}^n \mid \mathbf{m}(p_{i+1}) \leq 1\}$, $i = 1, 2$. For the indicated state $\mathbf{m}_0$ it is clear that $\mathbf{m}_0 \in (\mathcal{A}_1 \cup \mathcal{A}_2)^*$ and that $\mathbf{m}_0 \notin \mathcal{A}_1^* \cup \mathcal{A}_2^*$. This is explained intuitively as follows. The token in p_1 can make the state of p_2 or the state of p_3 greater than one, however not both at the same time. △

From (6.2) we conclude that decomposition for unions of legal sets provides states which satisfy the control goal; all states which are reachable via uncontrollable transitions from states in $\cup_{i=1}^{\nu} \mathcal{A}_i^*$ belong to $\cup_{i=1}^{\nu} \mathcal{A}_i$. However, when the sets in (6.2) are not equal, then $\cup_{i=1}^{\nu} \mathcal{A}_i^*$ does not contain all states for which all uncontrollably reachable states are legal. This means that the control law designed by using this modularly calculated set of states $\cup_{i=1}^{\nu} \mathcal{A}_i^*$ as control goal is permissive, but in general not maximally permissive. This control law then limits the behaviour of the system more than necessary.

Therefore, in this chapter we look for conditions under which there is equality in (6.2). Then, modular construction of the supremal controllable subset does not lead to too restrictive control laws. We formulate a result for ν equal to two. Decomposition for greater ν is possible when the presented result can be applied successively. The legal sets we consider here are general. No conditions are imposed on their structure or on their invariance properties.

2. Unions of legal sets

Consider two subsets $\mathcal{A}_1$ and $\mathcal{A}_2$ of $\mathbb{N}^n$. In this section we formulate sufficient conditions for

$$\mathcal{A}_1^* \cup \mathcal{A}_2^* = (\mathcal{A}_1 \cup \mathcal{A}_2)^*. \tag{6.3}$$

Define for $j = 1, 2$ the sets $P_j^{\downarrow}$, $P_j^{\uparrow}$, P_j^{d}, $\bar{I}_j$ and $\bar{U}_j$ as shorthand notation for $P^{\downarrow}(\mathcal{A}_j)$, $P^{\uparrow}(\mathcal{A}_j)$, $P^{\mathrm{d}}(\mathcal{A}_j)$, $\bar{I}(\mathcal{A}_j)$ and $\bar{U}(\mathcal{A}_j)$, respectively (see Chapter 4). Further, define the sets

$$\bar{I}_{1,2} := \bar{I}_1 \cap \bar{I}_2 \quad \text{and} \quad \bar{U}_{1,2} := \bar{U}_1 \cap \bar{U}_2. \tag{6.4}$$

Define the Petri net

$$\mathcal{N}_{1,2}^{\mathrm{uc}} := (\bar{I}_{1,2}, \bar{U}_{1,2}, f^+|_{\bar{U}_{1,2} \times \bar{I}_{1,2}}, f^-|_{\bar{I}_{1,2} \times \bar{U}_{1,2}}).$$

This uncontrollable Petri can be seen as the intersection of the influencing nets

$$\begin{aligned} \mathcal{N}_{\mathcal{A}_1}^{\mathrm{uc}} &= (\bar{I}_1, \bar{U}_1, f^+|_{\bar{U}_1 \times \bar{I}_1}, f^-|_{\bar{I}_1 \times \bar{U}_1}) \quad \text{and} \\ \mathcal{N}_{\mathcal{A}_2}^{\mathrm{uc}} &= (\bar{I}_2, \bar{U}_2, f^+|_{\bar{U}_2 \times \bar{I}_2}, f^-|_{\bar{I}_2 \times \bar{U}_2}); \end{aligned}$$

it consists of the set of common places $\bar{I}_{1,2}$ and the set of common transitions $\bar{U}_{1,2}$ (see (6.4)). Introduce the eight following conditions:

(C1) the Petri net $\mathcal{N}_{1,2}^{\mathrm{uc}}$ is acyclic,

(C2) $\sharp(p^\bullet \cap (\bar{U}_1 \cup \bar{U}_2)) \leq 1$ for all $p \in \bar{I}_{1,2}$,

(C3) $(P_1^{\downarrow})^\bullet \cap \bar{U}_2 = \emptyset$,

(C4) $(P_2^{\downarrow})^\bullet \cap \bar{U}_1 = \emptyset$,

(C5) ${}^\bullet P_1^{\uparrow} \cap \bar{U}_2 = \emptyset$,

(C6) ${}^\bullet P_2^{\uparrow} \cap \bar{U}_1 = \emptyset$,

(C7) $((P_1^{\mathrm{d}})^\bullet \cup {}^\bullet P_1^{\mathrm{d}}) \cap \bar{U}_2 = \emptyset$,

(C8) $((P_2^{\mathrm{d}})^\bullet \cup {}^\bullet P_2^{\mathrm{d}}) \cap \bar{U}_1 = \emptyset$.

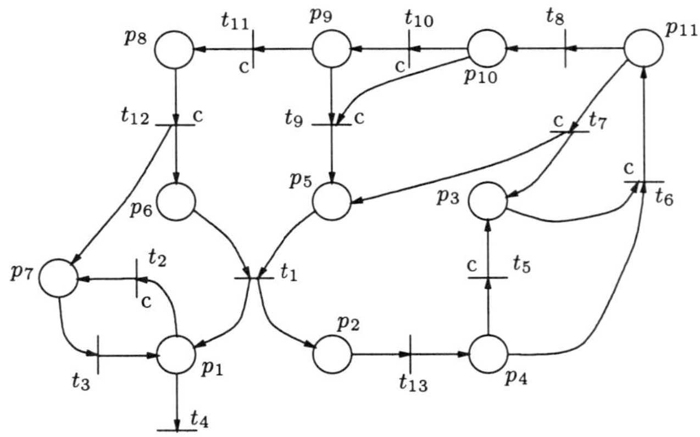

Figure 6.2. The Petri net of Example 6.2 as an illustration of Theorem 6.1.

From (C2) we have that $\sharp(p^\bullet \cap \bar{U}_{1,2}) \leq 1$ for all $p \in \bar{I}_{1,2}$. Together with (C1) this implies that $\mathcal{N}^{\mathrm{uc}}_{1,2}$ is an acyclic Petri net free of choice places. This class of Petri nets was studied in Section 5 of Chapter 5.

Theorem 6.1 contains the main result of this chapter.

Theorem 6.1. *Let $\mathcal{A}_1$, $\mathcal{A}_2 \subseteq \mathbb{N}^n$. If conditions (C1) to (C8) are satisfied, then it holds that*

$$(\mathcal{A}_1 \cup \mathcal{A}_2)^* = \mathcal{A}_1^* \cup \mathcal{A}_2^*.$$

When conditions (C1)–(C8) are satisfied, then the supremal controllable subset of the union of $\mathcal{A}_1$ and $\mathcal{A}_2$ is equal to the union of the supremal controllable subsets of both sets. After formulating a number of auxiliary results in the next section, the proof of Theorem 6.1 is given in Section 4.

Example 6.2. Consider the Petri net in Figure 6.2 and the sets $\mathcal{A}_1 = \{\mathbf{m} \in \mathbb{N}^{11} \mid \mathbf{m}(p_1) + 2\mathbf{m}(p_2) - \mathbf{m}(p_{11}) \leq 4\}$ and $\mathcal{A}_2 = \{\mathbf{m} \in \mathbb{N}^{11} \mid \mathbf{m}(p_2) + 2\mathbf{m}(p_3) \leq 5\}$. We know from Chapter 4 that $P_1^{\downarrow} = \{p_1, p_2\}$, $P_1^{\uparrow} = \{p_{11}\}$, $P_1^{\mathrm{d}} = \emptyset$, $P_2^{\downarrow} = \{p_2, p_3\}$ and $P_2^{\uparrow} = P_2^{\mathrm{d}} = \emptyset$. Thus is

$$\begin{aligned}
\bar{I}_1 &= \alpha(p_1) \cup \alpha(p_2) \cup \{p_{11}\} = \{p_1, p_2, p_5, p_6, p_7, p_{11}\}, \\
\bar{U}_1 &= \beta(p_1) \cup \beta(p_2) \cup \{t_8\} = \{t_1, t_3, t_8\}, \\
\bar{I}_2 &= \alpha(p_2) \cup \alpha(p_3) = \{p_2, p_3, p_5, p_6\}, \\
\bar{U}_2 &= \beta(p_2) \cup \beta(p_3) = \{t_1\}.
\end{aligned}$$

Herewith it can be verified that conditions (C1)–(C8) hold and thus that $(\mathcal{A}_1 \cup \mathcal{A}_2)^* = \mathcal{A}_1^* \cup \mathcal{A}_2^*$ (Theorem 6.1). △

Theorem 6.1 is formulated for the case $\nu = 2$ in (6.2). For greater ν this particular result can be used when it can be applied successively. If for example $\nu = 3$, i.e. $\mathcal{A} = \mathcal{A}_1 \cup \mathcal{A}_2 \cup \mathcal{A}_3$, then decomposition into three subproblems is possible when Theorem 6.1 can be applied to one of the three following cases:

1. for $\mathcal{A}_1 \cup \mathcal{A}_2$ and $\mathcal{A}_3$, and for $\mathcal{A}_1$ and $\mathcal{A}_2$,
2. for $\mathcal{A}_1 \cup \mathcal{A}_3$ and $\mathcal{A}_2$, and for $\mathcal{A}_1$ and $\mathcal{A}_3$,
3. for $\mathcal{A}_2 \cup \mathcal{A}_3$ and $\mathcal{A}_1$, and for $\mathcal{A}_2$ and $\mathcal{A}_3$.

3. Auxiliary results

In this section we formulate and prove a number of auxiliary results which are used in the proof of Theorem 6.1. The next two properties follow from the definition of ${}^\bullet(\cdot)$ and $(\cdot)^\bullet$ in Definition 1.6 (p. 16).

Proposition 6.1. *Let X_1, $X_2 \subseteq P$ and Y_1, $Y_2 \subseteq T$. It holds that ${}^\bullet(X_1 \cap X_2) \subseteq {}^\bullet X_1 \cap {}^\bullet X_2$ and that ${}^\bullet(Y_1 \cap Y_2) \subseteq {}^\bullet Y_1 \cap {}^\bullet Y_2$.*

Proposition 6.2. *Let $X \subseteq P$ and $Y \subseteq T$. It is true that $X^\bullet \cap Y = \emptyset$ if and only if $X \cap {}^\bullet Y = \emptyset$, and that ${}^\bullet X \cap Y = \emptyset$ if and only if $X \cap Y^\bullet = \emptyset$.*

Proposition 6.3 gives conditions under which the order of two successive transition vectors in a sequence can be changed. These conditions are structural, i.e. they are state independent.

Proposition 6.3. *Let Y_1, $Y_2 \subseteq T$, $\boldsymbol{\delta}_1$, $\boldsymbol{\delta}_2 \in \Delta$ and $\mathbf{m}_1$, $\mathbf{m}_2$, $\mathbf{m}_3 \in \mathbb{N}^n$. If $\mathbf{m}_1 \xrightarrow{\boldsymbol{\delta}_1} \mathbf{m}_2 \xrightarrow{\boldsymbol{\delta}_2} \mathbf{m}_3$, $\boldsymbol{\delta}_1|_{T\setminus Y_1} = \mathbf{0}$, $\boldsymbol{\delta}_2|_{T\setminus Y_2} = \mathbf{0}$, ${}^\bullet Y_1 \cap {}^\bullet Y_2 = \emptyset$ and $Y_1^\bullet \cap {}^\bullet Y_2 = \emptyset$, then it is true that $\mathbf{m}_1 \xrightarrow{\boldsymbol{\delta}_2} \mathbf{m}_2' \xrightarrow{\boldsymbol{\delta}_1} \mathbf{m}_3$.*

Proof. It suffices to prove that

$$\mathbf{m}_1(p) \geq \sum_{t \in p^\bullet \cap Y_2} f^-(p,t)\boldsymbol{\delta}_2(t) \qquad (p \in {}^\bullet Y_2), \tag{6.5}$$

and that

$$\mathbf{m}_2'(p) \geq \sum_{t \in p^\bullet \cap Y_1} f^-(p,t)\boldsymbol{\delta}_1(t) \qquad (p \in {}^\bullet Y_1). \tag{6.6}$$

From $\mathbf{m}_1 \xrightarrow{\boldsymbol{\delta}_1} \mathbf{m}_2 \xrightarrow{\boldsymbol{\delta}_2} \mathbf{m}_3$ it follows that

$$\mathbf{m}_1(p) \geq \sum_{t \in p^\bullet \cap Y_1} f^-(p,t)\boldsymbol{\delta}_1(t) \qquad (p \in {}^\bullet Y_1), \tag{6.7}$$

and that

$$\mathbf{m}_2(p) \geq \sum_{t \in p^\bullet \cap Y_2} f^-(p,t)\boldsymbol{\delta}_2(t) \qquad (p \in {}^\bullet Y_2). \tag{6.8}$$

For (6.8) to imply the inequalities in (6.5), it should hold that $\mathbf{m}_2(p) \leq \mathbf{m}_1(p)$ for all $p \in {}^\bullet Y_2$. As

$$\mathbf{m}_2(p) = \mathbf{m}_1(p) + \sum_{t \in {}^\bullet p \cap Y_1} f^+(t,p)\boldsymbol{\delta}_1(t) - \sum_{t \in p^\bullet \cap Y_1} f^-(p,t)\boldsymbol{\delta}_1(t),$$

a sufficient condition for this is that ${}^\bullet({}^\bullet Y_2) \cap Y_1 = \emptyset$. With Proposition 6.2 this is equivalent to $Y_1^\bullet \cap {}^\bullet Y_2 = \emptyset$.

For (6.7) to imply the inequalities in (6.6), it should hold that $\mathbf{m}_2'(p) \geq \mathbf{m}_1(p)$ for all $p \in {}^\bullet Y_1$. As

$$\mathbf{m}_2'(p) = \mathbf{m}_1(p) + \sum_{t \in {}^\bullet p \cap Y_2} f^+(t,p)\boldsymbol{\delta}_2(t) - \sum_{t \in p^\bullet \cap Y_2} f^-(p,t)\boldsymbol{\delta}_2(t),$$

a sufficient condition for this is that $({}^\bullet Y_1)^\bullet \cap Y_2 = \emptyset$. With Proposition 6.2 this is equivalent to ${}^\bullet Y_1 \cap {}^\bullet Y_2 = \emptyset$. □

Proposition 6.4. *Let $\mathcal{A} \subseteq \mathbb{N}^n$, $p \in P$ and $t \in {}^\bullet p \cap T^{uc}$. If $p \in \bar{I}(\mathcal{A})$ and $t \notin \bar{U}(\mathcal{A})$, then p belongs to $P^\uparrow(\mathcal{A})$.*

Proof. This follows from Definition 4.5 and the definition of $\bar{I}(\mathcal{A})$ and $\bar{U}(\mathcal{A})$ in (4.12) and (4.13). □

Proposition 6.5. *Let $\mathcal{A} \subseteq \mathbb{N}^n$ and $p \in P$. If $p \in \bar{I}(\mathcal{A})$, $p^\bullet \cap T^{uc} \neq \emptyset$ and $p^\bullet \cap \bar{U}(\mathcal{A}) = \emptyset$, then $p \in P^\downarrow(\mathcal{A})$.*

Proof. This follows from Definition 4.5 and the definition of $\bar{I}(\mathcal{A})$ and $\bar{U}(\mathcal{A})$ in (4.12) and (4.13). □

Proposition 6.6 contains properties of the sets $\bar{I}_j$ and $\bar{U}_j$ $(j = 1,2)$, and $\bar{I}_{1,2}$ and $\bar{U}_{1,2}$.

Proposition 6.6. *If conditions (C1) to (C8) are satisfied, then it holds that*

(a) ${}^\bullet\bar{U}_{1,2} \subseteq \bar{I}_{1,2}$,

(b) ${}^\bullet(\bar{U}_j \setminus \bar{U}_{1,2}) \cap \bar{I}_{1,2} = \emptyset$ $(j = 1,2)$,

(c) $(\bar{U}_j \setminus \bar{U}_{1,2})^\bullet \cap \bar{I}_{1,2} = \emptyset$ $(j = 1,2)$,

(d) ${}^\bullet(\bar{U}_j \setminus \bar{U}_{1,2}) \cap {}^\bullet\bar{U}_{1,2} = \emptyset$ $(j = 1,2)$,

(e) $(\bar{U}_j \setminus \bar{U}_{1,2})^\bullet \cap {}^\bullet\bar{U}_{1,2} = \emptyset$ $(j = 1,2)$,

(f) $^\bullet(\bar{U}_j \setminus \bar{U}_{1,2}) \subseteq \bar{I}_j \setminus \bar{I}_{1,2}$ $(j = 1, 2)$,

(g) $(\bar{I}_j \setminus \bar{I}_{1,2})^\bullet \cap \bar{U}_{1,2} = \emptyset$ $(j = 1, 2)$,

(h) $^\bullet(\bar{I}_2 \setminus \bar{I}_1) \cap (\bar{U}_1 \setminus \bar{U}_2) = \emptyset$ *and* $^\bullet(\bar{I}_1 \setminus \bar{I}_2) \cap (\bar{U}_2 \setminus \bar{U}_1) = \emptyset$,

(i) $(\bar{I}_2 \setminus \bar{I}_1)^\bullet \cap (\bar{U}_1 \setminus \bar{U}_2) = \emptyset$ *and* $(\bar{I}_1 \setminus \bar{I}_2)^\bullet \cap (\bar{U}_2 \setminus \bar{U}_1) = \emptyset$.

Proof. Because of symmetry, for items (b) to (g) we only prove the property for $j = 1$ and for items (h) and (i) we only prove the first of the two statements.

(a) This follows from Propositions 4.7 and 6.1, and (6.4):

$$^\bullet\bar{U}_{1,2} = {}^\bullet(\bar{U}_1 \cap \bar{U}_2) \subseteq {}^\bullet\bar{U}_1 \cap {}^\bullet\bar{U}_2 \subseteq \bar{I}_1 \cap \bar{I}_2 = \bar{I}_{1,2}.$$

(b) By contradiction. Let $p \in {}^\bullet(\bar{U}_1 \setminus \bar{U}_2) \cap \bar{I}_{1,2}$ and $t \in p^\bullet \cap (\bar{U}_1 \setminus \bar{U}_2)$. From (C2) it follows that $p^\bullet \cap (\bar{U}_1 \cup \bar{U}_2) = \{t\}$ and thus that $p^\bullet \cap \bar{U}_2 = \emptyset$. Together with $p \in \bar{I}_2$ it then follows from Proposition 6.5 that $p \in P_2^\downarrow$. This implies with condition (C4) that $p^\bullet \cap \bar{U}_1 = \emptyset$. This contradicts $t \in \bar{U}_1$.

(c) By contradiction. Let $p \in (\bar{U}_1 \setminus \bar{U}_2)^\bullet \cap \bar{I}_{1,2}$ and $t \in {}^\bullet p \cap (\bar{U}_1 \setminus \bar{U}_2)$. From $p \in \bar{I}_2$ and $t \notin \bar{U}_2$ it follows with Proposition 6.4 that $p \in P_2^\uparrow$. Together with condition (C6) this implies that ${}^\bullet p \cap \bar{U}_1 = \emptyset$. This contradicts $t \in \bar{U}_1$.

(d) This follows from (a) and (b).

(e) This follows from (a) and (c).

(f) This follows from ${}^\bullet\bar{U}(\mathcal{A}) \subseteq \bar{I}(\mathcal{A})$ (see Proposition 4.7) and (b).

(g) This follows from Proposition 6.2 and (a).

(h) By contradiction. Let $t \in {}^\bullet(\bar{I}_2 \setminus \bar{I}_1) \cap (\bar{U}_1 \setminus \bar{U}_2)$ and $p \in t^\bullet \cap (\bar{I}_2 \setminus \bar{I}_1)$. From $p \in \bar{I}_2$ and $t \notin \bar{U}_2$ it follows with Proposition 6.4 that $p \in P_2^\uparrow$. Together with (C6) this implies that ${}^\bullet p \cap \bar{U}_1 = \emptyset$. This contradicts $t \in \bar{U}_1$.

(i) This follows from Proposition 6.2 and (f). □

4. Proof of Theorem 6.1

This section is dedicated to the proof of the main decomposition result in Theorem 6.1. The proof consists of six parts. In part 1 we give an overview of the approach, in part 2 we construct a sequence of transition vectors and in parts 3 to 6 we show that this sequence is enabled.

Part 1. Consider the following diagram.

$$\begin{array}{ccc}
 & & \mathbf{m} \in (\mathcal{A}_1 \cup \mathcal{A}_2)^* \\
 & & \Updownarrow \text{ Def. 3.4} \\
\mathcal{R}_\infty^{\bar{U}(\mathcal{A}_1 \cup \mathcal{A}_2)}(\mathbf{m}) \subseteq \mathcal{A}_1 \cup \mathcal{A}_2 & \overset{\text{Th. 4.1}}{\Longleftrightarrow} & \mathcal{R}_\infty^{T^{\mathrm{uc}}}(\mathbf{m}) \subseteq \mathcal{A}_1 \cup \mathcal{A}_2 \\
\Downarrow \text{ Cor. 4.2} & & \\
\mathcal{R}_\infty^{\bar{U}_1 \cup \bar{U}_2}(\mathbf{m}) \subseteq \mathcal{A}_1 \cup \mathcal{A}_2 & & \Uparrow \\
(\Downarrow) \text{ (C1)–(C8)} & & \\
\begin{cases} \mathcal{R}_\infty^{\bar{U}_1}(\mathbf{m}) \subseteq \mathcal{A}_1 & \text{or} \\ \mathcal{R}_\infty^{\bar{U}_2}(\mathbf{m}) \subseteq \mathcal{A}_2 \end{cases} & \overset{\text{Th. 4.1}}{\Longleftrightarrow} & \begin{cases} \mathcal{R}_\infty^{T^{\mathrm{uc}}}(\mathbf{m}) \subseteq \mathcal{A}_1 & \text{or} \\ \mathcal{R}_\infty^{T^{\mathrm{uc}}}(\mathbf{m}) \subseteq \mathcal{A}_2 \end{cases} \\
 & & \Updownarrow \text{ Def. 3.4} \\
 & & \mathbf{m} \in \mathcal{A}_1^* \cup \mathcal{A}_2^*
\end{array}$$

As already mentioned in Section 1, it follows from the diagram that $\mathcal{A}_1^* \cup \mathcal{A}_2^* \subseteq (\mathcal{A}_1 \cup \mathcal{A}_2)^*$. To prove $\mathcal{A}_1^* \cup \mathcal{A}_2^* \supseteq (\mathcal{A}_1 \cup \mathcal{A}_2)^*$, we use reduction results from Chapter 4 (Theorem 4.1 and Corollary 4.2) and the conditions (C1) to (C8). By contraposition we show that the implication between brackets in the above diagram holds if (C1)–(C8) are satisfied. We show that, if

$$\mathcal{R}_\infty^{\bar{U}_1}(\mathbf{m}) \not\subseteq \mathcal{A}_1 \quad \text{and} \quad \mathcal{R}_\infty^{\bar{U}_2}(\mathbf{m}) \not\subseteq \mathcal{A}_2, \tag{6.9}$$

then

$$\mathcal{R}_\infty^{\bar{U}_1 \cup \bar{U}_2}(\mathbf{m}) \not\subseteq \mathcal{A}_1 \cup \mathcal{A}_2. \tag{6.10}$$

Throughout the proof the transition bag assumption is used, i.e. $\Omega = \Delta$. By definition, it follows from (6.9) that there exist transition vectors $\boldsymbol{\delta}_i^j$ $(i = 1, \ldots, k_j;\ j = 1, 2)$ with

$$\boldsymbol{\delta}_i^j|_{T \setminus \bar{U}_j} = \mathbf{0} \qquad (i = 1, \ldots, k_j; j = 1, 2) \tag{6.11}$$

such that

$$\mathbf{m} \xrightarrow{\boldsymbol{\delta}_1^j} \cdots \xrightarrow{\boldsymbol{\delta}_{k_j}^j} \mathbf{m}^j \notin \mathcal{A}_j \qquad (j = 1, 2). \tag{6.12}$$

Based on these two sequences and the conditions (C1)–(C8), we construct a state which belongs to $\mathcal{R}_\infty^{\bar{U}_1 \cup \bar{U}_2}(\mathbf{m})$ but not to $\mathcal{A}_1 \cup \mathcal{A}_2$. This state is reached by a sequence of transition vectors of which the sum is less than or equal to the sums of the two sequences of transition vectors in (6.12). In the rest of this section it is assumed implicitly that $j = 1, 2$, i.e. that all statements hold for both the values one and two for j.

Part 2. Define for $i = 1, \ldots, k_j$ the transition vectors $\tilde{\boldsymbol{\delta}}_i^j$ such that

$$\tilde{\boldsymbol{\delta}}_i^j|_{\bar{U}_{1,2}} = \boldsymbol{\delta}_i^j|_{\bar{U}_{1,2}} \quad \text{and} \quad \tilde{\boldsymbol{\delta}}_i^j|_{T \setminus \bar{U}_{1,2}} = \mathbf{0} \qquad (i = 1, \ldots, k_j), \tag{6.13}$$

and the transition vectors $\hat{\boldsymbol{\delta}}_i^j$ such that

$$\hat{\boldsymbol{\delta}}_i^j|_{\bar{U}_j \setminus \bar{U}_{1,2}} = \boldsymbol{\delta}_i^j|_{\bar{U}_j \setminus \bar{U}_{1,2}} \quad \text{and} \quad \hat{\boldsymbol{\delta}}_i^j|_{T \setminus (\bar{U}_j \setminus \bar{U}_{1,2})} = \mathbf{0} \qquad (i = 1, \ldots, k_j). \tag{6.14}$$

From this it follows that

$$\boldsymbol{\delta}_i^j = \tilde{\boldsymbol{\delta}}_i^j + \hat{\boldsymbol{\delta}}_i^j \qquad (i = 1, \ldots, k_j).$$

With the aid of Proposition 1.7 the sequences in (6.12) can be decomposed as:

$$\mathbf{m} \xrightarrow{\tilde{\boldsymbol{\delta}}_1^j} \cdot \xrightarrow{\hat{\boldsymbol{\delta}}_1^j} \cdots \xrightarrow{\tilde{\boldsymbol{\delta}}_{k_j}^j} \cdot \xrightarrow{\hat{\boldsymbol{\delta}}_{k_j}^j} \mathbf{m}^j. \tag{6.15}$$

Because of Proposition 6.6(d) and (e) the sequences in (6.15) can be reordered. Apply Proposition 6.3 multiple times for $Y_1 = \bar{U}_j \setminus \bar{U}_{1,2}$ and $Y_2 = \bar{U}_{1,2}$. One obtains:

$$\mathbf{m} \xrightarrow{\tilde{\boldsymbol{\delta}}_1^j} \cdots \xrightarrow{\tilde{\boldsymbol{\delta}}_{k_j}^j} \cdot \xrightarrow{\hat{\boldsymbol{\delta}}_1^j} \cdots \xrightarrow{\hat{\boldsymbol{\delta}}_{k_j}^j} \mathbf{m}^j. \tag{6.16}$$

Condition (C1) states that the Petri net

$$\mathcal{N}_{1,2}^{\text{uc}} = (\bar{I}_{1,2}, \bar{U}_{1,2}, f^+|_{\bar{U}_{1,2} \times \bar{I}_{1,2}}, f^-|_{\bar{I}_{1,2} \times \bar{U}_{1,2}})$$

is acyclic. Analogous to Proposition 5.10 and with Proposition 6.6(a), this is equivalent to $({}^\bullet\bar{U}_{1,2}, \bar{U}_{1,2}, f^+|_{\bar{U}_{1,2} \times {}^\bullet\bar{U}_{1,2}}, f^-|_{{}^\bullet\bar{U}_{1,2} \times \bar{U}_{1,2}})$ being acyclic. Consequently, the results of Chapter 5 can be applied. Partition, with the aid of Definition 5.1, the set of places $\bar{I}_{1,2}$ and the set of transitions $\bar{U}_{1,2}$ into sets $P_1, \ldots, P_{\lambda'}$ and $T_1, \ldots, T_\lambda$, respectively. Define for $i = 1, \ldots, \lambda$ transition vectors $\bar{\boldsymbol{\delta}}_i^j$ such that

$$\bar{\boldsymbol{\delta}}_i^j|_{T_i} = \sum_{v=1}^{k_j} \tilde{\boldsymbol{\delta}}_v^j|_{T_i} \quad \text{and} \quad \bar{\boldsymbol{\delta}}_i^j|_{T \setminus T_i} = \mathbf{0} \qquad (i = 1, \ldots, \lambda). \tag{6.17}$$

The definition of $\bar{\boldsymbol{\delta}}_i^j$ in (6.13) implies that $\sum_{v=1}^{\lambda} \bar{\boldsymbol{\delta}}_v^j = \sum_{v=1}^{k_j} \tilde{\boldsymbol{\delta}}_v^j$. As $\mathcal{N}_{1,2}^{\text{uc}}$ is acyclic, Corollary 5.3 can be applied to (6.16). We obtain:

$$\mathbf{m} \equiv \bar{\mathbf{m}}_0^j \xrightarrow{\bar{\boldsymbol{\delta}}_1^j} \cdots \xrightarrow{\bar{\boldsymbol{\delta}}_\lambda^j} \bar{\mathbf{m}}_\lambda^j \xrightarrow{\hat{\boldsymbol{\delta}}_1^j} \cdots \xrightarrow{\hat{\boldsymbol{\delta}}_{k_j}^j} \bar{\mathbf{m}}_{\lambda+k_j}^j \equiv \mathbf{m}^j. \tag{6.18}$$

With the definition of $\hat{\boldsymbol{\delta}}_i^j$ in (6.14), ${}^\bullet(\bar{U}_j \setminus \bar{U}_{1,2}) \cap \bar{I}_{1,2} = \emptyset$ (Proposition 6.6(d)) and $(\bar{U}_j \setminus \bar{U}_{1,2})^\bullet \cap \bar{I}_{1,2} = \emptyset$ (Proposition 6.6(e)) it follows from (6.18) that

$$\bar{\mathbf{m}}_\lambda^j|_{\bar{I}_{1,2}} = \bar{\mathbf{m}}_{\lambda+k_j}^j|_{\bar{I}_{1,2}}. \tag{6.19}$$

As $(\Delta \cup \{\mathbf{0}\}, \leq)$ is a lattice, the transition vectors

$$\boldsymbol{\delta}_i := \sup(\bar{\boldsymbol{\delta}}_i^1, \bar{\boldsymbol{\delta}}_i^2) \qquad (i = 1, \ldots, \lambda), \tag{6.20}$$

are well defined. In the rest of this section we show that

$$\mathbf{m} \equiv \hat{\mathbf{m}}_0 \xrightarrow{\boldsymbol{\delta}_1} \cdots \xrightarrow{\boldsymbol{\delta}_\lambda} \hat{\mathbf{m}}_\lambda \xrightarrow{\hat{\boldsymbol{\delta}}_1^1} \cdots \xrightarrow{\hat{\boldsymbol{\delta}}_{k_1}^1} \hat{\mathbf{m}}_{\lambda+k_1} \xrightarrow{\hat{\boldsymbol{\delta}}_1^2} \cdots \xrightarrow{\hat{\boldsymbol{\delta}}_{k_2}^2} \hat{\mathbf{m}}_{\lambda+k_1+k_2} \tag{6.21}$$

holds. The first λ steps in (6.21) are proven in part 3, the next k_1 in part 4 and the last k_2 in part 5. Finally, in part 6 we show that $\hat{\mathbf{m}}_{\lambda+k_1+k_2} \notin \mathcal{A}_1 \cup \mathcal{A}_2$, from which (6.10) follows.

Part 3. Because of the definition of $\bar{\boldsymbol{\delta}}_i^j$ in (6.17), the sequences in (6.18), condition (C2) and the definition of $\boldsymbol{\delta}_i$ in (6.20), one obtains from Proposition 5.9 (p. 121) that the sequence $(\boldsymbol{\delta}_1, \ldots, \boldsymbol{\delta}_\lambda)$ is enabled under $\mathbf{m}$ in (6.21). From (6.18) and (6.21) one finds that

$$\begin{aligned} \hat{\mathbf{m}}_\lambda(p) - \bar{\mathbf{m}}_\lambda^j(p) = & \sum_{t \in {}^\bullet p \cap \bar{U}_{1,2}} \sum_{v=1}^{\lambda} f^+(t,p) \left(\boldsymbol{\delta}_v(t) - \bar{\boldsymbol{\delta}}_v^j(t)\right) \\ & + \sum_{t \in p^\bullet \cap \bar{U}_{1,2}} \sum_{v=1}^{\lambda} f^-(p,t) \left(\bar{\boldsymbol{\delta}}_v^j(t) - \boldsymbol{\delta}_v(t)\right) \qquad (p \in P). \end{aligned} \tag{6.22}$$

From $(\bar{I}_j \setminus \bar{I}_{1,2})^\bullet \cap \bar{U}_{1,2}$ (Proposition 6.6(g)) it follows that the second term on the right hand side of (6.22) is equal to zero when $p \in \bar{I}_j \setminus \bar{I}_{1,2}$. Further it follows from (6.20) that $\boldsymbol{\delta}_i \geq \bar{\boldsymbol{\delta}}_i^j$ for $i = 1, \ldots, \lambda$ and thus that

$$\hat{\mathbf{m}}_\lambda|_{\bar{I}_j \setminus \bar{I}_{1,2}} \geq \bar{\mathbf{m}}_\lambda^j|_{\bar{I}_j \setminus \bar{I}_{1,2}}. \tag{6.23}$$

On the other hand, with conditions (C3) and (C4), it holds for each $p \in P_1^\downarrow \cup P_2^\downarrow$ that $p^\bullet \cap \bar{U}_{1,2} = \emptyset$ and thus, from (6.22), that

$$\hat{\mathbf{m}}_\lambda|_{P_1^\downarrow \cup P_2^\downarrow} \geq \bar{\mathbf{m}}_\lambda^j|_{P_1^\downarrow \cup P_2^\downarrow}. \tag{6.24}$$

For each $p \in P_1^\uparrow \cup P_2^\uparrow$ it holds with conditions (C5) and (C6) that ${}^\bullet p \cap \bar{U}_{1,2} = \emptyset$ and thus, from (6.22), that

$$\hat{\mathbf{m}}_\lambda|_{P_1^\uparrow \cup P_2^\uparrow} \leq \bar{\mathbf{m}}_\lambda^j|_{P_1^\uparrow \cup P_2^\uparrow}. \tag{6.25}$$

For $p \in P_1^{\mathrm{d}} \cup P_2^{\mathrm{d}}$ we have from conditions (C7) and (C8) that $(p^\bullet \cup {}^\bullet p) \cap \bar{U}_{1,2} = \emptyset$ and thus, from (6.22), that

$$\hat{\mathbf{m}}_\lambda|_{P_1^{\mathrm{d}} \cup P_2^{\mathrm{d}}} = \bar{\mathbf{m}}_\lambda^j|_{P_1^{\mathrm{d}} \cup P_2^{\mathrm{d}}}. \tag{6.26}$$

Part 4. From (6.18) we know that the sequence $(\hat{\boldsymbol{\delta}}^1_1, \ldots, \hat{\boldsymbol{\delta}}^1_{k_1})$ is enabled under $\bar{\mathbf{m}}^1_\lambda$. Because of $\hat{\boldsymbol{\delta}}^1_i|_{T\setminus(\bar{U}_1\setminus\bar{U}_2)} = \mathbf{0}$ (see (6.14)), ${}^\bullet(\bar{U}_1 \setminus \bar{U}_2) \subseteq \bar{I}_1 \setminus \bar{I}_2$ (Proposition 6.6(f)) and (6.23) for $j = 1$ this implies that $(\hat{\boldsymbol{\delta}}^1_1, \ldots, \hat{\boldsymbol{\delta}}^1_{k_1})$ is enabled under $\hat{\mathbf{m}}_\lambda$ in (6.21) (see Proposition 1.12). The resulting state is $\hat{\mathbf{m}}_{\lambda+k_1}$. From (6.18) and (6.21) it follows that $\hat{\mathbf{m}}_{\lambda+k_1} - \hat{\mathbf{m}}_\lambda = \bar{\mathbf{m}}^1_{\lambda+k_1} - \bar{\mathbf{m}}^1_\lambda$ or that $\hat{\mathbf{m}}_{\lambda+k_1} - \bar{\mathbf{m}}^1_{\lambda+k_1} = \hat{\mathbf{m}}_\lambda - \bar{\mathbf{m}}^1_\lambda$. Together with (6.23)–(6.26) we obtain that

$$\hat{\mathbf{m}}_{\lambda+k_1}|_{\bar{I}_1\setminus\bar{I}_2} \geq \bar{\mathbf{m}}^1_{\lambda+k_1}|_{\bar{I}_1\setminus\bar{I}_2}, \tag{6.27}$$

$$\hat{\mathbf{m}}_{\lambda+k_1}|_{P_1^\downarrow\cup P_2^\downarrow} \geq \bar{\mathbf{m}}^1_{\lambda+k_1}|_{P_1^\downarrow\cup P_2^\downarrow}, \tag{6.28}$$

$$\hat{\mathbf{m}}_{\lambda+k_1}|_{P_1^\uparrow\cup P_2^\uparrow} \leq \bar{\mathbf{m}}^1_{\lambda+k_1}|_{P_1^\uparrow\cup P_2^\uparrow}, \tag{6.29}$$

$$\hat{\mathbf{m}}_{\lambda+k_1}|_{P_1^{\mathrm{d}}\cup P_2^{\mathrm{d}}} = \bar{\mathbf{m}}^1_{\lambda+k_1}|_{P_1^{\mathrm{d}}\cup P_2^{\mathrm{d}}}. \tag{6.30}$$

Moreover, it follows from $\hat{\boldsymbol{\delta}}^1_i|_{T\setminus(\bar{U}_1\setminus\bar{U}_2)} = \mathbf{0}$ (see (6.14)), ${}^\bullet(\bar{I}_2 \setminus \bar{I}_1) \cap (\bar{U}_1 \setminus \bar{U}_2) = \emptyset$ (Proposition 6.6(h)) and $(\bar{I}_2 \setminus \bar{I}_1)^\bullet \cap (\bar{U}_1 \setminus \bar{U}_2) = \emptyset$ (Proposition 6.6(i)) that

$$\hat{\mathbf{m}}_{\lambda+k_1}|_{\bar{I}_2\setminus\bar{I}_1} = \hat{\mathbf{m}}_\lambda|_{\bar{I}_2\setminus\bar{I}_1}, \tag{6.31}$$

and from ${}^\bullet(\bar{U}_1\setminus\bar{U}_2)\cap\bar{I}_{1,2} = \emptyset$ (Proposition 6.6(d)) and $(\bar{U}_1\setminus\bar{U}_2)^\bullet\cap\bar{I}_{1,2} = \emptyset$ (Proposition 6.6(e)) that

$$\hat{\mathbf{m}}_{\lambda+k_1}|_{\bar{I}_{1,2}} = \hat{\mathbf{m}}_\lambda|_{\bar{I}_{1,2}}. \tag{6.32}$$

Part 5. We know from (6.18) that the sequence $(\hat{\boldsymbol{\delta}}^2_1, \ldots, \hat{\boldsymbol{\delta}}^2_{k_2})$ is enabled under $\bar{\mathbf{m}}^2_\lambda$. From (6.23) and (6.31) we find that

$$\hat{\mathbf{m}}_{\lambda+k_1}|_{\bar{I}_2\setminus\bar{I}_1} \geq \bar{\mathbf{m}}^2_\lambda|_{\bar{I}_2\setminus\bar{I}_1}. \tag{6.33}$$

Together with $\hat{\boldsymbol{\delta}}^2_i|_{T\setminus(\bar{U}\setminus\bar{U}_1)} = \mathbf{0}$ (see (6.14)) and ${}^\bullet(\bar{U}_2\setminus\bar{U}_1) \subseteq \bar{I}_2\setminus\bar{I}_1$ (Proposition 6.6(f)) this implies that $(\hat{\boldsymbol{\delta}}^2_1, \ldots, \hat{\boldsymbol{\delta}}^2_{k_2})$ is enabled under $\hat{\mathbf{m}}_{\lambda+k_1}$ in (6.21) (see Proposition 1.12). The resulting state is $\hat{\mathbf{m}}_{\lambda+k_1+k_2}$. From (6.18) and (6.21) we get that $\hat{\mathbf{m}}_{\lambda+k_1+k_2} - \hat{\mathbf{m}}_{\lambda+k_1} = \bar{\mathbf{m}}^2_{\lambda+k_2} - \bar{\mathbf{m}}^2_\lambda$, or that $\hat{\mathbf{m}}_{\lambda+k_1+k_2} - \bar{\mathbf{m}}^2_{\lambda+k_2} = \hat{\mathbf{m}}_{\lambda+k_1} - \bar{\mathbf{m}}^2_\lambda$. Together with (6.33) we then find that

$$\hat{\mathbf{m}}_{\lambda+k_1+k_2}|_{\bar{I}_2\setminus\bar{I}_1} \geq \bar{\mathbf{m}}^2_{\lambda+k_2}|_{\bar{I}_2\setminus\bar{I}_1}, \tag{6.34}$$

from (6.25) and (6.31) that

$$\hat{\mathbf{m}}_{\lambda+k_1+k_2}|_{P_2^\uparrow\setminus\bar{I}_1} \leq \bar{\mathbf{m}}^2_{\lambda+k_2}|_{P_2^\uparrow\setminus\bar{I}_1}, \tag{6.35}$$

and from (6.26) and (6.31) that

$$\hat{\mathbf{m}}_{\lambda+k_1+k_2}|_{P_2^{\mathrm{d}}\setminus\bar{I}_1} = \bar{\mathbf{m}}^2_{\lambda+k_2}|_{P_2^{\mathrm{d}}\setminus\bar{I}_1}. \tag{6.36}$$

On the other hand, it follows from $\hat{\boldsymbol{\delta}}^2_i|_{T\setminus(\bar{U}_2\setminus\bar{U}_1)} = \mathbf{0}$ (see (6.14)) and Proposition 6.6(d), (e), (h) and (i) that

$$\hat{\mathbf{m}}_{\lambda+k_1+k_2}|_{\bar{I}_1} = \hat{\mathbf{m}}_{\lambda+k_1}|_{\bar{I}_1}. \tag{6.37}$$

Part 6. In the final part of this proof we show that $\hat{\mathbf{m}}_{\lambda+k_1+k_2} \notin \mathcal{A}_1 \cup \mathcal{A}_2$. Note that $\hat{\mathbf{m}}_{\lambda+k_1+k_2} \in \mathcal{R}_\infty^{\bar{U}_1\cup\bar{U}_2}(\mathbf{m})$. From (6.28) and (6.37) we obtain:

$$\hat{\mathbf{m}}_{\lambda+k_1+k_2}|_{P_1^{\downarrow}} \geq \bar{\mathbf{m}}^1_{\lambda+k_1}|_{P_1^{\downarrow}}. \tag{6.38}$$

From (6.29) and (6.37) we find:

$$\hat{\mathbf{m}}_{\lambda+k_1+k_2}|_{P_1^{\uparrow}} \leq \bar{\mathbf{m}}^1_{\lambda+k_1}|_{P_1^{\uparrow}}. \tag{6.39}$$

From (6.30) and (6.37) we get:

$$\hat{\mathbf{m}}_{\lambda+k_1+k_2}|_{P_1^{\mathrm{d}}} = \bar{\mathbf{m}}^1_{\lambda+k_1}|_{P_1^{\mathrm{d}}}. \tag{6.40}$$

From (6.38)–(6.40) and $\bar{\mathbf{m}}^1_{\lambda+k_1} \equiv \mathbf{m}^1 \notin \mathcal{A}_1$ (see (6.18) for $j = 1$) it follows with Proposition 4.4 (p. 74) that $\hat{\mathbf{m}}_{\lambda+k_1+k_2} \notin \mathcal{A}_1$.

From (6.34) we find:

$$\hat{\mathbf{m}}_{\lambda+k_1+k_2}|_{P_2^{\downarrow}\setminus\bar{I}_1} \geq \bar{\mathbf{m}}^2_{\lambda+k_1}|_{P_2^{\downarrow}\setminus\bar{I}_1}. \tag{6.41}$$

From (6.19) for $j = 2$, (6.24), (6.32) and (6.37) we obtain:

$$\hat{\mathbf{m}}_{\lambda+k_1+k_2}|_{P_2^{\downarrow}\cap\bar{I}_1} \geq \bar{\mathbf{m}}^2_{\lambda+k_1}|_{P_2^{\downarrow}\cap\bar{I}_1}. \tag{6.42}$$

From (6.41) and (6.42) we thus have that

$$\hat{\mathbf{m}}_{\lambda+k_1+k_2}|_{P_2^{\downarrow}} \geq \bar{\mathbf{m}}^2_{\lambda+k_1}|_{P_2^{\downarrow}}. \tag{6.43}$$

From (6.19) for $j = 2$, (6.25) for $j = 2$, (6.32) and (6.37) we find:

$$\hat{\mathbf{m}}_{\lambda+k_1+k_2}|_{P_2^{\uparrow}\cap\bar{I}_1} \leq \bar{\mathbf{m}}^2_{\lambda+k_1}|_{P_2^{\uparrow}\cap\bar{I}_1}. \tag{6.44}$$

Together with (6.35) we obtain that

$$\hat{\mathbf{m}}_{\lambda+k_1+k_2}|_{P_2^{\uparrow}} \leq \bar{\mathbf{m}}^2_{\lambda+k_1}|_{P_2^{\uparrow}}. \tag{6.45}$$

From (6.19) for $j = 2$, (6.26) for $j = 2$, (6.32) and (6.37) it follows that

$$\hat{\mathbf{m}}_{\lambda+k_1+k_2}|_{P_2^{\mathrm{d}}\cap \bar{I}_1} = \bar{\mathbf{m}}^2_{\lambda+k_1}|_{P_2^{\mathrm{d}}\cap \bar{I}_1}. \tag{6.46}$$

Together with (6.36) we obtain that

$$\hat{\mathbf{m}}_{\lambda+k_1+k_2}|_{P_2^{\mathrm{d}}} = \bar{\mathbf{m}}^2_{\lambda+k_1}|_{P_2^{\mathrm{d}}}. \tag{6.47}$$

From (6.43), (6.45), (6.47) and $\bar{\mathbf{m}}^2_{\lambda+k_2} \equiv \mathbf{m}^2 \notin \mathcal{A}_2$ (see (6.18) for $j = 2$) it follows with Proposition 4.4 that $\hat{\mathbf{m}}_{\lambda+k_1+k_2} \notin \mathcal{A}_2$. Together with $\hat{\mathbf{m}}_{\lambda+k_1+k_2} \notin \mathcal{A}_1$, this implies (6.10).

5. Discussion

To determine $\mathcal{A}_1^*$ and $\mathcal{A}_2^*$, we know from Chapter 4 that it suffices to study the influencing nets $\mathcal{N}^{\mathrm{uc}}_{\mathcal{A}_1}$ and $\mathcal{N}^{\mathrm{uc}}_{\mathcal{A}_2}$, respectively. For the construction of $(\mathcal{A}_1 \cup \mathcal{A}_2)^*$ this is $\mathcal{N}^{\mathrm{uc}}_{\mathcal{A}_1\cup\mathcal{A}_2}$. In this section we discuss the relationship between the Petri nets $\mathcal{N}^{\mathrm{uc}}_{\mathcal{A}_1}$ and $\mathcal{N}^{\mathrm{uc}}_{\mathcal{A}_2}$ on the one hand and $\mathcal{N}^{\mathrm{uc}}_{\mathcal{A}_1\cup\mathcal{A}_2}$ on the other hand. We show that applying Theorem 6.1, and thus decomposing the construction of the supremal controllable subset of $\mathcal{A}_1 \cup \mathcal{A}_2$ into two subproblems, is not always beneficial.

Proposition 6.7. *Let $\mathcal{A}_1$, $\mathcal{A}_2 \subseteq \mathbb{N}^n$. It is true that*

$$\begin{aligned} &P^{\downarrow}(\mathcal{A}_1 \cup \mathcal{A}_2) \subseteq P^{\downarrow}(\mathcal{A}_1) \cup P^{\downarrow}(\mathcal{A}_2) \cup P^{\mathrm{d}}(\mathcal{A}_1) \cup P^{\mathrm{d}}(\mathcal{A}_2),\\ &P^{\uparrow}(\mathcal{A}_1 \cup \mathcal{A}_2) \subseteq P^{\uparrow}(\mathcal{A}_1) \cup P^{\uparrow}(\mathcal{A}_2) \cup P^{\mathrm{d}}(\mathcal{A}_1) \cup P^{\mathrm{d}}(\mathcal{A}_2),\\ &P^{\mathrm{d}}(\mathcal{A}_1 \cup \mathcal{A}_2) \subseteq P^{\downarrow}(\mathcal{A}_1) \cup P^{\downarrow}(\mathcal{A}_2) \cup P^{\uparrow}(\mathcal{A}_1) \cup P^{\uparrow}(\mathcal{A}_2)\\ &\qquad\qquad\qquad\qquad\qquad \cup P^{\mathrm{d}}(\mathcal{A}_1) \cup P^{\mathrm{d}}(\mathcal{A}_2). \end{aligned}$$

Proof. This follows from the definition of $P^{\downarrow}(\cdot)$, $P^{\uparrow}(\cdot)$ and $P^{\mathrm{d}}(\cdot)$ (Definition 4.2, p. 72) and Proposition 4.25 (p. 103). □

This proposition implies that $\bar{I}(\mathcal{A}_1 \cup \mathcal{A}_2) \subseteq \bar{I}(\mathcal{A}_1) \cup \bar{I}(\mathcal{A}_2)$ and $\bar{U}(\mathcal{A}_1 \cup \mathcal{A}_2) \subseteq \bar{U}(\mathcal{A}_1) \cup \bar{U}(\mathcal{A}_2)$. Consequently, $\mathcal{N}^{\mathrm{uc}}_{\mathcal{A}_1\cup\mathcal{A}_2}$ is always a sub-Petri net of the union of $\mathcal{N}^{\mathrm{uc}}_{\mathcal{A}_1}$ and $\mathcal{N}^{\mathrm{uc}}_{\mathcal{A}_2}$. It is even possible that $\mathcal{N}^{\mathrm{uc}}_{\mathcal{A}_1\cup\mathcal{A}_2}$ is a proper sub-Petri net of $\mathcal{N}^{\mathrm{uc}}_{\mathcal{A}_1}$ or of $\mathcal{N}^{\mathrm{uc}}_{\mathcal{A}_2}$, as it is illustrated next.

Example 6.3. Consider the uncontrollable Petri net $\mathcal{N}^{\mathrm{uc}}$ in Figure 6.3(a) and two legal sets $\mathcal{A}_1$ and $\mathcal{A}_2$ with

$$\begin{aligned} &P^{\mathrm{i}}(\mathcal{A}_1) = \{p_3, p_4, p_5, p_6, p_7\}, \quad && P^{\uparrow}(\mathcal{A}_1) = \{p_2\}, \quad P^{\mathrm{d}}(\mathcal{A}_1) = \{p_1\},\\ &P^{\mathrm{i}}(\mathcal{A}_2) = \{p_2, p_3, p_4, p_5, p_6, p_7\}, && P^{\uparrow}(\mathcal{A}_2) = \{p_1\}. \end{aligned}$$

The projection of $\mathcal{A}_1$ and $\mathcal{A}_2$ on the $(\mathbf{m}(p_1), \mathbf{m}(p_2))$-plane is given in Figure 6.3(b). For the union of these sets it holds that

$$P^{\mathrm{i}}(\mathcal{A}_1 \cup \mathcal{A}_2) = \{p_3, p_4, p_5, p_6, p_7\}, \qquad P^{\uparrow}(\mathcal{A}_1 \cup \mathcal{A}_2) = \{p_1, p_2\}.$$

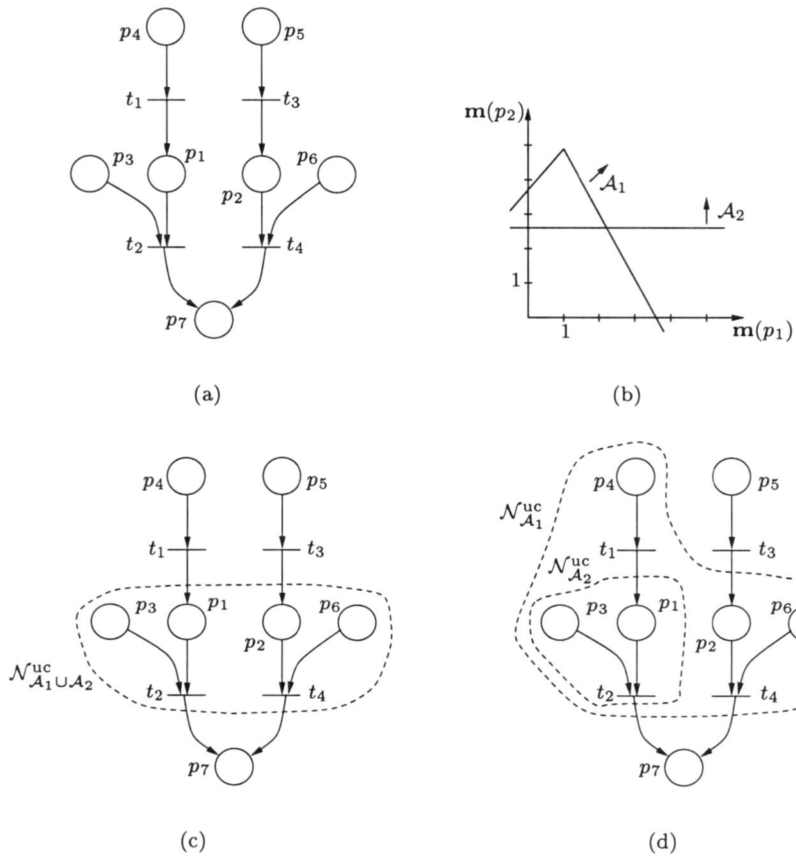

Figure 6.3. (a) The uncontrollable Petri net of Example 6.3, (b) The projection of $\mathcal{A}_1$ and $\mathcal{A}_2$ on the $(\mathbf{m}(p_1), \mathbf{m}(p_2))$-plane, (c) The influencing net of $\mathcal{A}_1 \cup \mathcal{A}_2$, (d) The influencing nets of $\mathcal{A}_1$ and $\mathcal{A}_2$.

The respective influencing nets are depicted in Figures 6.3(c) and 6.3(d). The influencing net of $\mathcal{A}_1 \cup \mathcal{A}_2$, i.e. $\mathcal{N}^{\mathrm{uc}}_{\mathcal{A}_1\cup\mathcal{A}_2}$, is a proper sub-Petri net of $\mathcal{N}^{\mathrm{uc}}_{\mathcal{A}_1}$, the influencing net of $\mathcal{A}_1$. $\triangle$

This example shows that the choice of $\mathcal{A}_1$ and $\mathcal{A}_2$ for a given legal set $\mathcal{A} = \mathcal{A}_1 \cup \mathcal{A}_2$ needs to be done with care, to make sure that the union of $\mathcal{N}^{\mathrm{uc}}_{\mathcal{A}_1}$ and $\mathcal{N}^{\mathrm{uc}}_{\mathcal{A}_2}$ is as small as possible or, preferably, equal to $\mathcal{N}^{\mathrm{uc}}_{\mathcal{A}_1\cup\mathcal{A}_2}$.

From Propositions 4.26, 4.27 and 4.29 it follows that there exist classes of legal sets for which this problem, i.e. the risk for larger influencing nets after decomposition, does not exist:

1. Both $\mathcal{A}_1$ and $\mathcal{A}_2$ are given by a finite number of disjunctions of linear inequalities which represent down-sets, and each of these sets is not redundant in $\mathcal{A}_1 \cup \mathcal{A}_2$.
2. Both $\mathcal{A}_1$ and $\mathcal{A}_2$ are given by a finite number of disjunctions of linear inequalities which represent up-sets, and $\mathcal{A}_1 \cup \mathcal{A}_2 \neq \mathbb{N}^n$.
3. Both $\mathcal{A}_1$ and $\mathcal{A}_2$ are given by a finite number of disjunctions of linear inequalites and the independence condition (4.81) on page 107 is satisfied.

6. Control design

In the preceding sections we have studied how the construction of the supremal controllable subset of a given legal set can be done in a modular way. In this section we look at the same problem for the construction of the set of permissive control values $\mathcal{U}_{\mathbf{m}}^{\mathcal{A}}$ in (3.14).

According to (3.22), this set can be written as

$$\mathcal{U}_{\mathbf{m}}^{\mathcal{A}} = \{\zeta \in \mathcal{U} \mid \mathcal{R}_{\infty}^{T^{uc}}(\mathcal{R}_{1,\Omega}^{T^c}(\zeta, \mathbf{m})) \subseteq \mathcal{A}\}. \tag{6.48}$$

From this, as in Propositions 2.9 and 2.10 (p. 38) for Petri nets without uncontrollable transitions, one obtains that

$$\mathcal{U}_{\mathbf{m}}^{\mathcal{A}} = \bigcap_{i=1}^{\nu} \mathcal{U}_{\mathbf{m}}^{\mathcal{A}_i} \quad \text{if } \mathcal{A} = \bigcap_{i=1}^{\nu} \mathcal{A}_i,$$

and that

$$\mathcal{U}_{\mathbf{m}}^{\mathcal{A}} \supseteq \bigcup_{i=1}^{\nu} \mathcal{U}_{\mathbf{m}}^{\mathcal{A}_i} \quad \text{if } \mathcal{A} = \bigcup_{i=1}^{\nu} \mathcal{A}_i. \tag{6.49}$$

From the latter it follows that by taking the union of $\mathcal{U}_{\mathbf{m}}^{\mathcal{A}_i}$ for $i = 1, \ldots, \nu$, one obtains permissive control values for $\mathcal{A} = \cup_{i=1}^{\nu} \mathcal{A}_i$. However, as the subset relation in (6.49) can be proper, one does not necessarily obtain all of them. Moreover, following Proposition 3.9 (p. 62), it holds for each

$$\zeta \in \mathcal{U}_{\mathbf{m}}^{\mathcal{A}} \setminus \left(\bigcup_{i=1}^{\nu} \mathcal{U}_{\mathbf{m}}^{\mathcal{A}_i} \right) \tag{6.50}$$

with $\mathcal{A} = \cup_{i=1}^{\nu} \mathcal{A}_i$ that

$$\forall \zeta' \in \mathcal{U}_{\mathbf{m}}^{\mathcal{A}}, \zeta' \supseteq \zeta : \zeta' \notin \bigcup_{i=1}^{\nu} \mathcal{U}_{\mathbf{m}}^{\mathcal{A}_i}. \tag{6.51}$$

Thus if a permissive control value is not found via decomposition, then each larger permissive control value is not found either. Because one needs the maximal elements in $(\mathcal{E}^{\mathcal{A}}_{\mathbf{m}}, \leq_{\mathbf{m}})$ to construct maximally permissive control laws, the importance of equality in (6.49) becomes clear.

Like in the first part of this chapter, we consider the case ν equal to two. Define for $j = 1, 2$ the sets $\bar{N}_j$ and $\bar{O}_j$ as shorthand notation for $\bar{N}(\mathcal{A}_j)$ and $\bar{O}(\mathcal{A}_j)$ (see Chapter 4), respectively. Further, define the sets

$$\bar{O}_{1,2} := \bar{O}_1 \cap \bar{O}_2 \quad \text{and} \quad \bar{N}_{1,2} := \bar{N}_1 \cap \bar{N}_2. \tag{6.52}$$

Introduce the following conditions:

(C9) $\sharp(p^\bullet \cap (\bar{N}_1 \cup \bar{N}_2)) \leq 1$ for all $p \in \bar{O}_{1,2}$,
(C10) $(P_1^{\downarrow})^\bullet \cap \bar{N}_2 = \emptyset$,
(C11) $(P_2^{\downarrow})^\bullet \cap \bar{N}_1 = \emptyset$,
(C12) ${}^\bullet P_1^{\uparrow} \cap \bar{N}_2 = \emptyset$,
(C13) ${}^\bullet P_2^{\uparrow} \cap \bar{N}_1 = \emptyset$,
(C14) $((P_1^{\mathrm{d}})^\bullet \cup {}^\bullet P_1^{\mathrm{d}}) \cap \bar{N}_2 = \emptyset$,
(C15) $((P_2^{\mathrm{d}})^\bullet \cup {}^\bullet P_2^{\mathrm{d}}) \cap \bar{N}_1 = \emptyset$,
(C16) ${}^\bullet\bar{U}_1 \cap O_2 = \emptyset$,
(C17) ${}^\bullet\bar{U}_2 \cap O_1 = \emptyset$.

Conditions (C9) to (C15) are similar to (C2) to (C8). Theorem 6.2 formulates conditions under which decomposition of the control design is possible for the union of two sets. We use a general concurrency assumption Ω. Recall that a partially ordered set is sup-closed if the supremum of each pair of elements belongs to this set. Consequently, a partially ordered set $(V, \leq)$ with $V \in \mathcal{O}^{\downarrow}(\Omega, \leq)$ is sup-closed if and only if $(V \cup \{\mathbf{0}\}, \leq)$ is a lattice.

Theorem 6.2. *Consider a Petri net $\mathcal{N} = (P, T, f^+, f^-, T^{\mathrm{uc}})$ with concurrency assumption Ω and control set $\mathcal{U}$. Let $\mathcal{A}_1$, $\mathcal{A}_2 \subseteq \mathbb{N}^n$ and $\mathbf{m} \in \mathbb{N}^n$. If all elements of $\mathcal{U}$ are sup-closed under the componentwise partial order relation and the conditions (C1) to (C17) are satisfied, then*

$$\mathcal{U}^{\mathcal{A}_1 \cup \mathcal{A}_2}_{\mathbf{m}} = \mathcal{U}^{\mathcal{A}_1}_{\mathbf{m}} \cup \mathcal{U}^{\mathcal{A}_2}_{\mathbf{m}}.$$

Proof. The proof is analogous to that of Theorem 6.1. Because of the expression for $\mathcal{U}^{\mathcal{A}}_{\mathbf{m}}$ in (6.48), Theorem 4.2 and Corollary 4.3 it suffices to prove that

$$\mathcal{R}^{\bar{U}_1}_{\infty}(\mathcal{R}^{\bar{N}_1}_{1,\Omega}(\zeta, \mathbf{m})) \not\subseteq \mathcal{A}_1 \quad \text{and} \quad \mathcal{R}^{\bar{U}_2}_{\infty}(\mathcal{R}^{\bar{N}_2}_{1,\Omega}(\zeta, \mathbf{m})) \not\subseteq \mathcal{A}_2 \tag{6.53}$$

implies that

$$\mathcal{R}^{\bar{U}_1 \cup \bar{U}_2}_{\infty}(\mathcal{R}^{\bar{N}_1 \cup \bar{N}_2}_{1,\Omega}(\zeta, \mathbf{m})) \not\subseteq \mathcal{A}_1 \cup \mathcal{A}_2. \tag{6.54}$$

By definition, it follows from (6.53) that there exist transition vectors $\boldsymbol{\delta}_0^j \in \Omega$ $(j = 1, 2)$ with $\boldsymbol{\delta}_0^j|_{T\setminus\bar{N}_j} = \mathbf{0}$ and $\boldsymbol{\delta}_i^j \in \Delta$ $(i = 1, \ldots, k_j; j = 1, 2)$ with

$$\boldsymbol{\delta}_i^j|_{T\setminus\bar{U}_j} = \mathbf{0} \qquad (i = 1, \ldots, k_j; j = 1, 2) \tag{6.55}$$

such that

$$\mathbf{m} \xrightarrow{\boldsymbol{\delta}_0^j, \zeta} \mathbf{m}_0^j \xrightarrow{(\boldsymbol{\delta}_1^j, \ldots, \boldsymbol{\delta}_{k_j}^j)} \mathbf{m}^j \notin \mathcal{A}_j \qquad (j = 1, 2). \tag{6.56}$$

As $(\zeta, \leq)$ is sup-closed, the transition vector $\boldsymbol{\delta}_0 := \sup(\boldsymbol{\delta}_0^1, \boldsymbol{\delta}_0^2)$ is well defined and belongs to ζ. We show that

$$\mathbf{m} \xrightarrow{\boldsymbol{\delta}_0, \zeta} \hat{\mathbf{m}}_0 \xrightarrow{(\boldsymbol{\delta}_1, \ldots, \boldsymbol{\delta}_\lambda, \hat{\boldsymbol{\delta}}_1^1, \ldots, \hat{\boldsymbol{\delta}}_{k_1}^1, \hat{\boldsymbol{\delta}}_1^2, \ldots, \hat{\boldsymbol{\delta}}_{k_2}^2)} \hat{\mathbf{m}}_{\lambda+k_1+k_2} \notin \mathcal{A}_1 \cup \mathcal{A}_2, \tag{6.57}$$

where $\boldsymbol{\delta}_i$ $(i = 1, \ldots, \lambda)$ and $\hat{\boldsymbol{\delta}}_i^j$ $(i = 1, \ldots, k_j;\ j = 1, 2)$ are defined as in the proof of Theorem 6.1 (Section 4). As $\boldsymbol{\delta}_0$ only contains transitions in $\bar{N}_1 \cup \bar{N}_2$, it is true that $\boldsymbol{\delta}_0 \in \Omega_{\mathbf{m}}$ if and only if

$$\begin{aligned} \mathbf{m}(p) &\geq \sum_{t \in p^\bullet \cap (\bar{N}_1 \cup \bar{N}_2)} f^-(p, t)\boldsymbol{\delta}_0(t) \\ &= \sum_{t \in p^\bullet \cap (\bar{N}_1 \cup \bar{N}_2)} f^-(p, t) \max(\boldsymbol{\delta}_0^1(t), \boldsymbol{\delta}_0^2(t)) \qquad (p \in {}^\bullet(\bar{N}_1 \cup \bar{N}_2)). \end{aligned} \tag{6.58}$$

For $p \in {}^\bullet\bar{N}_1 \setminus {}^\bullet\bar{N}_2$ we have that $p^\bullet \cap \bar{N}_2 = \emptyset$. As $\boldsymbol{\delta}_0^j|_{T\setminus\bar{N}_j} = \mathbf{0}$ $(j = 1, 2)$, the condition in (6.58) becomes $\mathbf{m}(p) \geq \sum_{t \in p^\bullet \cap \bar{N}_1} f^-(p, t)\boldsymbol{\delta}_0^1(t)$, which follows from $\boldsymbol{\delta}_0^1 \in \Omega_{\mathbf{m}}$. The proof for $p \in {}^\bullet\bar{N}_2 \setminus {}^\bullet\bar{N}_1$ is analogous. If $p \in {}^\bullet\bar{N}_1 \cap {}^\bullet\bar{N}_2 = \bar{O}_{1,2}$, then (6.58) is, because of (C9), equivalent to

$$\begin{aligned} \mathbf{m}(p) \geq &\sum_{t \in p^\bullet \cap (\bar{N}_1 \cap \bar{N}_2)} f^-(p, t)\boldsymbol{\delta}_0(t), \\ &= f^-(p, t_p)\boldsymbol{\delta}_0(t_p), \\ &= f^-(p, t_p) \max(\boldsymbol{\delta}_0^1(t_p), \boldsymbol{\delta}_0^2(t_p)), \end{aligned}$$

where t_p is the unique output transition of p in $\bar{N}_1 \cup \bar{N}_2$ (because of $p \in \bar{O}_{1,2}$, is $t_p \in \bar{N}_{1,2}$). This is equivalent to the conjunction of $\mathbf{m}(p) \geq f^-(p, t_p)\boldsymbol{\delta}_0^1(t_p)$ and $\mathbf{m}(p) \geq f^-(p, t_p)\boldsymbol{\delta}_0^2(t_p)$, which follows from $\boldsymbol{\delta}_0^j \in \Omega_{\mathbf{m}}$ $(j = 1, 2)$. The state after the execution of $\boldsymbol{\delta}_0$ is $\hat{\mathbf{m}}_0$.

For each $p \in {}^\bullet\bar{U}_1$ we have from the first step in (6.56) and (6.57) that

$$\begin{aligned}\hat{\mathbf{m}}_0(p) - \mathbf{m}_0^1(p) = & \sum_{t \in {}^\bullet p \cap (\bar{N}_1 \setminus \bar{N}_2)} \big(\boldsymbol{\delta}_0(t) - \boldsymbol{\delta}_0^1(t)\big) + \sum_{t \in {}^\bullet p \cap \bar{N}_{1,2}} \big(\boldsymbol{\delta}_0(t) - \boldsymbol{\delta}_0^1(t)\big) \\ & + \sum_{t \in {}^\bullet p \cap (\bar{N}_2 \setminus \bar{N}_1)} \boldsymbol{\delta}_0(t) - \sum_{t \in p^\bullet \cap (\bar{N}_1 \setminus \bar{N}_2)} \big(\boldsymbol{\delta}_0(t) - \boldsymbol{\delta}_0^1(t)\big) \\ & - \sum_{t \in p^\bullet \cap \bar{N}_{1,2}} \big(\boldsymbol{\delta}_0(t) - \boldsymbol{\delta}_0^1(t)\big) - \sum_{t \in p^\bullet \cap (\bar{N}_2 \setminus \bar{N}_1)} \boldsymbol{\delta}_0(t). \end{aligned} \quad (6.59)$$

From the definition of $\boldsymbol{\delta}_0$ it follows that the first three terms on the right hand side are nonnegative and that the fourth term equals zero. From (C16) it follows that the fifth and the sixth term are equal to zero. With an analogous argument for $\hat{\mathbf{m}}_0(p) - \mathbf{m}_0^2(p)$ with $p \in {}^\bullet\bar{U}_2$ we obtain that

$$\hat{\mathbf{m}}_0|_{{}^\bullet\bar{U}_j} \geq \mathbf{m}_0^j(p)|_{{}^\bullet\bar{U}_j} \qquad (j = 1, 2). \quad (6.60)$$

Analogously, one finds from (C10) and (C11) that

$$\hat{\mathbf{m}}_0|_{P_j^\downarrow} \geq \mathbf{m}_0^j(p)|_{P_j^\downarrow} \qquad (j = 1, 2), \quad (6.61)$$

from (C12) and (C13) that

$$\hat{\mathbf{m}}_0|_{P_j^\uparrow} \leq \mathbf{m}_0^j(p)|_{P_j^\uparrow} \qquad (j = 1, 2), \quad (6.62)$$

and from (C14) and (C15) that

$$\hat{\mathbf{m}}_0|_{P_j^{\mathrm{d}}} = \mathbf{m}_0^j(p)|_{P_j^{\mathrm{d}}} \qquad (j = 1, 2). \quad (6.63)$$

As in the proof of Theorem 6.1 it can be shown from (6.56) and (6.60) that the sequence of transition vectors in (6.57) is enabled under $\hat{\mathbf{m}}_0$. This also implies, together with (6.61)–(6.63), that $\hat{\mathbf{m}}_{\lambda+k_1+k_2} \notin \mathcal{A}_1 \cup \mathcal{A}_2$. □

The fact that this result is structural is very important. The conditions are all independent of the state $\mathbf{m}$. As Ω always belongs to $\mathcal{U}$, a necessary condition for this theorem to be applicable is that $(\Omega \cup \{\mathbf{0}\}, \leq)$ is a lattice. This is the case for the transition bag assumption ($\Omega = \Delta$) and the concurrency assumption ($\Omega = \Delta_{\mathrm{N}}$). Under on/off control and the transition bag assumption or the concurrency assumption, all control values are sup-closed. In general this is not true for fine control.

The conditions of the theorem can be made weaker at the cost of making them state dependent, i.e. no longer structural. Namely, if all control values which belong to $\mathcal{U} \setminus (\mathcal{U}_{\mathbf{m}}^{\mathcal{A}_1} \cup \mathcal{U}_{\mathbf{m}}^{\mathcal{A}_2})$ are sup-closed and conditions (C1) to (C17) are satisfied, then $\mathcal{U}_{\mathbf{m}}^{\mathcal{A}_1 \cup \mathcal{A}_2} = \mathcal{U}_{\mathbf{m}}^{\mathcal{A}_1} \cup \mathcal{U}_{\mathbf{m}}^{\mathcal{A}_2}$.

7. Notes and references

Decomposition of the control design has hardly been studied. The results of this chapter generalise results for legal down-sets and 1-safe Petri nets in [18] and [21]. A summary of the results of this chapter specialised to legal down-sets, appears in [38].

Chapter 7

CONTINUOUS VERSUS DISCRETE EVENTS

When a legal set $\mathcal{A}$ is described by a finite disjunction of linear inequalities and its influencing net $\mathcal{N}_{\mathcal{A}}^{\mathrm{uc}}$ is acyclic, the problem of deciding whether a given state belongs to $\mathcal{A}^*$ can be written as an integer linear programming problem (see (5.32) on page 126). In this chapter we construct a closed-form solution to the corresponding linear programming problem (take $\boldsymbol{\delta} \in \mathbb{Q}_+^{\sharp \bar{U}}$ and $z \in \mathbb{Q}$ in (5.32)). We do this by introducing *continuous Petri nets* in Section 1. The difference with the Petri net definition of Chapter 1 is that in the enabling condition (1.3) the transition vectors $\boldsymbol{\delta}$ can belong to $\mathbb{Q}_+^m \setminus \{\mathbf{0}\}$ and are not restricted to $\Delta = \mathbb{N}^m \setminus \{\mathbf{0}\}$. In Section 2, this approach allows us, by using reachability results for acyclic Petri nets, to construct a subset of $\mathcal{A}^*$. With this subset as the control goal, permissive control laws can be constructed.

In Section 3 we discuss this technique for the construction of $\mathcal{A}^*$ itself. We give in Sections 4 to 6 a number of classes of Petri nets and legal sets for which this is possible. Further we study the difficulties which obstruct the general application of this method. The general structure of the supremal controllable subset is discussed in Section 7.

1. Continuous Petri nets

1.1. Definition

A *continuous Petri net* $\tilde{\mathcal{N}} = (P, T, f^+, f^-, \mathbf{x}_0)$ is defined in the same manner as a Petri net in Definition 1.5 (p. 15). Namely, P and T are disjoint, finite sets of places and transitions, respectively, and $f^+ : T \times P \to \mathbb{N}$ and $f^- : P \times T \to \mathbb{N}$ are the incidence functions from transitions to places and from places to transition, respectively. There are two differences between continuous Petri nets and Petri nets. First, the initial state $\mathbf{x}_0$ of a continuous Petri net belongs to $\mathbb{Q}_+^n$ and thus not necessarily to $\mathbb{N}^n$. The second difference lies in the state evolution. For a given state $\mathbf{x}_k \in \mathbb{Q}_+^n$, all possible next states $\mathbf{x}_{k+1}$ satisfy $\mathbf{x}_{k+1} = \mathbf{x}_k + \mathbf{F}\mathbf{y}$, where $\mathbf{y} \in \mathbb{Q}_{+,0}^m := \mathbb{Q}_+^m \setminus \{\mathbf{0}\}$ and $\mathbf{x}_k \geq \mathbf{F}^-\mathbf{y}$. The continuous Petri net with the same set of places, set of transitions and incidence functions as the Petri net $\mathcal{N} = (P, T, f^+, f^-)$ is denoted $\tilde{\mathcal{N}} = (P, T, f^+, f^-)$. The structure of both nets can be represented by the same graph.

The state of a continuous Petri net evolves, like that of a Petri net, stepwise. The term 'continuous' indicates that the vectors $\mathbf{y}$, which determine the state change, belong to $\mathbb{Q}_{+,0}^m$ and are not limited to $\Delta = \mathbb{N}^m \setminus \{\mathbf{0}\}$. In this chapter the (continuous equivalent of the) transition bag assumption is adopted.

1.2. Reachability

Reachable sets in a continuous Petri net $\tilde{\mathcal{N}}$ are defined like reachable sets for Petri nets in Chapter 1. They are denoted $\tilde{\mathcal{R}}_k(\mathbf{x})$ with $k \in \mathbb{N}$ and $\mathbf{x} \in \mathbb{Q}_+^n$, to distinguish them from the corresponding sets $\mathcal{R}_k(\mathbf{m})$ for a Petri net.

Definition 7.1. *Consider a continuous Petri net $\tilde{\mathcal{N}}$. Define for all $T' \subseteq T$, $\mathbf{x} \in \mathbb{Q}_+^n$ and $k \in \mathbb{N}$ the set of states*

$$\begin{aligned} \tilde{\mathcal{R}}_k^{T'}(\mathbf{x}) := \{\mathbf{x}' \in \mathbb{Q}_+^n \mid \exists \mathbf{x}'' \in \tilde{\mathcal{R}}_{k-1}^{T'}(\mathbf{x}), \\ \exists \mathbf{y} \in \mathbb{Q}_{+,0}^m, \mathbf{x}'' \geq \mathbf{F}^-\mathbf{y}, \mathbf{y}|_{T \setminus T'} = \mathbf{0} : \mathbf{x}' = \mathbf{x}'' + \mathbf{F}\mathbf{y}\} \cup \tilde{\mathcal{R}}_{k-1}^{T'}(\mathbf{x}), \end{aligned}$$

where $\mathcal{R}_0^{T'}(\mathbf{x}) := \{\mathbf{x}\}$.

For the same reason as in (1.9),

$$\tilde{\mathcal{R}}_\infty(\mathbf{x}) := \lim_{k \to \infty} \tilde{\mathcal{R}}_k(\mathbf{x})$$

is well defined for all $\mathbf{x} \in \mathbb{Q}_+^n$. The following property immediately follows from the definition of continuous Petri nets.

Proposition 7.1. *Consider a Petri net $\mathcal{N}$ and the corresponding continuous Petri net $\tilde{\mathcal{N}}$. Then it is true for all $\mathbf{m} \in \mathbb{N}^n$ that*

$$\mathcal{R}_k(\mathbf{m}) \subseteq \tilde{\mathcal{R}}_k(\mathbf{m}) \cap \mathbb{N}^n \qquad (k \in \mathbb{N} \cup \{\infty\}).$$

In Example 7.1 we illustrate that $\mathcal{R}_\infty(\mathbf{m})$ can be a proper subset of $\tilde{\mathcal{R}}_\infty(\mathbf{m}) \cap \mathbb{N}^n$.

Example 7.1. For the continuous Petri and the state $\mathbf{m}_0 = (0, 0, 0, 1)$ in Figure 7.1(a) it is true that

$$\mathbf{m}_\mathrm{f} = (1, 0, 0, 0) \in \tilde{\mathcal{R}}_2(\mathbf{m}_0).$$

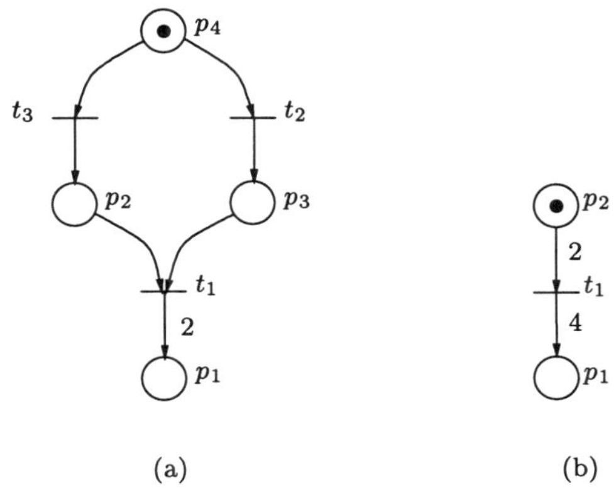

Figure 7.1. Two continuous Petri nets with initial states $\mathbf{m}_0 = (0,0,0,1)$ and $\mathbf{m}_0 = (1,0)$, respectively, for which $\mathcal{R}_\infty(\mathbf{m}_0) \subset \tilde{\mathcal{R}}_\infty(\mathbf{m}_0) \cap \mathbb{N}^n$.

This state is reached via $\mathbf{y}_1 = \left(0, \frac{1}{2}, \frac{1}{2}\right)$ and $\mathbf{y}_2 = \left(\frac{1}{2}, 0, 0\right)$. On the other hand, in the Petri net we find that $\mathbf{m}_f \notin \mathcal{R}_\infty(\mathbf{m}_0)$.

For the continuous Petri net and the state $\mathbf{m}_0 = (0,1)$ in Figure 7.1(b) it holds that $\mathbf{m}_f = (2,0) \in \tilde{\mathcal{R}}_1(\mathbf{m}_0)$ (reached by executing t_1 '$\frac{1}{2}$ times') and that $\mathbf{m}_f \notin \mathcal{R}_\infty(\mathbf{m}_0) = \{\mathbf{m}_0\}$ in the Petri net. △

1.3. Acyclic continuous Petri nets

A number of the results of Chapter 5 also hold for acyclic continuous Petri nets. Their proof does not depend on the integer nature of the transition vector. Proposition 7.2 is the continuous counterpart of Proposition 5.5 (p. 118).

Proposition 7.2. *Consider an acyclic continuous Petri net $\tilde{\mathcal{N}}$ with initial state $\mathbf{x}_0$. It holds that*

$$\tilde{\mathcal{R}}_\infty(\mathbf{x}_0) = \tilde{\mathcal{R}}_1^{t_m}(\cdots(\tilde{\mathcal{R}}_1^{t_1}(\mathbf{x}_0))\cdots). \tag{7.1}$$

We refer to Proposition 5.5 for the notation $t_1, \ldots, t_m$. Proposition 7.3 is the continuous equivalent of Theorem 5.1 (p. 114).

Proposition 7.3. *Consider an acyclic continuous Petri net $\tilde{\mathcal{N}}$ with initial state $\mathbf{x}_0$. Let $\mathbf{x}_f \in \mathbb{Q}_+^n$. It is true that $\mathbf{x}_f \in \tilde{\mathcal{R}}_\infty(\mathbf{x}_0)$ if and only if there exists a $\mathbf{y} \in \mathbb{Q}_+^m$ such that $\mathbf{x}_f = \mathbf{x}_0 + \mathbf{F}\mathbf{y}$.*

Reachability results for acyclic continuous Petri nets which are embedded in a larger continuous Petri net can be obtained immediately and are therefore not given explicitly.

2. A subset of the supremal controllable subset

In this section we construct, for legal sets which are described by a finite disjunction of linear inequalities and for which the influencing net is acyclic, a subset of their supremal controllable subset.

2.1. The approach

Consider a Petri net $\mathcal{N}$ and a legal set

$$\mathcal{A} = \bigcup_{i=1}^{\nu} \{\mathbf{m} \in \mathbb{N}^n \mid \mathbf{A}_i^T \mathbf{m} \leq b_i\} \quad \text{with } \mathbf{A}_i \in \mathbb{Z}^n, b_i \in \mathbb{Z} \; (i = 1, \ldots, \nu). \tag{7.2}$$

Finite intersections of sets of this type can be handled by using the decomposition result of Proposition 3.13 (p. 66). For the continuous Petri net $\tilde{\mathcal{N}}$ that corresponds to $\mathcal{N}$, we consider a legal set $\tilde{\mathcal{A}}$ which is described by the same equations as $\mathcal{A}$ in (7.2):

$$\tilde{\mathcal{A}} = \bigcup_{i=1}^{\nu} \{\mathbf{x} \in \mathbb{Q}_+^n \mid \mathbf{A}_i^T \mathbf{x} \leq b_i\}. \tag{7.3}$$

Hence,

$$\mathcal{A} = \tilde{\mathcal{A}} \cap \mathbb{N}^n. \tag{7.4}$$

Analogously to the supremal controllable subset

$$\mathcal{A}^* = \{\mathbf{m} \in \mathbb{N}^n \mid \mathcal{R}_\infty^{\bar{U}}(\mathbf{m}) \subseteq \mathcal{A}\}, \tag{7.5}$$

define the set

$$\tilde{\mathcal{A}}^* := \{\mathbf{x} \in \mathbb{Q}_+^n \mid \tilde{\mathcal{R}}_\infty^{\bar{U}}(\mathbf{x}) \subseteq \tilde{\mathcal{A}}\} \tag{7.6}$$

for the continuous case. It is clear that $\tilde{\mathcal{A}}^* \subseteq \tilde{\mathcal{A}}$.

Proposition 7.4. *Let $\mathcal{A} \subseteq \mathbb{N}^n$ and $\tilde{\mathcal{A}} \subseteq \mathbb{Q}_+^n$. If $\mathcal{A} = \tilde{\mathcal{A}} \cap \mathbb{N}^n$, then $\tilde{\mathcal{A}}^* \cap \mathbb{N}^n \subseteq \mathcal{A}^*$.*

Proof. From (7.6) we find that

$$\tilde{\mathcal{A}}^* \cap \mathbb{N}^n = \{\mathbf{m} \in \mathbb{N}^n \mid \tilde{\mathcal{R}}_\infty^{\bar{U}}(\mathbf{m}) \subseteq \tilde{\mathcal{A}}\}.$$

Moreover, we have for each $\mathbf{m} \in \mathbb{N}^n$ that

$$\tilde{\mathcal{R}}_\infty^{\bar{U}}(\mathbf{m}) \subseteq \tilde{\mathcal{A}} \quad \Rightarrow \quad \tilde{\mathcal{R}}_\infty^{\bar{U}}(\mathbf{m}) \cap \mathbb{N}^n \subseteq \tilde{\mathcal{A}} \cap \mathbb{N}^n,$$

$$\Leftrightarrow \quad \tilde{\mathcal{R}}_{\infty}^{\bar{U}}(\mathbf{m}) \cap \mathbb{N}^n \subseteq \mathcal{A} \qquad (7.4),$$

$$\Rightarrow \quad \mathcal{R}_{\infty}^{\bar{U}}(\mathbf{m}) \subseteq \mathcal{A} \qquad \text{Proposition 7.1.}$$

From (7.5) and (7.6) it then follows that $\tilde{\mathcal{A}}^* \cap \mathbb{N}^n \subseteq \mathcal{A}^*$. □

Thus, $\tilde{\mathcal{A}}^* \cap \mathbb{N}^n$ is a subset of the supremal controllable subset. As is shown later, $\tilde{\mathcal{A}}^* \cap \mathbb{N}^n$ mostly is a good, in a lot of cases exact, approximation of $\mathcal{A}^*$.

Note that the result of Proposition 7.4 is independent of the form of $\mathcal{A}$ and $\tilde{\mathcal{A}}$ in (7.2) and (7.3). It is based exclusively on the relationship (7.4). In the next subsection we construct $\tilde{\mathcal{A}}^*$ for $\tilde{\mathcal{A}}$ in (7.3) if the influencing net $\mathcal{N}_{\mathcal{A}}^{\mathrm{uc}}$, and thus also $\tilde{\mathcal{N}}_{\mathcal{A}}^{\mathrm{uc}}$, is acyclic.

2.2. Construction of $\tilde{\mathcal{A}}^*$

In this subsection we show in a constructive manner that $\tilde{\mathcal{A}}^*$ in (7.6) can be described with a finite disjunction of linear inequalities if $\tilde{\mathcal{A}}$ is as in (7.3) and $\mathcal{N}_{\mathcal{A}}^{\mathrm{uc}}$ is acyclic. Lemma 7.1 is essential for the algorithm.

Lemma 7.1. *Define* $K_l := \{1, \ldots, k_l\}$ *for* $l = 1, 2, 3$. *Let* $a_1, \ldots, a_{k_1}$, $b_1, \ldots, b_{k_2}$, $c_1, \ldots, c_{k_3} \in \mathbb{Q}$ *and* $d \in \mathbb{Q}_+$. *Then,*

$$\bigvee_{u \in K_1} (y \leq a_u) \vee \bigvee_{v \in K_2} (b_v \leq y) \vee \bigvee_{w \in K_3} (0 \leq c_w) \qquad (y \in \mathbb{Q}_+, y \leq d),$$

if and only if

$$\bigvee_{u \in K_1} (d \leq a_u) \vee \bigvee_{v \in K_2} (b_v \leq 0) \vee \bigvee_{u \in K_1, v \in K_2} (b_v \leq a_u) \vee \bigvee_{w \in K_3} (0 \leq c_w).$$

Proof. For each a and b in $\mathbb{Q}$ and for each d in $\mathbb{Q}_+$ one easily verifies that

$$y \leq a \vee b \leq y \qquad (y \in \mathbb{Q}_+, y \leq d),$$

if and only if

$$d \leq a \vee b \leq 0 \vee b \leq a.$$

When we apply this result for

$$a = \max_{u \in K_1}(a_u) \quad \text{and} \quad b = \min_{v \in K_2}(b_v),$$

then we obtain the statement of the lemma. □

For the construction of the algorithm, assume first that there are no source transitions in $\mathcal{N}_{\mathcal{A}}^{\mathrm{uc}}$, i.e. $T^{\mathrm{so}} \cap \bar{U} = \emptyset$. The case $T^{\mathrm{so}} \cap \bar{U} \neq \emptyset$ is discussed in Section 2.3. As the influencing net

$$\mathcal{N}_{\mathcal{A}}^{\mathrm{uc}} = (\bar{I}, \bar{U}, f^+|_{\bar{U}\times\bar{I}}, f^-|_{\bar{I}\times\bar{U}})$$

is acyclic, the sets $\bar{I}$ and $\bar{U}$ can be partitioned as in Chapter 5 (Definition 5.1) into sets $P_1, \ldots, P_{\lambda'}$ and sets $T_2, \ldots, T_\lambda$, respectively. As $T^{\mathrm{so}} \cap \bar{U} = \emptyset$, we have $T_1 = \emptyset$. Analogously to the notation that precedes Proposition 5.5 (p. 118), define

$$\tau_i := \sharp T_i \quad \text{and} \quad \tau_i' := \sum_{k=2}^{i} \tau_k \quad \text{for } i = 2, \ldots, \lambda.$$

Let $\tau := \bar{U}$. Thus, $\tau = \tau_\lambda'$. Introduce the notation $\bar{U} = \{t_1, \ldots, t_\tau\}$ such that

$$t_{\tau_{i-1}'+1}, \ldots, t_{\tau_i'} \in T_i \quad \text{for } i = 2, \ldots, \lambda, \tag{7.7}$$

where $\tau_1' := 0$.

For $i = 1, \ldots, \tau$ recursively define the sets

$$\tilde{\mathcal{A}}_i^* := \{\mathbf{x} \in \mathbb{Q}_+^n \mid \tilde{\mathcal{R}}_1^{t_{\tau-i+1}}(\mathbf{x}) \subseteq \tilde{\mathcal{A}}_{i-1}^*\}, \tag{7.8}$$

where

$$\tilde{\mathcal{A}}_0^* := \tilde{\mathcal{A}}. \tag{7.9}$$

From the embedded version of Proposition 7.2 and (7.6) it follows that $\tilde{\mathcal{A}}_\tau^* = \tilde{\mathcal{A}}^*$. Next, we show by induction that for each $i = 0, \ldots, \tau$ there exist vectors $\tilde{\mathbf{A}}_{i,j}^* \in \mathbb{Q}^n$ and numbers $\tilde{b}_{i,j}^* \in \mathbb{Q}$ ($j = 1, \ldots, \nu_i$) such that

$$\tilde{\mathcal{A}}_i^* = \bigcup_{j=1}^{\nu_i} \{\mathbf{x} \in \mathbb{Q}_+^n \mid \tilde{\mathbf{A}}_{i,j}^{*,T}\mathbf{x} \leq \tilde{b}_{i,j}^*\}. \tag{7.10}$$

From (7.3) and (7.9) it is clear that this holds for $i = 0$. The induction hypothesis is that (7.10) holds for i. The proof is constructive. We use the transition bag assumption.

From (7.8) it follows that $\mathbf{x} \in \mathbb{Q}_+^n$ belongs to $\tilde{\mathcal{A}}_{i+1}^*$ if and only if

$$\tilde{\mathcal{R}}_1^{t_{\tau-i}}(\mathbf{x}) \subseteq \tilde{\mathcal{A}}_i^*. \tag{7.11}$$

The Petri net

$$({}^\bullet t_{\tau-i}, \{t_{\tau-i}\}, f^+|_{\{t_{\tau-i}\}\times{}^\bullet t_{\tau-i}}, f^-|_{{}^\bullet t_{\tau-i}\times\{t_{\tau-i}\}})$$

is a sub-Petri net of $\tilde{\mathcal{N}}_{\mathcal{A}}^{\mathrm{uc}}$ and is thus acyclic. Because of the embedded version of Proposition 7.3, the induction hypothesis (7.10) and the absence of source transitions, condition (7.11) is equivalent to

$$\bigvee_{j=1}^{\nu_i} \tilde{\mathbf{A}}_{i,j}^{*,T}(\mathbf{x} + \mathbf{F}(t_{\tau-i})y) \leq \tilde{b}_{i,j}^{*} \qquad \left(y \in \mathbb{Q}_+, y \leq \min_{p \in {}^\bullet t_{\tau-i}} \frac{\mathbf{x}(p)}{f^-(p, t_{\tau-i})}\right). \tag{7.12}$$

Herein, $\mathbf{F}(t_{\tau-i})$ is the column of $\mathbf{F}$ which corresponds to the transition $t_{\tau-i}$. Partition $\{1, \ldots, \nu_i\}$ into three sets:

$$K_1^i := \{j \in \{1, \ldots, \nu_i\} \mid \tilde{\mathbf{A}}_{i,j}^{*,T}\mathbf{F}(t_{\tau-i}) > 0\},$$
$$K_2^i := \{j \in \{1, \ldots, \nu_i\} \mid \tilde{\mathbf{A}}_{i,j}^{*,T}\mathbf{F}(t_{\tau-i}) < 0\},$$
$$K_3^i := \{j \in \{1, \ldots, \nu_i\} \mid \tilde{\mathbf{A}}_{i,j}^{*,T}\mathbf{F}(t_{\tau-i}) = 0\}.$$

Then (7.12) is equivalent to

$$\bigvee_{j \in K_1^i} \left(y \leq \frac{\tilde{b}_{i,j}^{*} - \tilde{\mathbf{A}}_{i,j}^{*,T}\mathbf{x}}{\tilde{\mathbf{A}}_{i,j}^{*,T}\mathbf{F}(t_{\tau-i})}\right) \vee \bigvee_{j \in K_2^i} \left(\frac{\tilde{b}_{i,j}^{*} - \tilde{\mathbf{A}}_{i,j}^{*,T}\mathbf{x}}{\tilde{\mathbf{A}}_{i,j}^{*,T}\mathbf{F}(t_{\tau-i})} \leq y\right)$$
$$\vee \bigvee_{j \in K_3^i} \left(0 \leq \tilde{b}_{i,j}^{*} - \tilde{\mathbf{A}}_{i,j}^{*,T}\mathbf{x}\right) \qquad \left(y \in \mathbb{Q}_+, y \leq \min_{p \in {}^\bullet t_{\tau-i}} \frac{\mathbf{x}(p)}{f^-(p, t_{\tau-i})}\right).$$

By applying Lemma 7.1, this is equivalent to

$$\bigvee_{j \in K_1^i} \left(\min_{p \in {}^\bullet t_{\tau-i}} \frac{\mathbf{x}(p)}{f^-(p, t_{\tau-i})} \leq \frac{\tilde{b}_{i,j}^{*} - \tilde{\mathbf{A}}_{i,j}^{*,T}\mathbf{x}}{\tilde{\mathbf{A}}_{i,j}^{*,T}\mathbf{F}(t_{\tau-i})}\right) \vee \bigvee_{j \in K_2^i} \left(\frac{\tilde{b}_{i,j}^{*} - \tilde{\mathbf{A}}_{i,j}^{*,T}\mathbf{x}}{\tilde{\mathbf{A}}_{i,j}^{*,T}\mathbf{F}(t_{\tau-i})} \leq 0\right)$$
$$\vee \bigvee_{j_1 \in K_1^i, j_2 \in K_2^i} \left(\frac{\tilde{b}_{i,j_2}^{*} - \tilde{\mathbf{A}}_{i,j_2}^{*,T}\mathbf{x}}{\tilde{\mathbf{A}}_{i,j_2}^{*,T}\mathbf{F}(t_{\tau-i})} \leq \frac{\tilde{b}_{i,j_1}^{*} - \tilde{\mathbf{A}}_{i,j_1}^{*,T}\mathbf{x}}{\tilde{\mathbf{A}}_{i,j_1}^{*,T}\mathbf{F}(t_{\tau-i})}\right)$$
$$\vee \bigvee_{j \in K_3^i} \left(0 \leq \tilde{b}_{i,j}^{*} - \tilde{\mathbf{A}}_{i,j}^{*,T}\mathbf{x}\right). \tag{7.13}$$

Lemma 7.2. *Consider a finite subset I of $\mathbb{N}$, and numbers $r_i \in \mathbb{R}$ ($i \in I$) and $r \in \mathbb{R}$. Then, $\min_{i \in I}(r_i) \leq r$ if and only if $\vee_{i \in I}(r_i \leq r)$.*

By applying this lemma to (7.13), we find vectors $\tilde{\mathbf{A}}_{i+1,j}^{*} \in \mathbb{Q}^n$ and numbers $\tilde{b}_{i+1,j}^{*} \in \mathbb{Q}$ ($j = 1, \ldots, \nu_{i+1}$) such that

$$\tilde{\mathcal{A}}_{i+1}^{*} = \bigcup_{j=1}^{\nu_{i+1}} \{\mathbf{x} \in \mathbb{Q}_+^n \mid \tilde{\mathbf{A}}_{i+1,j}^{*,T}\mathbf{x} \leq \tilde{b}_{i+1,j}^{*}\}.$$

This completes the proof of the induction step. Thus, in $\tau = \sharp\bar{U}$ of these steps the set $\tilde{\mathcal{A}}^*$ can be constructed.

Theorem 7.1. *If $\tilde{\mathcal{N}}^{\text{uc}}_{\mathcal{A}}$ is acyclic and does not contain any source transitions, and $\tilde{A}$ is as in* (7.3), *then the above algorithm constructs vectors $\tilde{\mathbf{A}}_j^* \in \mathbb{Q}^n$ and numbers $\tilde{b}_j^* \in \mathbb{Q}$ $(j = 1, \ldots, \nu)$ such that*

$$\tilde{\mathcal{A}}^* = \bigcup_{j=1}^{\nu} \{\mathbf{x} \in \mathbb{Q}_+^n \mid \tilde{\mathbf{A}}_j^{*,T}\mathbf{x} \leq \tilde{b}_j^*\}. \tag{7.14}$$

From Proposition 7.4 it follows that we have constructed the subset

$$\tilde{\mathcal{A}}^* \cap \mathbb{N}^n = \bigcup_{j=1}^{\nu} \{\mathbf{m} \in \mathbb{N}^n \mid \tilde{\mathbf{A}}_j^{*,T}\mathbf{m} \leq \tilde{b}_j^*\}$$

of $\mathcal{A}^*$. This subset is nontrivial as is illustrated next.

Example 7.2. Consider the acyclic influencing net $\tilde{\mathcal{N}}^{\text{uc}}_{\mathcal{A}}$ in Figure 7.1(a). The legal set is $\mathcal{A} = \{\mathbf{m} \in \mathbb{N}^n \mid \mathbf{m}(p_1) \leq b\}$ with $b \in \mathbb{N}$. For the partitions of Definition 5.1 we have that $T_1 = \emptyset$, $T_2 = \{t_2, t_3\}$, $T_3 = \{t_1\}$, $P_1 = \{p_4\}$, $P_2 = \{p_2, p_3\}$ and $P_3 = \{p_1\}$. Thus is $\tilde{\mathcal{R}}_\infty(\mathbf{x}) = \tilde{\mathcal{R}}_1^{t_1}(\tilde{\mathcal{R}}_1^{t_2}(\tilde{R}_1^{t_3}(\mathbf{x})))$. Note that the transition names in Figure 7.1(a) do not correspond to the order in (7.7). Recall from (7.8) the definition of $\tilde{\mathcal{A}}_1^*$, $\tilde{\mathcal{A}}_2^*$ and $\tilde{\mathcal{A}}_3^*$. We construct $\tilde{\mathcal{A}}_3^* = \tilde{\mathcal{A}}^*$ with the aid of Lemma 7.1.

We have that $\mathbf{x} \in \tilde{\mathcal{A}}_1^*$ if and only if $\tilde{\mathcal{R}}_1^{t_1}(\mathbf{x}) \subseteq \tilde{\mathcal{A}}_0^*$, or if and only if

$$\begin{aligned} & \mathbf{m}(p_1) + y \leq b \qquad (y \in \mathbb{Q}_+, y \leq \min(\mathbf{x}(p_2), \mathbf{x}(p_3))), \\ \Leftrightarrow\ & \mathbf{x}(p_1) + \mathbf{x}(p_2) \leq b \vee \mathbf{x}(p_1) + \mathbf{x}(p_3) \leq b, \end{aligned}$$

and that $\mathbf{x} \in \tilde{\mathcal{A}}_2^*$ if and only if $\tilde{\mathcal{R}}_1^{t_2}(\mathbf{x}) \subseteq \tilde{\mathcal{A}}_1^*$, or if and only if

$$\begin{aligned} & \mathbf{x}(p_1) + \mathbf{x}(p_2) \leq b \vee \mathbf{x}(p_1) + \mathbf{x}(p_3) + y \leq b \quad (y \in \mathbb{Q}_+, y \leq \mathbf{x}(p_4)), \\ \Leftrightarrow\ & \mathbf{x}(p_1) + \mathbf{x}(p_2) \leq b \vee \mathbf{x}(p_1) + \mathbf{x}(p_3) + \mathbf{x}(p_4) \leq b, \end{aligned}$$

and, finally, $\mathbf{x} \in \tilde{\mathcal{A}}_3^* = \tilde{\mathcal{A}}^*$ if and only if $\tilde{\mathcal{R}}_1^{t_3}(\mathbf{x}) \subseteq \tilde{\mathcal{A}}_2^*$, or if and only if

$$\begin{aligned} & \mathbf{x}(p_1) + \mathbf{x}(p_2) + y \leq b \vee \mathbf{x}(p_1) + \mathbf{x}(p_3) + \mathbf{x}(p_4) - y \leq b \\ & \qquad\qquad (y \in \mathbb{Q}_+, y \leq \mathbf{x}(p_4)), \\ \Leftrightarrow\ & \mathbf{x}(p_1) + \mathbf{x}(p_2) + \mathbf{x}(p_4) \leq b \vee \mathbf{x}(p_1) + \mathbf{x}(p_3) + \mathbf{x}(p_4) \leq b \\ & \qquad \vee 2\mathbf{x}(p_1) + \mathbf{x}(p_2) + \mathbf{x}(p_3) + \mathbf{x}(p_4) \leq 2b. \end{aligned}$$

Thus, $\tilde{\mathcal{A}}^* \cap \mathbb{N}^n = \{\mathbf{m} \in \mathbb{N}^n \mid \mathbf{m}(p_1) + \mathbf{m}(p_2) + \mathbf{m}(p_4) \leq b \vee \mathbf{m}(p_1) + \mathbf{m}(p_3) + \mathbf{m}(p_4) \leq b \vee 2\mathbf{m}(p_1) + \mathbf{m}(p_2) + \mathbf{m}(p_3) + \mathbf{m}(p_4) \leq 2b\}$. △

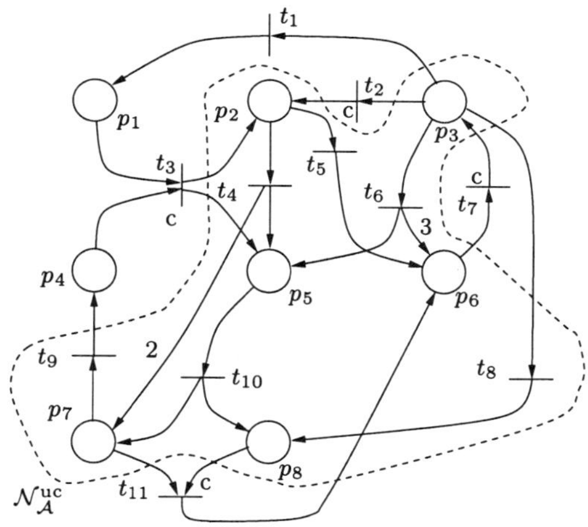

Figure 7.2. The Petri net of Example 7.3. The influencing net $\mathcal{N}_{\mathcal{A}}^{\mathrm{uc}}$ is indicated with the dashed line.

Example 7.3. For the Petri net in Figure 7.2 and the legal set

$$\mathcal{A} = \{\mathbf{m} \in \mathbb{N}^8 \mid -2\mathbf{m}(p_2) + 3\mathbf{m}(p_3) - \mathbf{m}(p_5) \\ + \mathbf{m}(p_6) - \mathbf{m}(p_7) + 2\mathbf{m}(p_8) \leq 6\},$$

one finds that

$$\tilde{\mathcal{A}}^* \cap \mathbb{N}^8 = \{\mathbf{m} \in \mathbb{N}^8 \mid 2\mathbf{m}(p_2) + 5\mathbf{m}(p_3) + 2\mathbf{m}(p_5) \\ + \mathbf{m}(p_6) + 2\mathbf{m}(p_8) \leq 6\}.$$

In Section 4 it is shown that in this example $\mathcal{A}^* = \tilde{\mathcal{A}}^* \cap \mathbb{N}^n$. △

Recall from (5.32) that when $\mathcal{A}$ is described by a finite disjunction of linear inequalities and the influencing net $\mathcal{N}_{\mathcal{A}}^{\mathrm{uc}}$ is acyclic, then the problem of deciding whether a given state belongs to $\mathcal{A}^*$ can be written as an integer linear programming inequality. With Proposition 7.3 one verifies that in Theorem 7.1 we have constructed a closed-form solution to the corresponding linear programming inequality. Let $z \in \mathbb{Q}$. Then $\mathbf{x} \in \mathbb{Q}_+^n$ belongs to $\tilde{\mathcal{A}}^*$ if and only if

$$\max z \leq 0 \tag{7.15}$$

$$\text{subject to} \begin{cases} z \leq \mathbf{A}_i^T|_{\bar{I}}\,(\mathbf{x}|_{\bar{I}} + \mathbf{F}_{\mathcal{A}}\mathbf{y}) - b_i & (i = 1, \ldots, \nu) \\ \mathbf{y} \in \mathbb{Q}_+^{\sharp\bar{U}} \\ \mathbf{0} \leq \mathbf{x}|_{\bar{I}} + \mathbf{F}_{\mathcal{A}}\mathbf{y}. \end{cases}$$

From Corollary 4.1 (p. 87) and the definition of $\mathcal{A}^*$ it follows that

$$\mathcal{A}^* = \{\mathbf{m} \in \mathbb{N}^n \mid \mathcal{R}_\infty^{T'}(\mathbf{m}) \subseteq \mathcal{A}\}$$

for all subsets T' of T^{uc} with $\bar{U} \subseteq T'$. Because of the possible difficulty to determine the sets $P^\uparrow$, $P^\downarrow$ and P^{d}, the set of uncontrollable transitions $\bar{U}$ is not always exactly known. In this case the above algorithm can be applied for a set $T' \supseteq \bar{U}$ instead of for $\bar{U}$, if the uncontrollable sub-Petri net

$$({}^\bullet T', T', f^+|_{T' \times {}^\bullet T'}, f^-|_{{}^\bullet T' \times T'})$$

of $\mathcal{N}^{\mathrm{uc}}$ is acyclic. The construction of $\tilde{\mathcal{A}}^*$ is then more involved. It takes $\sharp T'$, instead of $\tau = \sharp \bar{U}$, steps.

2.3. Presence of source transitions

If there are source transitions in $\mathcal{N}_{\mathcal{A}}^{\mathrm{uc}}$, i.e. $T^{\mathrm{so}} \cap \bar{U} \neq \emptyset$, then the above algorithm can also be used to construct $\tilde{\mathcal{A}}^*$.

Proposition 7.5. *In a Petri net with initial state $\mathbf{m}_0 \in \mathbb{N}^n$, it holds that*

$$\mathcal{R}_\infty^{T'}(\mathbf{m}_0) = \mathcal{R}_\infty^{T' \setminus T^{\mathrm{so}}}(\mathcal{R}_\infty^{T^{\mathrm{so}} \cap T'}(\mathbf{m}_0))$$

for each $T' \subseteq T$.

Proof. As ${}^\bullet T^{\mathrm{so}} = \emptyset$, this follows from Proposition 6.3 (p. 133) for $Y_1 = T' \setminus T^{\mathrm{so}}$ and $Y_2 = T^{\mathrm{so}}$. □

In words, each reachable state can be reached by first executing source transitions and then transitions which are not source transitions. This property also holds for continuous Petri nets.

If $\tilde{\mathcal{N}}_{\mathcal{A}}^{\mathrm{uc}}$ contains source transitions, then first consider the set

$$\{\mathbf{x} \in \mathbb{Q}_+^n \mid \tilde{\mathcal{R}}_\infty^{\bar{U} \setminus T^{\mathrm{so}}}(\mathbf{x}) \subseteq \tilde{\mathcal{A}}\}.$$

As the corresponding uncontrollable Petri net does not contain any source transitions, Theorem 7.1 can be used to construct a representation

$$\bigcup_{j=1,\ldots,\nu} \{\mathbf{x} \in \mathbb{Q}_+^n \mid \tilde{\mathbf{A}}_j^{*,T} \mathbf{x} \leq \tilde{b}_j^*\}$$

for this set. As source transitions can be executed any number of times, with Proposition 7.5 it is clear that

$$\tilde{\mathcal{A}}^* = \{\mathbf{x} \in \mathbb{Q}_+^n \mid \forall \mathbf{x}' \in \tilde{\mathcal{R}}_\infty^{T^{\mathrm{so}} \cap \bar{U}}(\mathbf{x}) : \vee_{j=1,\ldots,\nu} \tilde{\mathbf{A}}_j^{*,T} \mathbf{x}' \leq \tilde{b}_j^*\},$$

$$= \bigcup_{j=1,\ldots,\nu} \{\mathbf{x} \in \mathbb{Q}_+^n \mid \forall \mathbf{y} \in \mathbb{Q}_+^m, \mathbf{y}|_{T\setminus(T^{\mathrm{so}}\cap\bar{U})} = \mathbf{0} : \tilde{\mathbf{A}}_j^{*,T}(\mathbf{x}+\mathbf{F}\mathbf{y}) \leq \tilde{b}_j^*\},$$

$$= \bigcup_{\substack{j=1,\ldots,\nu \\ \tilde{\mathbf{A}}_j^{*,T}\mathbf{F}(t)\leq 0, \forall t\in T^{\mathrm{so}}\cap\bar{U}}} \{\mathbf{x} \in \mathbb{Q}_+^n \mid \tilde{\mathbf{A}}_j^{*,T}\mathbf{x} \leq \tilde{b}_j^*\}.$$

3. Construction of the supremal controllable subset

In this section we investigate whether the approach of Section 2 can be used to construct the supremal controllable subset of a given legal set. Therefore, we consider Petri nets instead of continuous Petri nets. The legal set $\mathcal{A}$ is as in (7.2) and its influencing net is assumed acyclic.

3.1. Auxiliary lemmas

We formulate the discrete equivalent of Lemma 7.1 in Lemma 7.4. Lemma 7.3 is used in its proof.

Lemma 7.3. *Consider a finite subset I of $\mathbb{N}$ and rational numbers a_i $(i \in I)$. Then it holds that*

$$\left\lceil \min_{i\in I}(a_i) \right\rceil = \min_{i\in I}(\lceil a_i \rceil), \tag{7.16}$$

and that

$$\left\lfloor \max_{i\in I}(a_i) \right\rfloor = \max_{i\in I}(\lfloor a_i \rfloor). \tag{7.17}$$

Proof. We only prove (7.16). The proof of (7.17) is analogous. For rational numbers a and b it holds that

$$a \leq b \quad \Rightarrow \quad \lceil a \rceil \leq \lceil b \rceil.$$

Because of $\min_{i\in I}(a_i) \leq a_j$ for all $j \in I$, we thus have that

$$\left\lceil \min_{i\in I}(a_i) \right\rceil \leq \lceil a_j \rceil \quad \text{for } j \in I.$$

This is equivalent to

$$\left\lceil \min_{i\in I}(a_i) \right\rceil \leq \min_{i\in I}(\lceil a_i \rceil).$$

On the other hand,

$$\left\lceil \min_{i\in I}(a_i) \right\rceil \geq \min_{i\in I}(\lceil a_i \rceil) \tag{7.18}$$

is equivalent to

$$\exists j \in I : \left\lceil \min_{i \in I}(a_i) \right\rceil \geq \lceil a_j \rceil. \tag{7.19}$$

Because there exists an $i^* \in I$ for which $a_{i^*} = \min_{i \in I}(a_i)$, (7.18) is proven because one can take $j = i^*$ in (7.19). □

Lemma 7.4 is the discrete counterpart of Lemma 7.1.

Lemma 7.4. *Define $K_l := \{1, \ldots, k_l\}$ for $l = 1, 2, 3$. Let $a_1, \ldots, a_{k_1}$, $b_1, \ldots, b_{k_2}$, $c_1, \ldots, c_{k_3} \in \mathbb{Q}$ and $d \in \mathbb{N}$. Then,*

$$\bigvee_{u \in K_1} (k \leq a_u) \vee \bigvee_{v \in K_2} (b_v \leq k) \vee \bigvee_{w \in K_3} (0 \leq c_w) \qquad (k \in \mathbb{N}, k \leq d),$$

if and only if

$$\bigvee_{u \in K_1} (d \leq a_u) \vee \bigvee_{v \in K_2} (b_v \leq 0) \vee \bigvee_{u \in K_1, v \in K_2} (\lceil b_v \rceil \leq \lfloor a_u \rfloor + 1) \vee \bigvee_{w \in K_3} (0 \leq c_w). \tag{7.20}$$

Proof. One easily verifies that for all a and b in $\mathbb{Q}$ it holds that

$$k \leq a \vee b \leq k \qquad (k \in \mathbb{N}, k \leq d),$$

if and only if

$$d \leq a \vee b \leq 0 \vee \lceil b \rceil \leq \lfloor a \rfloor + 1.$$

When we apply this result together with Lemma 7.3 for

$$a = \max_{u \in K_1}(a_u) \quad \text{and} \quad b = \min_{v \in K_2}(b_v),$$

we obtain the result of the lemma. □

3.2. Discussion

Next, we discuss the applicability of Lemma 7.4 to the construction of $\mathcal{A}^*$. The approach is similar to the one in Section 2 for the construction of $\tilde{\mathcal{A}}^*$. Lemma 7.4 is very similar to Lemma 7.1, its continuous counterpart. Nevertheless, it is not similar enough to be able to construct $\mathcal{A}^*$ in all cases. The nonlinear character of the third term in (7.20), i.e. the presence of the operators $\lceil \cdot \rceil$ and $\lfloor \cdot \rfloor$, does not always allow to apply Lemma 7.4 multiple times in succession.

Consider a legal set $\mathcal{A}$ as in (7.2). Assume $\mathcal{N}_{\mathcal{A}}^{\text{uc}}$ acyclic and $T^{\text{so}} \cap \bar{U} = \emptyset$. Analogously to (7.8), define for $i = 1, \ldots, \tau$ the sets

$$\mathcal{A}_i^* := \{\mathbf{m} \in \mathbb{N}^n \mid \mathcal{R}_1^{t_{\tau-i+1}}(\mathbf{m}) \subseteq \mathcal{A}_{i-1}^*\}, \tag{7.21}$$

where $\mathcal{A}_0^* := \mathcal{A}$. Thus, $\mathcal{A}^* = \mathcal{A}_\tau^*$. Suppose that for an $i \in \{0, \ldots, \tau\}$ there exist vectors $\mathbf{A}_{i,j}^* \in \mathbb{Q}^n$ and numbers $b_{i,j}^* \in \mathbb{Q}$ $(j = 1, \ldots, \nu)$ such that

$$\mathcal{A}_i^* = \bigcup_{j=1}^{\nu_i} \{\mathbf{m} \in \mathbb{N}^n \mid \mathbf{A}_{i,j}^{*,T} \mathbf{m} \leq b_{i,j}^*\}. \tag{7.22}$$

Then, $\mathbf{m} \in \mathcal{A}_{i+1}^*$ if and only if $\mathcal{R}_1^{t_{\tau-i}}(\mathbf{m}) \subseteq \mathcal{A}_i^*$, or if and only if

$$\bigvee_{j=1}^{\nu_i} \mathbf{A}_{i,j}^{*,T}(\mathbf{m} + \mathbf{F}(t_{\tau-i})k) \leq b_{i,j}^* \qquad \left(k \in \mathbb{N}, k \leq \min_{p \in {}^\bullet t_{\tau-i}} \left\lfloor \frac{\mathbf{m}(p)}{f^-(p, t_{\tau-i})} \right\rfloor\right). \tag{7.23}$$

Lemma 7.4 can be applied here. As the upper limit of k is nonlinear in the state, in general this lemma cannot be used in the next step. Therefore, we assume that $f^-|_{\bar{I} \times \bar{U}} : \bar{I} \times \bar{U} \to \{0, 1\}$, i.e. each arc in $\mathcal{N}_{\mathcal{A}}^{\text{uc}}$ from a place to a transition has weight one. Then (7.23) becomes

$$\bigvee_{j=1}^{\nu_i} \mathbf{A}_{i,j}^{*,T}(\mathbf{m} + \mathbf{F}(t_{\tau-i})k) \leq b_{i,j}^* \qquad \left(k \in \mathbb{N}, k \leq \min_{p \in {}^\bullet t_{\tau-i}} \mathbf{m}(p)\right). \tag{7.24}$$

When one applies Lemma 7.4 to (7.24), a nonlinear term appears if $K_1^i \neq \emptyset$, $K_2^i \neq \emptyset$ and there exist $j \in K_1^i \cup K_2^i$ and $p \in P$ such that

$$\frac{\tilde{\mathbf{A}}_{i,j}^*(p)}{\tilde{\mathbf{A}}_{i,j}^{*,T} \mathbf{F}(t_{\tau-i})} \notin \mathbb{Z}.$$

This implies that due to the limitations of Lemma 7.4 this procedure does not lead to $\mathcal{A}^*$ when a nonlinear term is dependent of k in one of the following steps $i + 2, \ldots, \tau$.

Nevertheless, this method allows us to construct $\mathcal{A}^*$ for a number of classes of systems and legal sets. Existing results can be proven in a different, uniform manner and new results can be formulated. In the three following sections we study these results in more detail. The above analysis shows that the final form of $\mathcal{A}^*$ strongly depends on the coefficients in the linear inequalities which describe $\mathcal{A}$.

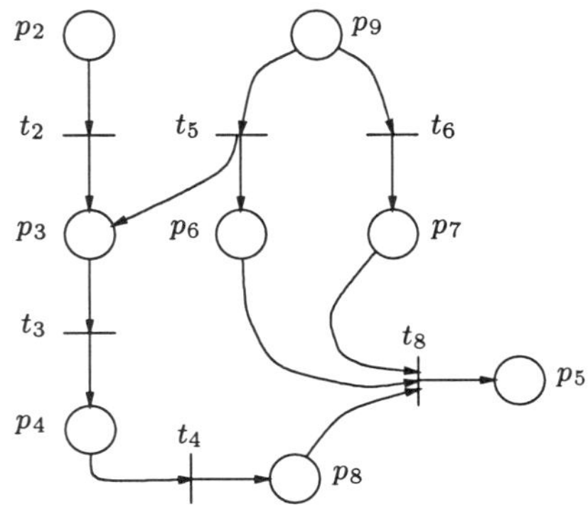

Figure 7.3. The influencing net $\mathcal{N}_{\mathcal{A}}^{\mathrm{uc}}$ of Example 7.5.

Example 7.4. With Lemma 7.4 we go through the calculations of Example 7.2 (p. 156) for the Petri net in Figure 7.1(a). Only the final stage differs from the continuous case. The result is that

$$\begin{aligned}\mathcal{A}^* = \{\mathbf{m} \in \mathbb{N}^n \mid \mathbf{m}(p_1) + \mathbf{m}(p_2) + \mathbf{m}(p_4) \leq b \vee \mathbf{m}(p_1) + \mathbf{m}(p_3) \\ + \mathbf{m}(p_4) \leq b \vee 2\mathbf{m}(p_1) + \mathbf{m}(p_2) + \mathbf{m}(p_3) + \mathbf{m}(p_4) \leq 2b + 1\}.\end{aligned}$$

Thus, $\tilde{\mathcal{A}}^* \cap \mathbb{N}^n$ is a proper subset of $\mathcal{A}^*$. △

Our approach also allows to treat examples which were obtained by Holloway *et al.* with the so-called *path algebra* [17]. The following example can be found in [17] and is solved here in a different way.

Example 7.5. Consider the influencing net $\mathcal{N}_{\mathcal{A}}^{\mathrm{uc}}$ in Figure 7.3 with the legal set

$$\mathcal{A} = \{\mathbf{m} \in \mathbb{N}^n \mid \mathbf{m}(p_4) = 0 \vee \mathbf{m}(p_5) = 0\}. \tag{7.25}$$

Following Chapter 5, the partitions of the set of places and the set of transitions are:

$$\begin{aligned} T_1 &= \emptyset, & P_1 &= \{p_2, p_9\}, \\ T_2 &= \{t_2, t_5, t_6\}, & P_2 &= \{p_3, p_6, p_7\}, \\ T_3 &= \{t_3\}, & P_3 &= \{p_4\}, \\ T_4 &= \{t_4\}, & P_4 &= \{p_8\}, \\ T_5 &= \{t_8\}, & P_5 &= \{p_5\}. \end{aligned}$$

By putting the transitions in the order $t_8, t_4, t_3, t_2, t_6, t_5$ it is possible to apply the above approach to construct $\mathcal{A}^*$. By eliminating redundant

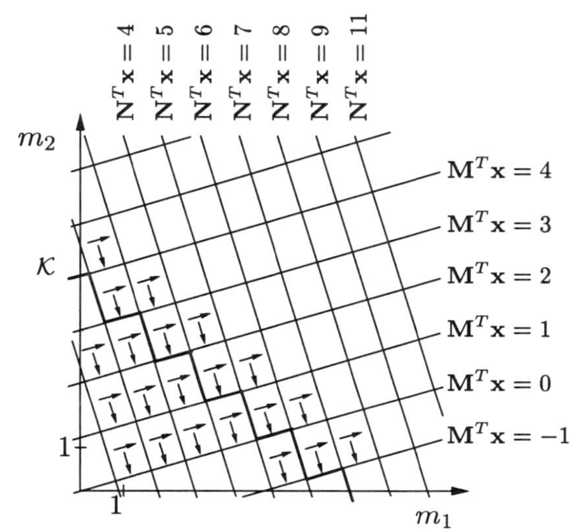

Figure 7.4. The set in (7.26).

inequalities in each step, we obtain that $\mathbf{m}$ belongs to $\mathcal{A}^*$ if and only if

$$\begin{gathered}\mathbf{m}(p_5) + \mathbf{m}(p_6) + \mathbf{m}(p_9) \leq 0 \vee \mathbf{m}(p_2) + \mathbf{m}(p_3) + \mathbf{m}(p_4) + \mathbf{m}(p_9) \leq 0 \\ \vee\, \mathbf{m}(p_2) + \mathbf{m}(p_3) + \mathbf{m}(p_4) + \mathbf{m}(p_5) + \mathbf{m}(p_8) + \mathbf{m}(p_9) \leq 1 \\ \vee\, \mathbf{m}(p_5) + \mathbf{m}(p_7) + \mathbf{m}(p_9) \leq 0 \vee 2\mathbf{m}(p_5) + \mathbf{m}(p_6) + \mathbf{m}(p_7) + \mathbf{m}(p_9) \leq 1 \\ \vee\, \mathbf{m}(p_2) + \mathbf{m}(p_3) + \mathbf{m}(p_4) + \mathbf{m}(p_5) + \mathbf{m}(p_7) + \mathbf{m}(p_9) \leq 1.\end{gathered}$$

From this reasoning it also follows that $\tilde{\mathcal{A}}^* \cap \mathbb{N}^n \subset \mathcal{A}^*$. △

To give some intuition for the 'nonlinear' sets which are possibly obtained when constructing the supremal controllable subset, the set

$$\{\mathbf{m} \in \mathbb{N}^2 \mid \lfloor \mathbf{M}^T\mathbf{m} \rfloor + \lceil \mathbf{N}^T\mathbf{m} \rceil \leq 5\} \tag{7.26}$$

is depicted in Figure 7.4. Each $\mathbf{m} \in \mathbb{N}^2$ that lies on or beneath the piecewise linear bold line $\mathcal{K}$ and does not belong to the intersection of $\mathcal{K}$ and a line $\mathbf{M}^T\mathbf{m} = k$ ($k \in \mathbb{N}$), belongs to this set. The arrows indicate the influence of the operators $\lfloor\cdot\rfloor$ and $\lceil\cdot\rceil$. One can conclude that the saw shape edge is typical for this kind of sets.

In the three following sections we discuss three general classes of Petri nets and legal sets for which the procedure of this section leads to the supremal controllable subset.

4. No synchronising transitions in $\mathcal{N}_{\mathcal{A}}^{\text{uc}}$

As shown in Chapter 2 (Section 6), under fine control all maximally permissive control laws are equivalent. When, in addition, the legal set $\mathcal{A}$ is given by a finite conjunction of linear inequalities and all transitions in the Petri net are controllable, then at least one of these maximally permissive supervisors can be realised by a Petri net. This is done by extending the original Petri net model by as many control places as there are linear inequalities in the description of $\mathcal{A}$. As the closed-loop system is a Petri net, it can be studied by using theoretical results and algorithms from Petri net theory.

For this reason it is important to know the classes of Petri nets and legal sets $\mathcal{A}$ for which the above property holds in the presence of uncontrollable transitions, i.e. for which $\mathcal{A}^*$ can be written as a finite intersection of linear inequalities. This is the subject of this section.

We consider a legal set which is described by one linear inequality:

$$\mathcal{A} = \{\mathbf{m} \in \mathbb{N}^n \mid \mathbf{A}^T\mathbf{m} \leq b\} \quad \text{with } \mathbf{A} \in \mathbb{Z}^n, b \in \mathbb{Z}. \tag{7.27}$$

Assume $\mathcal{A} \neq \emptyset$ and $\mathcal{A} \neq \mathbb{N}^n$. As usual, the presented results can also be used for intersections of sets of this type. Introduce the following conditions:

(C1) $\mathcal{N}_{\mathcal{A}}^{\text{uc}}$, the influencing net of $\mathcal{A}$, is acyclic,
(C2) $\forall t \in \bar{U} : \sharp({}^{\bullet}t) = 1$,
(C3) $\forall p \in \bar{I}, \forall t \in \bar{U} : f^-(p,t) \leq 1$.

Conditions (C1) and (C3) were already satisfied in Section 3. Therefore, it suffices to study the consequences of the additional condition (C2) for the discussion in that section.

From Proposition 4.24 (p. 102) it follows for $\mathcal{A}$ in (7.27) that

$$\begin{aligned} P^{\text{i}} &= \{p \in P \mid \mathbf{A}(p) = 0\}, \\ P^{\downarrow} &= \{p \in P \mid \mathbf{A}(p) > 0\}, \\ P^{\uparrow} &= \{p \in P \mid \mathbf{A}(p) < 0\}, \\ P^{\text{d}} &= \emptyset. \end{aligned}$$

From (C3) it follows for the sets of Definition 4.4 that $P_{\text{d}}^{\uparrow} = \emptyset$ and thus that $P^{\uparrow} = P_{\text{u}}^{\uparrow}$. Because of $P^{\text{d}} = \emptyset$ and $P_{\text{d}}^{\uparrow} = \emptyset$, it is true that $\bar{P}^{\text{d}} = \emptyset$. Therefore, $I^{\text{d}} = U^{\text{d}} = \emptyset$ (see Definition 4.5), $\bar{I} = I^{\downarrow} \cup I^{\uparrow}$ and $\bar{U} = U^{\downarrow} \cup U^{\uparrow}$ (see (4.12) and (4.13)). From (C2) it follows that $\sigma'(p) = \emptyset$ for all $p \in P^{\uparrow}$. With Definition 4.5 this implies that $I^{\uparrow} = P_{\text{u}}^{\uparrow}$ and that $U^{\uparrow} = (P_{\text{u}}^{\uparrow})^{\bullet} \cap T^{\text{uc}}$.

As explained next, from the discussion in Section 2.2 it follows that the supremal controllable subset of $\mathcal{A}$ in (7.27) can be described by one linear inequality $\mathbf{A}^{*,T}\mathbf{m} \leq b^*$ when conditions (C1) to (C3) are satisfied. To construct $\mathcal{A}^*$, we go again through Section 3.2.

In the first step of the procedure, $\mathcal{A}_1^*$ is constructed from $\mathcal{A}_0^* = \mathcal{A}^*$ by applying Lemma 7.4. As $\mathcal{A}$ is described by one linear inequality, in Lemma 7.4 two of the sets K_1, K_2 and K_3 are empty and the nonempty one is a singleton. Moreover, d in this lemma is equal to $\mathbf{m}(p)$, where, because of (C2), p is the unique input place of t_τ. Consequently, $\mathcal{A}_1^*$ is given by one linear inequality. This reasoning can be repeated for each of the other transitions in $\bar{U}$. This leads to the conclusion that the algorithm in Section 3 constructs a vector $\mathbf{A}^*$ such that $\mathcal{A}^* = \{\mathbf{m} \in \mathbb{N}^n \mid \mathbf{A}^{*,T}\mathbf{m} \leq b^* = b\}$.

For this construction of $\mathcal{A}^*$, it is possible to avoid going through the algorithm step by step. In the rest of this section we deduce a closed-form expression for $\mathbf{A}^*$. This approach makes the structure of $\mathbf{A}^*$ intuitively clear.

Because of (C1) we can, with the aid of Definition 5.1, partition the set of places $\bar{I}$ into nonempty sets $P_1, \ldots, P_{\lambda'}$ and the set of transitions $\bar{U}$ into nonempty sets $T_2, \ldots, T_\lambda$. Because of (C2), there are no source transitions in $\mathcal{N}_{\mathcal{A}}^{\text{uc}}$, i.e. $T^{\text{so}} \cap \bar{U} = \emptyset$, and thus $T_1 = \emptyset$. Order and rename places in $\bar{I}$ such that places in P_1 come first, then places in P_2, and so forth until $P_{\lambda'}$:

$$p_1, p_2, \ldots, p_{\sharp\bar{I}}.$$

Conditions (C2) and (C3) imply that the unique input place of a transition which belongs to T_{i+1}, belongs to P_i. Moreover, all output transitions of a place in P_i which belong to $\bar{U}$, belong to T_{i+1}. Hence, $P_i^\bullet \cap \bar{U} = T_{i+1}$. As there are no synchronising transitions in $\mathcal{N}_{\mathcal{A}}^{\text{uc}}$, places in $\bar{I}$ do not share any output transitions in $\bar{U}$. Because of these findings, one can order and rename transitions in $\bar{U}$ as follows:

$$\underbrace{t_1, \ldots,}_{p_1^\bullet \cap \bar{U}} \cdots \underbrace{, \ldots, t_\tau,}_{p_{\sharp I}^\bullet \cap \bar{U}} \tag{7.28}$$

where $\tau = \sharp\bar{U}$. From the above it is clear that in this order transitions in T_2 come first, then transitions in T_3, etc.

To construct the supremal controllable subset of $\mathcal{A}$ in (7.27) under conditions (C1) to (C3), we go again through the reasoning of Section 3.2 with the transitions order in (7.28).

We reuse the definition of $\mathcal{A}_i^*$ $(i = 0, \ldots, \tau)$ in (7.21). From the embedded version of Proposition 5.5 it follows that $\mathcal{A}_\tau^* = \mathcal{A}^*$. Next, we construct for all $i = 0, \ldots, \tau$ a vector $\mathbf{A}_i^* \in \mathbb{Z}^n$ and an integer b_i^* such that

$$\mathcal{A}_i^* = \{\mathbf{m} \in \mathbb{N}^n \mid \mathbf{A}_i^{*,T}\mathbf{m} \leq b_i^*\}. \tag{7.29}$$

From (7.27) and $\mathcal{A}_0^* = \mathcal{A}$, it follows that $\mathbf{A}_0^* = \mathbf{A}$ and $b_0^* = b_0$. Suppose that $\mathbf{A}_j^*$ and b_j^* are known for all indices j in $\{1, \ldots, i\}$. Next, we construct $\mathbf{A}_{i+1}^*$ and b_{i+1}^*.

From (7.21) it follows that $\mathbf{m} \in \mathbb{N}^n$ belongs to $\mathcal{A}_{i+1}^*$ if and only if

$$\mathcal{R}_1^{t_{\tau-i}}(\mathbf{m}) \subseteq \mathcal{A}_i^*. \tag{7.30}$$

Because of Theorem 5.1, (7.29) and the induction hypothesis, condition (7.30) is equivalent to

$$\mathbf{A}_i^{*,T}(\mathbf{m} + \mathbf{F}(t_{\tau-i})k) \leq b_i^* \qquad \left(k \in \mathbb{N}, k \leq \min_{p \in {}^\bullet t_{\tau-i}} \left\lfloor \frac{\mathbf{m}(p)}{f^-(p, t_{\tau-i})} \right\rfloor \right). \tag{7.31}$$

Because of (C2), ${}^\bullet t_{\tau-i}$ is a singleton $\{\bar{p}_{\tau-i}\}$ and the arc from this place to $t_{\tau-i}$ has weight one, i.e. $f^-(\bar{p}_{\tau-i}, t_{\tau-i}) = 1$. Therefore, (7.31) is equivalent to

$$\mathbf{A}_i^{*,T}(\mathbf{m} + \mathbf{F}(t_{\tau-i})k) \leq b_i^* \qquad (k \in \mathbb{N}, k \leq \mathbf{m}(\bar{p}_{\tau-i})). \tag{7.32}$$

Lemma 7.5. *Let $a, b \in \mathbb{Z}$ and $c \in \mathbb{N}$. Then,*

$$an \leq b \quad \textit{for all } n \in \mathbb{N} \textit{ with } n \leq c, \tag{7.33}$$

if and only if

$$\max(ac, 0) \leq b.$$

Proof. Because an is linear in n, (7.33) is equivalent to

$$0 \leq b \quad \wedge \quad ac \leq b. \tag{7.34}$$

These conditions are found by substituting the lower and upper bounds 0 and c for n in $an \leq b$, respectively. □

With the aid of this lemma, (7.32) is equivalent to

$$\begin{aligned} & \mathbf{A}_i^{*,T}\mathbf{m} + \max\left(\mathbf{m}(\bar{p}_{\tau-i})\mathbf{A}_i^{*,T}\mathbf{F}(t_{\tau-i}), 0\right) \leq b_i^*, \\ \Leftrightarrow \quad & \mathbf{A}_i^{*,T}\mathbf{m} + \mathbf{m}(\bar{p}_{\tau-i}) \max\left(\mathbf{A}_i^{*,T}\mathbf{F}(t_{\tau-i}), 0\right) \leq b_i^*, \\ \Leftrightarrow \quad & \mathbf{A}_i^{*,T}\mathbf{m} + \mathbf{m}(\bar{p}_{\tau-i}) \max\left(\mathbf{A}_i^{*,T}\mathbf{F}^+(t_{\tau-i}) - \mathbf{A}_i^*(\bar{p}_{\tau-i}), 0\right) \leq b_i^*. \end{aligned}$$

The last equivalence is motivated by the fact that $\mathbf{F}^{-}(t_{\tau-i})$, the column of $\mathbf{F}^{-}$ which corresponds to $t_{\tau-i}$, is zero apart from the element $\mathbf{F}^{-}(\bar{p}_{\tau-i}, t_{\tau-i})$ which is equal to one (see (C3)).

Consequently, we have constructed a vector $\mathbf{A}^{*}_{i+1} \in \mathbb{Z}^{n}$ and an integer b^{*}_{i+1} such that $\mathcal{A}^{*}_{i+1} = \{\mathbf{m} \in \mathbb{N}^{n} \mid \mathbf{A}^{*,T}_{i+1}\mathbf{m} \leq b^{*}_{i+1}\}$. It holds that $b^{*}_{i+1} = b^{*}_{i}$, that

$$\mathbf{A}^{*}_{i+1}(p) = \mathbf{A}^{*}_{i}(p) \quad \text{for all } p \neq \bar{p}_{\tau-i}, \tag{7.35}$$

and that

$$\begin{aligned}\mathbf{A}^{*}_{i+1}(\bar{p}_{\tau-i}) &= \mathbf{A}^{*}_{i}(\bar{p}_{\tau-i}) + \max\left(\mathbf{A}^{*,T}_{i}\mathbf{F}^{+}(t_{\tau-i}) - \mathbf{A}^{*}_{i}(\bar{p}_{\tau-i}), 0\right), \\ &= \max\left(\mathbf{A}^{*,T}_{i}\mathbf{F}^{+}(t_{\tau-i}), \mathbf{A}^{*}_{i}(\bar{p}_{\tau-i})\right).\end{aligned} \tag{7.36}$$

Consequently, when comparing the vectors $\mathbf{A}^{*}_{i}$ and $\mathbf{A}^{*}_{i+1}$, we see that only their $\bar{p}_{\tau-i}$-component can differ. This finishes the proof of the induction step.

Next, we summarise the $\tau = \sharp\bar{U}$ steps in the construction of the inequality $\mathbf{A}^{*,T}\mathbf{m} \leq b^{*}$ which describes $\mathcal{A}^{*}$, in order to obtain a closed-form expression for it. It is clear that

$$b^{*} = b^{*}_{\tau} = \cdots = b^{*}_{0} = b.$$

For the construction of $\mathbf{A}^{*}$, recall that transitions in $\bar{U}$ are ordered as in (7.28). When going through the above construction for all output transitions of a place p (these transitions are grouped in (7.28)) only the component of this place changes. After all these transitions it is equal to the maximum of $\mathbf{A}^{*,T}\mathbf{F}^{+}(t)$ over all output transitions t of p in $\bar{U}$, and $\mathbf{A}(p)$. Because of the structure of $\mathbf{F}^{+}$ and $\mathbf{F}^{-}$ with the above ordering of transitions and places in $\bar{U}$ and $\bar{I}$, respectively, we obtain the following important result.

Theorem 7.2. *Consider a Petri net $\mathcal{N}$ and a legal set $\mathcal{A}$ as in (7.27). If conditions (C1) to (C3) are satisfied, then it holds for the vector $\mathbf{A}^{*} \in \mathbb{Z}^{n}$ with*

$$\mathbf{A}^{*}(p) = \begin{cases} \max\left[\max_{t \in p^{\bullet} \cap \bar{U}}\left(\mathbf{A}^{*,T}\mathbf{F}^{+}(t)\right), \mathbf{A}(p)\right] & p \in \bar{I}, p^{\bullet} \cap \bar{U} \neq \emptyset, \\ \mathbf{A}(p) & p \in \bar{I}, p^{\bullet} \cap \bar{U} = \emptyset, \\ 0 & p \notin \bar{I}, \end{cases}$$

that $\mathcal{A}^{*} = \{\mathbf{m} \in \mathbb{N}^{n} \mid \mathbf{A}^{*,T}\mathbf{m} \leq b\}$.

Example 7.6. Consider again the Petri net in Figure 7.2 (see also Example 7.3, p. 157). By applying the result of Theorem 7.2, we obtain that

$$\begin{aligned}
\mathbf{A}^*(p_7) &= \max(0, \mathbf{A}(p_7)) = 0,\\
\mathbf{A}^*(p_8) &= \mathbf{A}(p_8) = 2,\\
\mathbf{A}^*(p_5) &= \max(\mathbf{A}^*(p_7) + \mathbf{A}^*(p_8), \mathbf{A}(p_5)) = 2,\\
\mathbf{A}^*(p_6) &= \mathbf{A}(p_6) = 1,\\
\mathbf{A}^*(p_2) &= \max(2\mathbf{A}^*(p_7) + \mathbf{A}^*(p_5), \mathbf{A}^*(p_6), \mathbf{A}(p_2)) = 2,\\
\mathbf{A}^*(p_3) &= \max(\mathbf{A}^*(p_5) + 3\mathbf{A}^*(p_6), \mathbf{A}^*(p_2), \mathbf{A}(p_3)) = 5,\\
\mathbf{A}^*(p_1) &= \mathbf{A}^*(p_4) = 0.
\end{aligned}$$

Consequently,

$$\mathbf{A}^* = (2, 5, 2, 1, 0, 2, 0, 0).$$

For the case of fine control (cfr. Chapter 2), at least one maximally permissive supervisor for $\mathcal{A}^*$ can be represented by a Petri net. It is given by

$$-\mathbf{A}^{*,T}\mathbf{F} = (0, 4, 2, 3, 0, 0, 5, 3, -4, -4, 1).$$

The structure of the closed-loop system is depicted in Figure 7.5. The supervisor consists of one control place p^c and the corresponding arcs. Note that none of the uncontrollable transitions has p^c as an input place. When comparing $\mathcal{A}^*$ with $\tilde{\mathcal{A}}^*$ in Example 7.3 we see that these sets are equal. The above approximation is exact. △

Theorem 7.2 has an instructive intuitive explanation. As there are no synchronising transitions in the influencing net of $\mathcal{A}$, tokens can move 'freely' through it. The enabling condition of a transition is determined by the state of one place —its unique input place— only. Consider a state $\mathbf{m}$. To decide whether $\mathbf{m} \in \mathcal{A}^*$, one looks for the maximum attainable value of $\mathbf{A}^T\mathbf{m}'$ over all $\mathbf{m}' \in \mathcal{R}_\infty^{\bar{U}}(\mathbf{m})$. For each place p one can choose between 'keeping' the $\mathbf{m}(p)$ tokens in p (resulting in a contribution $\mathbf{A}(p)\mathbf{m}(p)$) and 'sending' them to one of the uncontrollable output transitions t of p in $\bar{U}$ (resulting in a contribution $\mathbf{A}^{*,T}\mathbf{F}^+(t)\mathbf{m}(p)$). Therefore, to decide which option is the worst with respect to the legal set, in Theorem 7.2 the maximum among the different options is taken by backtracking through the acyclic Petri net from bottom to top.

From Theorem 7.2 it follows that $\mathbf{A}^* \geq \mathbf{A}$. Consequently, $\mathcal{A}^* \subseteq \mathcal{A}$. Condition (C2) can be weakened to $\sharp({}^\bullet t) \leq 1$ for all $t \in \bar{U}$. Theorem 7.2 can be extended to any uncontrollable super-Petri of $\mathcal{N}_\mathcal{A}^{\mathrm{uc}}$ if the conditions (C1) to (C3) hold for this net.

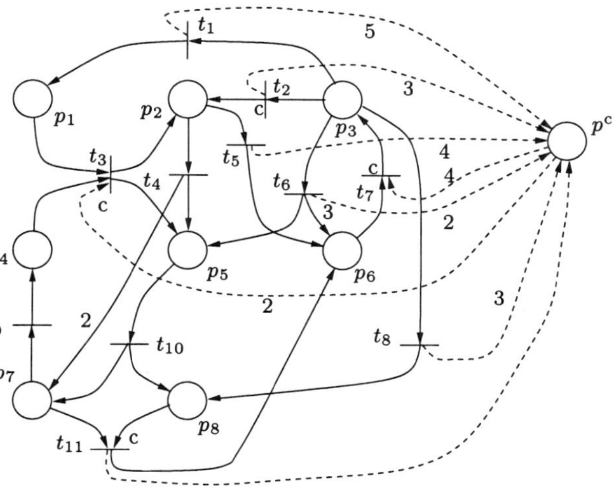

Figure 7.5. Closed-loop system of Example 7.6.

5. No choice places in $\mathcal{N}_{\mathcal{A}}^{\text{uc}}$

Assume that $\mathcal{N}_{\mathcal{A}}^{\text{uc}}$ is acyclic and free of choice places:

$$p^{\bullet} \cap \bar{U} \leq 1 \quad \text{for all } p \in \bar{I}. \tag{7.37}$$

In Chapter 5 (Section 5) these Petri nets have been studied in detail. Note that, following (7.37), it is possible that a place in $\bar{I}$ has more than one output transition. However, at most one belongs to $\bar{U}$.

In first instance, the legal set $\mathcal{A}$ can be any subset of $\mathbb{N}^n$. Further, assume that

$$p^{\bullet} \cap \bar{U} = \emptyset \quad \text{for } p \in P^{\downarrow} \cup P^{\text{d}}, \tag{7.38}$$

$$^{\bullet}p \cap \bar{U} = \emptyset \quad \text{for } p \in P^{\uparrow} \cup P^{\text{d}}. \tag{7.39}$$

Proposition 7.6. *Consider a Petri net $\mathcal{N}$ and a legal set $\mathcal{A}$ such that $\mathcal{N}_{\mathcal{A}}^{\text{uc}}$ is acyclic, and* (7.37), (7.38) *and* (7.39) *are satisfied. Let $\mathbf{m}_0 \in \mathbb{N}^n$ and $\boldsymbol{\delta}_1, \boldsymbol{\delta}_2 \in \Delta$ such that $\boldsymbol{\delta}_1|_{T\setminus\bar{U}} = \boldsymbol{\delta}_2|_{T\setminus\bar{U}} = \mathbf{0}$ and $\boldsymbol{\delta}_1 \leq \boldsymbol{\delta}_2$. If $\mathbf{m}_1 = \mathbf{m}_0 + \mathbf{F}\boldsymbol{\delta}_1 \geq \mathbf{0}$ and $\mathbf{m}_2 = \mathbf{m}_0 + \mathbf{F}\boldsymbol{\delta}_2 \geq \mathbf{0}$, then*

$$\mathbf{m}_1|_{P^{\downarrow}} \leq \mathbf{m}_2|_{P^{\downarrow}}, \tag{7.40}$$

$$\mathbf{m}_1|_{P^{\uparrow}} \geq \mathbf{m}_2|_{P^{\uparrow}}, \tag{7.41}$$

$$\mathbf{m}_1|_{P^{\text{d}}} = \mathbf{m}_2|_{P^{\text{d}}}. \tag{7.42}$$

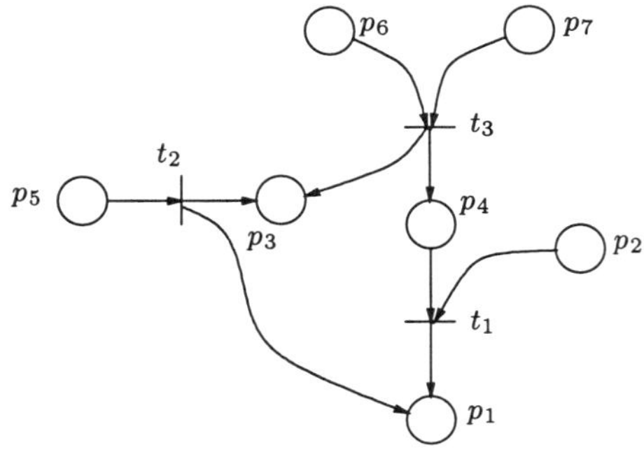

Figure 7.6. An acyclic influencing net $\mathcal{N}_{\mathcal{A}}^{\text{uc}}$ that is free of choice places.

Proof. We have for $p \in P^{\downarrow}$ that

$$\mathbf{m}_2(p) - \mathbf{m}_1(p) = \sum_{t \in {}^\bullet p \cap \bar{U}} f^+(t,p)(\boldsymbol{\delta}_2(t) - \boldsymbol{\delta}_1(t)) - \sum_{t \in p^\bullet \cap \bar{U}} f^-(p,t)(\boldsymbol{\delta}_2(t) - \boldsymbol{\delta}_1(t)).$$

The first term on the right hand side is greater than or equal to zero because of $\boldsymbol{\delta}_1 \leq \boldsymbol{\delta}_2$. The second term is zero because of (7.38). The proof of (7.41) and (7.42) is similar. □

For the states $\mathbf{m}_1$ and $\mathbf{m}_2$ in Proposition 7.6 we have, with Proposition 4.4 (p. 74), that $\mathbf{m}_1 \in \mathcal{A}$ if $\mathbf{m}_2 \in \mathcal{A}$. From Chapter 5 (Section 5) we know that in an acyclic Petri that is free of choice places, for each initial state $\mathbf{m}_0$ there exists a maximum enabled sequence of transition vectors with sum $\mathbf{y}^{\text{M}}$. For the value of this vector we refer to Chapter 5 (Section 5). This allows to formulate the following proposition.

Proposition 7.7. *Consider a Petri net $\mathcal{N}$ and a legal set $\mathcal{A}$ such that $\mathcal{N}_{\mathcal{A}}^{\text{uc}}$ is acyclic, and* (7.37), (7.38) *and* (7.39) *are satisfied. If $\mathbf{m}_0 \in \mathbb{N}^n$, then*

$$\mathbf{m}_0 \in \mathcal{A}^* \quad \Leftrightarrow \quad \mathbf{m}_0 + \mathbf{F}\mathbf{y}^{\text{M}} \in \mathcal{A}.$$

If $\mathcal{A}$ is described by a union of linear inequalities as in (7.2), then we obtain from the expression for $\mathbf{y}^{\text{M}}$ the set $\mathcal{A}^*$. The set $\mathcal{A}^*$ is described by linear inequalities if in addition $f^-|_{\bar{I}\times\bar{U}} : \bar{I} \times \bar{U} \to \{0,1\}$.

To obtain this result we have not used the technique of Section 3. However, it is clear that it can be applied here too to construct $\mathcal{A}^*$.

Example 7.7. Consider the acyclic influencing net $\mathcal{N}_{\mathcal{A}}^{\mathrm{uc}}$ of $\mathcal{A}$ in Figure 7.6. It is free of choice places. The legal set is

$$\mathcal{A} = \{\mathbf{m} \in \mathbb{N}^n \mid 2\mathbf{m}(p_1) + \mathbf{m}(p_3) - 3\mathbf{m}(p_5) - 2\mathbf{m}(p_6) \le b\},$$

where $b \in \mathbb{Z}$. We obtain with Proposition 4.24 that $P^{\downarrow} = \{p_1, p_3\}$, $P^{\uparrow} = \{p_5, p_6\}$ and $P^{\mathrm{d}} = \emptyset$. With the aid of the results in Chapter 5 (Section 5) it follows that

$$\begin{aligned} \mathbf{y}^{\mathrm{M}}(t_3) &= \min(\mathbf{m}_0(p_6), \mathbf{m}_0(p_7)), \\ \mathbf{y}^{\mathrm{M}}(t_2) &= \mathbf{m}_0(p_5), \\ \mathbf{y}^{\mathrm{M}}(t_1) &= \min(\mathbf{m}_0(p_2), \mathbf{m}_0(p_4) + \mathbf{y}^{\mathrm{M}}(t_3)), \end{aligned}$$

and thus that

$$\mathbf{m}_0|_{\bar{I}} + \mathbf{F}_{\mathcal{A}}\mathbf{y}^{\mathrm{M}}|_{\bar{U}} = \begin{bmatrix} \mathbf{m}_0(p_1) + \mathbf{y}^{\mathrm{M}}(t_1) + \mathbf{y}^{\mathrm{M}}(t_2) \\ \mathbf{m}_0(p_2) - \mathbf{y}^{\mathrm{M}}(t_1) \\ \mathbf{m}_0(p_3) + \mathbf{y}^{\mathrm{M}}(t_2) + \mathbf{y}^{\mathrm{M}}(t_3) \\ \mathbf{m}_0(p_4) - \mathbf{y}^{\mathrm{M}}(t_1) + \mathbf{y}^{\mathrm{M}}(t_3) \\ \mathbf{m}_0(p_5) - \mathbf{y}^{\mathrm{M}}(t_2) \\ \mathbf{m}_0(p_6) - \mathbf{y}^{\mathrm{M}}(t_3) \\ \mathbf{m}_0(p_7) - \mathbf{y}^{\mathrm{M}}(t_3) \end{bmatrix}.$$

From Proposition 7.7 we then find that $\mathbf{m} \in \mathcal{A}^*$ if and only if

$$\begin{aligned} 2\mathbf{m}(p_1) + \mathbf{m}(p_3) + \mathbf{m}(p_5) - 2\mathbf{m}(p_6) + 2\min(\mathbf{m}(p_2), \mathbf{m}(p_4) + \mathbf{m}(p_6), \\ \mathbf{m}(p_4) + \mathbf{m}(p_7)) + 3\min(\mathbf{m}(p_6), \mathbf{m}(p_7)) \le b. \end{aligned}$$

△

6. A third class

Finally, we consider a third class of systems for which the procedure of the Section 3 leads to a description with linear inequalities of $\mathcal{A}^*$. Assume that $f^-|_{\bar{I}\times\bar{U}} : \bar{I} \times \bar{U} \to \{0, 1\}$.

Proposition 7.8. *Consider a Petri net $\mathcal{N}$ and a legal set $\mathcal{A}$ which is described by one linear inequality. Moreover, let $\mathcal{N}_{\mathcal{A}}^{\mathrm{uc}}$ be acyclic and $\bar{U} \cap T^{\mathrm{so}} = \emptyset$. Partition $\bar{U}$ in sets $T_2, \ldots, T_\lambda$ as in Definition 5.1. If, apart from at most one transition $\bar{t} \in T_2$, each transition in $\bar{U}$ has precisely one input place, then there exist vectors $\mathbf{A}_j^* \in \mathbb{Z}^n$ and numbers $b_j^* \in \mathbb{Z}$ $(j = 1, \ldots, \nu)$ such that*

$$\mathcal{A}^* = \bigcup_{j=1}^{\nu} \{\mathbf{m} \in \mathbb{N}^n \mid \mathbf{A}_j^{*,T}\mathbf{m} \le b_j^*\}.$$

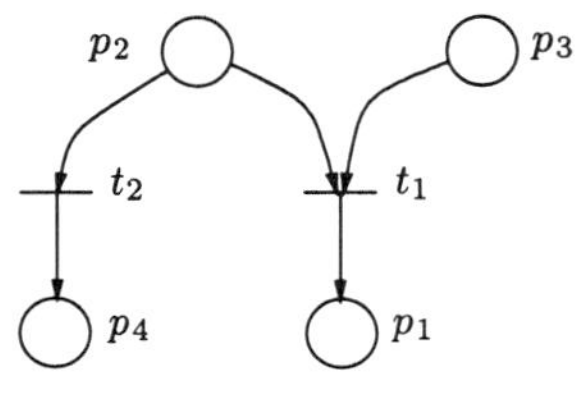

Figure 7.7. An influencing net $\mathcal{N}_{\mathcal{A}}^{\mathrm{uc}}$ illustrating Proposition 7.8.

Proof. Because of the freedom one has for the order within one partition of transitions T_i in Proposition 7.2, we have that $\mathcal{R}_{\infty}^{\bar{U}}(\mathbf{m}) = \mathcal{R}_1^{t_r}(\cdots(\mathcal{R}_1^{t_2}(\mathcal{R}_1^{\bar{t}}(\mathbf{m})))\cdots)$. Then the statement of the proof follows in a similar way as in Section 4. □

As was shown before, this result can immediately be extended to influencing nets $\mathcal{N}_{\mathcal{A}}^{\mathrm{uc}}$ which do contain any source transitions.

Example 7.8. Consider $\mathcal{N}_{\mathcal{A}}^{\mathrm{uc}}$ in Figure 7.7 with $\mathcal{A} = \{\mathbf{m} \in \mathbb{N}^n \mid 6\mathbf{m}(p_1) + \mathbf{m}(p_2) + 4\mathbf{m}(p_4) \leq b\}$ and $b \in \mathbb{N}$. For the partitions of Chapter 4 we have that $T_1 = \emptyset$, $T_2 = \{t_1, t_2\}$, $P_1 = \{p_2, p_3\}$ and $P_2 = \{p_1, p_4\}$. In $\mathcal{N}_{\mathcal{A}}^{\mathrm{uc}}$ it holds that $\mathcal{R}_{\infty}^{\bar{U}}(\mathbf{m}) = \mathcal{R}_1^{t_1}(\mathcal{R}_1^{t_2}(\mathbf{m}))$. Note that this order is not as specified in the proof of Proposition 7.8. If we apply for this order the procedure of Section 3, then the state $\mathbf{m}$ belongs to $\mathcal{A}^*$ if and only if

$$\begin{aligned} 6\mathbf{m}(p_1)+6\mathbf{m}(p_2)+4\mathbf{m}(p_4) \leq b \vee 6\mathbf{m}(p_1)+4\mathbf{m}(p_2)+5\mathbf{m}(p_3)+4\mathbf{m}(p_4) \leq b \\ \vee \left\lceil \frac{6\mathbf{m}(p_1) + 6\mathbf{m}(p_2) + 4\mathbf{m}(p_4) - b}{2} \right\rceil \\ \leq \left\lfloor \frac{b - 6\mathbf{m}(p_1) - \mathbf{m}(p_2) - 5\mathbf{m}(p_3) - 4\mathbf{m}(p_4)}{3} \right\rfloor + 1. \end{aligned}$$

However, it also holds that $\mathcal{R}_{\infty}^{\bar{U}}(\mathbf{m}) = \mathcal{R}_1^{t_2}(\mathcal{R}_1^{t_1}(\mathbf{m}))$. This order is as specified in the proof of Proposition 7.8. If we use this order, then $\mathbf{m}$ belongs to $\mathcal{A}^*$ if and only if

$$\begin{aligned} 6\mathbf{m}(p_1) + 6\mathbf{m}(p_2) + 4\mathbf{m}(p_4) \leq b \\ \vee\, 6\mathbf{m}(p_1) + 4\mathbf{m}(p_2) + 2\mathbf{m}(p_3) + 4\mathbf{m}(p_4) \leq b. \end{aligned}$$

Thus by changing the order of execution of the transitions t_1 and t_2, we obtain a much simpler expression for $\mathcal{A}^*$. △

It is clear that apart from the Petri nets and legal sets in Sections 4 to 6, there exist other classes of systems for which the supremal controllable subset can be constructed by the method of Section 3.

7. Structure of $\mathcal{A}^*$

The analysis of the preceding sections allows us to study the general structure of $\mathcal{A}^*$. As mentioned in Chapter 2, this is important for the structure, i.e. enabling condition and thus reachability properties, of the closed-loop system.

Proposition 7.9. *If $\mathcal{A} \in \mathcal{O}^{\downarrow}(\mathbb{N}^n, \leq)$, then $\mathcal{A}^* \in \mathcal{O}^{\downarrow}(\mathbb{N}^n, \leq)$.*

In words, the supremal controllable subset of a down-set is also a down-set.

Proof. If $\mathcal{A}^*$ is empty, it is a down-set in $(\mathbb{N}^n, \leq)$. Therefore, in the rest of the proof we assume $\mathcal{A}^*$ nonempty.

Let $\mathbf{m} \in \mathcal{A}^*$ and $\mathbf{m}' \leq \mathbf{m}$ with $\mathbf{m}' \in \mathbb{N}^n$. We show that $\mathbf{m}' \in \mathcal{A}^*$. By definition, this is equivalent to $\mathcal{R}_\infty^{T^{\text{uc}}}(\mathbf{m}') \subseteq \mathcal{A}$. First we prove that

$$\forall \bar{\mathbf{m}}' \in \mathcal{R}_\infty^{T^{\text{uc}}}(\mathbf{m}'), \exists \bar{\mathbf{m}} \in \mathcal{R}_\infty^{T^{\text{uc}}}(\mathbf{m}) : \bar{\mathbf{m}}' \leq \bar{\mathbf{m}}. \tag{7.43}$$

By definition, for each $\bar{\mathbf{m}}' \in \mathcal{R}_\infty^{T^{\text{uc}}}(\mathbf{m}')$ there exist transition vectors $\boldsymbol{\delta}_1, \ldots, \boldsymbol{\delta}_k \in \Delta$ with $\boldsymbol{\delta}_i|_{T^c} = \mathbf{0}$ such that $\mathbf{m}' \xrightarrow{(\boldsymbol{\delta}_1,\ldots,\boldsymbol{\delta}_k)} \bar{\mathbf{m}}'$. As $\mathbf{m}' \leq \mathbf{m}$, we obtain with Proposition 1.12 that $\mathbf{m} \xrightarrow{(\boldsymbol{\delta}_1,\ldots,\boldsymbol{\delta}_k)} \bar{\mathbf{m}}$ and that $\bar{\mathbf{m}}' \leq \bar{\mathbf{m}}$. This proves (7.43). From (7.43), $\mathcal{A} \in \mathcal{O}^{\downarrow}(\mathbb{N}^n, \leq)$ and $\mathcal{R}_\infty^{T^{\text{uc}}}(\mathbf{m}) \subseteq \mathcal{A}$ it follows that $\mathcal{R}_\infty^{T^{\text{uc}}}(\mathbf{m}') \subseteq \mathcal{A}$ for all $\mathbf{m}' \leq \mathbf{m}$. Thus, $\mathcal{A}^* \in \mathcal{O}^{\downarrow}(\mathbb{N}^n, \leq)$. □

Because of Theorem 4.3 (p. 99), Proposition 7.9 implies that for each legal down-set $\mathcal{A}$ there exist vectors $\mathbf{A}^*_{i,j} \in \mathbb{Z}^n$ and integers $b^*_{i,j}$ ($i = 1, \ldots, \nu; j = 1, \ldots, \nu_i$) such that

$$\mathcal{A}^* = \bigcap_{i=1}^{\nu} \bigcup_{j=1}^{\nu_i} \{\mathbf{m} \in \mathbb{N}^n \mid \mathbf{A}^{*,T}_{i,j} \mathbf{m} \leq b^*_{i,j}\}. \tag{7.44}$$

To construct this representation, the procedure in Section 3 can be used partially. For this approach to lead to a representation for $\mathcal{A}^*$ as in (7.44) one needs in addition an algorithm that constructs a representation as a finite conjunction and disjunction of linear inequalities for a down-set

$$\{\mathbf{m} \in \mathbb{N}^n \mid \lfloor \mathbf{M}^T \mathbf{m} \rfloor + \lceil \mathbf{N}^T \mathbf{m} \rceil \leq k\}, \tag{7.45}$$

with $\mathbf{M}, \mathbf{N} \in \mathbb{Q}^n_+$ and $k \in \mathbb{Z}$. Because of Theorem 4.3 we know that it exists. A possible approach, like in the proof of Theorem 4.3, is to construct the set of minimal, with respect to the componentwise partial order, elements of the complement of the set in (7.45).

Note that Proposition 7.9 does not have a counterpart for legal up-sets in $(\mathbb{N}^n, \leq)$. If $\mathcal{A}$ is an up-set, then $\mathcal{A}^*$ is not necessarily an up-set, as it is illustrated next.

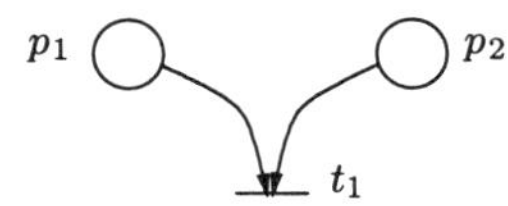

Figure 7.8. An influencing net $\mathcal{N}^{uc}_{\mathcal{A}}$ illustrating the fact that the supremal controllable subset of an up-set is not always an up-set.

Example 7.9. Consider the uncontrollable Petri net in Figure 7.8. The legal set is $\mathcal{A} = \{\mathbf{m} \in \mathbb{N}^2 \mid \mathbf{m}(p_1) \geq 1\}$. With the procedure of Section 3 it then follows that $\mathcal{A}^* = \{\mathbf{m} \in \mathbb{N}^2 \mid \mathbf{m}(p_1) \geq \mathbf{m}(p_2) + 1\}$. This result is intuitively clear. Consequently, $\mathcal{A}^*$ is not an up-set. We have $P^{\downarrow}(\mathcal{A}^*) = \{p_2\}$ and $P^{\uparrow}(\mathcal{A}^*) = \{p_1\}$. $\triangle$

As the supremal controllable subset $\mathcal{A}^*$ of a legal set $\mathcal{A}$ cannot be a down-set, nor an up-set, Theorem 4.3 is of no use and little can be said about the structure of $\mathcal{A}^*$. Thus it is not necessarily so that $\mathcal{A}^*$ can be written as in (7.44).

Note that if a Petri net $\mathcal{N}^{uc} = (P, T, f^+, f^-, \mathbf{m}_0, T^{uc})$ is bounded, then there exists a finite set Z such that $\mathcal{R}_\infty(\mathbf{m}_0) \subseteq Z$. The legal set $\mathcal{A}$ can then be replaced by the finite legal set $\mathcal{A}_1 = \mathcal{A} \cap Z$. Consequently, the supremal controllable subset $\mathcal{A}_1^*$ of $\mathcal{A}_1$ is a finite set and can thus be represented as in (7.44).

8. Notes and references

Note that the definition of a continuous Petri net should not be confused with that of [1]. There, 'continuous' refers to the fact that the state does not change in discrete steps, but continuously with respect to time.

Another general method that, like here in Section 2, also results in a subset of the supremal controllable subset is found in [8]. In [39] we have proven the result of Section 4 in a slightly different manner. The result of Section 5 generalises results of, among others, [18] and [24] to general acyclic Petri nets without choice places and to general legal sets.

Chapter 8

STRUCTURAL LINEAR ALGEBRAIC CONTROL DESIGN

1. Unobservable events

So far in this book, it was assumed that all events are observable. This means that the supervisory controller can observe each event which occurs in the studied discrete event system. Consequently, the state of the Petri net that models this system is completely known at any time.

In practice, not all events are observable, for instance because the controller has a local character and cannot oversee the entire system or because communication between measuring instruments and control unit is unreliable. As a consequence, control design becomes more difficult because the system's evolution via uncontrollable transitions should be anticipated while the system's state is not known exactly due to the presence of unobservable events. This problem has been studied in detail for automata (see [6, Section 3.7] for an overview). However, it received little attention in the Petri net literature so far.

The control laws which are constructed in this chapter, extending the terminology of Chapters 2 and 3 to Petri nets with unobservable events, are permissive but not necessarily maximally permissive. Their construction is based on the Petri net supervisors in Chapter 2 (Section 6). We replace the legal set $\mathcal{A}$ which is defined by a finite conjunction of linear inequalities, by an appropriate subset $\mathcal{A}'$. Define $T^{\mathrm{uo}} \subseteq T^{\mathrm{uc}}$ as the set of *unobservable transitions.* We assume that each unobservable transition is uncontrollable. The set of *observable transitions* is $T^{\mathrm{o}} := T \setminus T^{\mathrm{uo}}$. We use a linear-algebraic approach. The legal set is, as in the first part of Chapter 7, considered to be a subset of $\mathbb{Q}_+^n$.

In Section 2 we give an outline of our approach. Section 3 is dedicated to the study of the intersection of a linear halfspace with the first orthant in $\mathbb{Q}^n$. Conditions for candidate and optimal control laws are deduced in Sections 4 and 5, respectively. In Section 6 a more liberal control law is proposed by taking the disjunction of these optimal control laws. Finally, Section 7 presents an algorithm to obtain a subset of the supremal controllable subset.

2. Overview of the approach

In this section we give an overview of our approach to construct control laws for Petri nets with unobservable transitions. The no concurrency assumption is used, i.e. $\Omega = \Delta_{\mathrm{N}}$. Further, we assume that all Petri nets are free of self-loops. With these assumptions and following Proposition 1.15 (p. 23), the enabling condition $\mathbf{m} \geq \mathbf{F}^-\boldsymbol{\delta}$ is equivalent to $\mathbf{m} + \mathbf{F}\boldsymbol{\delta} \geq \mathbf{0}$.

We assume that the legal set $\mathcal{A}$ is described by one linear inequality:

$$\mathcal{A} = \{\mathbf{m} \in \mathbb{N}^n \mid \mathbf{A}^T\mathbf{m} \leq b\} \quad \text{with } \mathbf{A} \in \mathbb{Z}^n, b \in \mathbb{Z}, \tag{8.1}$$

and that

$$\mathcal{A} \neq \emptyset, \tag{8.2}$$

$$\mathcal{A} \neq \mathbb{N}^n. \tag{8.3}$$

As always, the results in this chapter are immediately generalisable to legal sets given by a finite conjunction of linear inequalities.

In Chapter 2 (Section 6) we have shown for fine control that, when all transitions are controllable, all maximally permissive control laws for $\mathcal{A}$ in (8.1) are equivalent and at least one of them can be realised by a Petri net without modifying the original Petri net model. The original Petri net, with n places and m transitions, is extended with a control place p^{c} with state $b - \mathbf{Am}$. This extra place is connected to the transitions of the original Petri net. The incidence matrices of the closed-loop Petri net are obtained by adding the vectors $\mathbf{B}^+$ and $\mathbf{B}^-$ with (see Section 6 of Chapter 2):

$$B_i^+ = \max(-(\mathbf{A}^T\mathbf{F})_i, 0) \qquad (i = 1, \ldots, m), \tag{8.4}$$

$$B_i^- = \max((\mathbf{A}^T\mathbf{F})_i, 0) \qquad (i = 1, \ldots, m), \tag{8.5}$$

to the incidence matrices $\mathbf{F}^+$ and $\mathbf{F}^-$. The enabling condition of the supervisory controller $b - \mathbf{A}^T\mathbf{m} \geq \mathbf{B}^{-,T}\boldsymbol{\delta}$ is —because of Proposition 1.15 (p. 23)— equivalent to $b - \mathbf{A}^T\mathbf{m} + \mathbf{B}^T\boldsymbol{\delta} \geq \mathbf{0}$, where $\mathbf{B} = \mathbf{B}^+ - \mathbf{B}^-$.

The added place and arcs act both as a controller and an observer. A transition $t \in T$ is not controlled if $\mathbf{B}^-(t) = 0$. In the Petri net graph of the closed-loop system there is no arc from p^{c} to t: the enabling condition of t does not depend on the state of $p^{\rm c}$ and is thus the same as in the original Petri net. A transition $t \in T$ with $\mathbf{B}(t) = 0$ is not observed by the supervisory controller. There is no arc from $p^{\rm c}$ to t and vice versa. The state of $p^{\rm c}$ does not change when t is executed.

Example 8.1. Revisit Figure 2.4 on page 46. The Petri net depicts the closed-loop system of the Petri net model for the railway example together with a maximally permissive supervisor which guarantees that the number of trains within Zone 5 is not greater than one. Transitions t_1, t_2, t_4, t_5, t_7 and t_8 are not controlled nor observed, and transition t_9 is not controlled. △

As this example illustrates, it follows from the structure of the Petri net supervisor that, if t is controlled, then $\mathbf{B}^-(t) \neq 0$. Together with (8.4)–(8.5) it follows that $\mathbf{B}(t) \neq 0$, and thus that t is also observed. For this reason, we assume in this chapter that each controllable transition is also observable, i.e. $T^{\rm c} \subseteq T^{\rm o}$ or $T^{\rm uo} \subseteq T^{\rm uc}$.

If the Petri net supervisor for $\mathcal{A}$ in (8.1) satisfies

$$\mathbf{B}^-(t) = 0 \quad \text{for all } t \in T^{\rm uc}, \tag{8.6}$$

$$\mathbf{B}(t) = 0 \quad \text{for all } t \in T^{\rm uo}, \tag{8.7}$$

then this supervisor is acceptable. None of the uncontrollable transitions is controlled, and none of the unobservable transitions is observed. Because of (8.4)–(8.5), the conditions (8.6)–(8.7) are equivalent to

$$(\mathbf{A}^T\mathbf{F})(t) \leq 0 \quad \text{for all } t \in T^{\rm uc} \setminus T^{\rm uo}, \tag{8.8}$$

$$(\mathbf{A}^T\mathbf{F})(t) = 0 \quad \text{for all } t \in T^{\rm uo}. \tag{8.9}$$

These conditions can be written more compactly as (note that $x = 0$ is equivalent to the conjunction of $x \leq 0$ and $-x \leq 0$):

$$\mathbf{A}^T\mathbf{D} \leq \mathbf{0}. \tag{8.10}$$

Herein $\mathbf{D}$ is an integer valued matrix with n rows and $\sharp(T^{\rm uc} \setminus T^{\rm uo}) + 2\sharp T^{\rm uo}$ columns, which —making abstraction of the sign— consists of columns of $\mathbf{F}$.

If the sufficient condition for an acceptable supervisor in (8.10) is not satisfied, then an acceptable supervisor can be constructed in a simple manner. This supervisor keeps the state within $\mathcal{A}$ and satisfies (8.10). Its construction is explained next.

When there exists a vector $\mathbf{A}'_{\mathbb{Q}} \in \mathbb{Q}^n$ such that

$$\mathbf{A}'_{\mathbb{Q}} \geq \mathbf{A} \quad \text{and} \quad (\mathbf{A}'_{\mathbb{Q}})^T\mathbf{D} \leq \mathbf{0}, \tag{8.11}$$

then a set of states

$$\mathcal{A}' = \{\mathbf{m} \in \mathbb{N}^n \mid (\mathbf{A}'_{\mathbb{Q}})^T\mathbf{m} \leq b\} = \{\mathbf{m} \in \mathbb{N}^n \mid (\mathbf{A}')^T\mathbf{m} \leq b'\} \qquad (8.12)$$

can be defined. Herein, $\mathbf{A}' \in \mathbb{Z}^n$ and $b' \in \mathbb{Z}$. They are obtained by multiplying both sides of the inequality $(\mathbf{A}'_{\mathbb{Q}})^T\mathbf{m} \leq b$ with the least common multiple of the denominators of the nonzero elements of $\mathbf{A}'_{\mathbb{Q}}$. As $\mathbf{A}'_{\mathbb{Q}} \geq \mathbf{A}$, $\mathcal{A}'$ is a subset of $\mathcal{A}$. Moreover, $(\mathbf{A}')^T\mathbf{D} \leq \mathbf{0}$ because of the homogeneous character of the inequality. Consequently, the Petri net supervisor for $\mathcal{A}'$ is acceptable.

Because of $\mathcal{A}' \subseteq \mathcal{A}$, all reachable states under this control law belong to $\mathcal{A}$ and because of $(\mathbf{A}')^T\mathbf{D} \leq \mathbf{0}$ only controllable transitions are controlled and only observable transitions are observed. As the vector $\mathbf{A}'_{\mathbb{Q}}$ can be chosen freely within the conditions (8.11), this control law can be very restrictive. For this reason, in this chapter we look for vectors $\mathbf{A}'_{\mathbb{Q}}$ which satisfy (8.11) and which in addition make the set $\mathcal{A}'$ in (8.12) as large as possible.

Because for integers there does not exist a result similar to Farkas' Lemma for rational numbers, we work in $\mathbb{Q}^n$. We consider the legal set $\mathcal{A}$ as a subset of $\mathbb{Q}^n_+$.

Define the set

$$\mathcal{A}_{\mathbb{Q}} := \{\mathbf{x} \in \mathbb{Q}^n_+ \mid \mathbf{A}^T\mathbf{x} \leq b\}. \qquad (8.13)$$

It is a subset of $\mathbb{Q}^n_+$ and it is defined by the same inequality as $\mathcal{A}$ in (8.1). If $\mathbf{A} \geq \mathbf{0}$ or $\mathbf{A} \leq \mathbf{0}$, we construct in this chapter all maximal —with respect to the subset relation— sets

$$\mathcal{A}'_{\mathbb{Q}} = \{\mathbf{x} \in \mathbb{Q}^n_+ \mid (\mathbf{A}'_{\mathbb{Q}})^T\mathbf{x} \leq b'_{\mathbb{Q}}\} \quad \text{with } \mathbf{A}'_{\mathbb{Q}} \in \mathbb{Q}^n, b'_{\mathbb{Q}} \in \mathbb{Q}, \qquad (8.14)$$

which satisfy

$$\mathcal{A}'_{\mathbb{Q}} \subseteq \mathcal{A}_{\mathbb{Q}}, \qquad (8.15)$$

$$(\mathbf{A}'_{\mathbb{Q}})^T\mathbf{D} \leq \mathbf{0}. \qquad (8.16)$$

Once optimal values for the vector $\mathbf{A}'_{\mathbb{Q}}$ and the number $b'_{\mathbb{Q}}$ are obtained, the most liberal —within the limitations of the present approach— control laws can be constructed. In general, there is more than one such a Petri net supervisor.

3. Intersection of a linear halfspace with the first orthant

In this section necessary and sufficient conditions are given under which in $\mathbb{Q}^n$ the intersection of a linear halfspace and $\mathbb{Q}^n_+$, the closed first orthant of $\mathbb{Q}^n$, is a subset of the intersection of another linear halfspace and $\mathbb{Q}^n_+$. Consider the sets

$$\mathcal{H}_1 = \{\mathbf{x} \in \mathbb{Q}^n \mid \mathbf{H}_1^T\mathbf{x} \le k_1\} \quad \text{with } \mathbf{H}_1 \in \mathbb{Q}^n, k_1 \in \mathbb{Q},$$
$$\mathcal{H}_2 = \{\mathbf{x} \in \mathbb{Q}^n \mid \mathbf{H}_2^T\mathbf{x} \le k_2\} \quad \text{with } \mathbf{H}_2 \in \mathbb{Q}^n, k_2 \in \mathbb{Q}.$$

If not empty nor equal to $\mathbb{Q}^n$, $\mathcal{H}_1$ and $\mathcal{H}_2$ are linear halfspaces in $\mathbb{Q}^n$. Next, necessary and sufficient conditions are formulated under which

$$\mathcal{H}_1 \cap \mathbb{Q}^n_+ \subseteq \mathcal{H}_2 \cap \mathbb{Q}^n_+. \tag{8.17}$$

This is equivalent to $\mathcal{H}_1 \cap \mathbb{Q}^n_+ \subseteq \mathcal{H}_2$, or to requiring that

$$\begin{cases} \mathbf{0} \le \mathbf{x}, \\ \mathbf{H}_1^T\mathbf{x} \le k_1, \\ \mathbf{H}_2^T\mathbf{x} > k_2, \end{cases} \tag{8.18}$$

has no solution $\mathbf{x} \in \mathbb{Q}^n$. Theorem 8.1 originally appeared in [27] and can also be found in [40].

Theorem 8.1. *If* $\mathbf{Z} \in \mathbb{Q}^{w \times v}, \mathbf{d} \in \mathbb{Q}^w, \mathbf{f} \in \mathbb{Q}^v$, *and* $c \in \mathbb{Q}$, *then either*

$$\begin{cases} \mathbf{Z}\mathbf{x} \ge \mathbf{d}, \\ \mathbf{f}^T\mathbf{x} < c, \end{cases}$$

has a solution $\mathbf{x} \in \mathbb{Q}^v$, *or*

$$\begin{cases} \mathbf{Z}^T\mathbf{y} = \mathbf{f}, \\ \mathbf{d}^T\mathbf{y} \ge c, \\ \mathbf{y} \ge \mathbf{0}, \end{cases}$$

has a solution $\mathbf{y} \in \mathbb{Q}^w$.

If $\mathbf{I}_n$ is the $n \times n$ unit matrix, then Theorem 8.1 can be applied to (8.18) with

$$\mathbf{Z} = \begin{bmatrix} \mathbf{I}_n \\ -\mathbf{H}_1^T \end{bmatrix}, \quad \mathbf{d} = \begin{bmatrix} 0 \\ \vdots \\ 0 \\ -k_1 \end{bmatrix}, \quad \mathbf{f} = -\mathbf{H}_2, \quad c = -k_2.$$

Consequently, for (8.17) to hold it is necessary and sufficient that there exists a $\mathbf{y} \in \mathbb{Q}^{n+1}$ such that

$$\begin{cases} \mathbf{y} \ge \mathbf{0}, \\ \begin{bmatrix} \mathbf{I}_n & -\mathbf{H}_1 \end{bmatrix} \mathbf{y} = -\mathbf{H}_2, \\ \begin{bmatrix} 0 & \cdots & 0 & -k_1 \end{bmatrix} \mathbf{y} \ge -k_2. \end{cases} \tag{8.19}$$

This can be simplified. With the notation

$$\mathbf{y} = \begin{bmatrix} y_1 \\ \vdots \\ y_{n+1} \end{bmatrix}, \quad \mathbf{H}_1 = \begin{bmatrix} H_{1,1} \\ \vdots \\ H_{1,n} \end{bmatrix}, \quad \mathbf{H}_2 = \begin{bmatrix} H_{2,1} \\ \vdots \\ H_{2,n} \end{bmatrix},$$

the second equation in (8.19) can be written as

$$y_i - H_{1,i} y_{n+1} = -H_{2,i} \qquad (i = 1, \ldots, n).$$

As the first n components of $\mathbf{y}$ do not appear in the third equation in (8.19), a simpler necessary and sufficient condition for (8.17) can be formulated.

Theorem 8.2. *The relation in* (8.17) *holds if and only if there exists an* $l \in \mathbb{Q}_+$ *such that*

$$\mathbf{H}_1 l \geq \mathbf{H}_2 \quad \textit{and} \quad k_1 l \leq k_2.$$

By choosing proper reference sets in Theorem 8.2, we obtain two properties which where stated in Chapter 4 without proof (see Propositions 4.20 and 4.21 on page 100).

Proposition 8.1. *Let* $\mathcal{H} = \{\mathbf{x} \in \mathbb{Q}^n \mid \mathbf{H}^T\mathbf{x} \leq k\}$ *with* $\mathbf{H} \in \mathbb{Q}^n$ *and* $k \in \mathbb{Q}$. *Then* $\mathcal{H} \cap \mathbb{Q}^n_+ = \emptyset$ *if and only if* $\mathbf{H} \geq \mathbf{0}$ *and* $k < 0$.

Proof. Take $\mathcal{H}_1 = \mathcal{H}$, $\mathbf{H}_2 = \mathbf{0}$ and $k_2 = -1$ in Theorem 8.2. Thus, $\mathcal{H}_2 = \emptyset$. Then, $\mathcal{H} \cap \mathbb{Q}^n_+ = \emptyset$ if and only if there exists a nonnegative l such that $\mathbf{H}l \geq 0$ and $kl \leq -1$. It follows that $l \neq 0$, from which the statement of the proposition follows. □

Proposition 8.2. *Let* $\mathcal{H} = \{\mathbf{x} \in \mathbb{Q}^n \mid \mathbf{H}^T\mathbf{x} \leq k\}$ *with* $\mathbf{H} \in \mathbb{Q}^n$ *and* $k \in \mathbb{Q}$. *Then* $\mathcal{H} \supseteq \mathbb{Q}^n_+$ *if and only if* $\mathbf{H} \leq \mathbf{0}$ *and* $k \geq 0$.

Proof. Take $\mathbf{H}_1 = \mathbf{0}$, $k_1 = 0$ and $\mathcal{H}_2 = \mathcal{H}$ in Theorem 8.2. □

4. Candidate sets $\mathcal{A}'_{\mathbb{Q}}$

Consider the sets $\mathcal{A}_{\mathbb{Q}}$ and $\mathcal{A}'_{\mathbb{Q}}$ in (8.13) and (8.14). In this section it is shown that if $\mathbf{A} \geq \mathbf{0}$ or $\mathbf{A} \leq \mathbf{0}$ in (8.1), then the necessary and sufficient condition given by Theorem 8.2 for $\mathcal{A}'_{\mathbb{Q}} \subseteq \mathcal{A}_{\mathbb{Q}}$ (see (8.15)) can be further simplified. Introduce the notations $\mathbf{A} = (a_1, \ldots, a_n)$ and $\mathbf{A}'_{\mathbb{Q}} = (q_1, \ldots, q_n)$.

4.1. The case $\mathrm{A} \geq 0$

Assume $\mathbf{A} \geq \mathbf{0}$ in (8.1), or equivalently $\mathbf{A} \in \mathbb{N}^n$. This implies that if $\mathbf{m} \in \mathcal{A}$, then each $\mathbf{m}' \leq \mathbf{m}$ ($\mathbf{m}' \in \mathbb{N}^n$) is also in $\mathcal{A}$. By Proposition 4.20, $\mathcal{A} \neq \emptyset$ (see (8.2)) is equivalent to $\mathbf{A} \not\geq \mathbf{0} \vee b \geq 0$, which is, by $\mathbf{A} \geq \mathbf{0}$, equivalent to $b \geq 0$. By Proposition 4.21 and $b \geq 0$, $\mathcal{A} \neq \mathbb{N}^n$ (see (8.3)) is equivalent to $\mathbf{A} \neq \mathbf{0}$. Further, it is clear that if $b = 0$ in (8.1) then for all $\epsilon_1, \epsilon_2 \in \mathbb{N} \setminus \{0\}$ with $\epsilon_1 < \epsilon_2$, it holds that

$$\mathcal{A} = \{\mathbf{m} \in \mathbb{N}^n \mid \mathbf{A}^T\mathbf{m} \leq 0\} = \{\mathbf{m} \in \mathbb{N}^n \mid \epsilon_2\mathbf{A}^T\mathbf{m} \leq \epsilon_1\}.$$

Therefore it is further assumed, without loss of generality, that $b > 0$. To summarise, we assume in this subsection that:

$$\mathbf{A} \in \mathbb{N}^n, \tag{8.20}$$

$$\mathbf{A} \neq \mathbf{0}, \tag{8.21}$$

$$b > 0. \tag{8.22}$$

Depending on $b'_{\mathbb{Q}}$ in (8.14), we consider three cases: $b'_{\mathbb{Q}} > 0$ (Proposition 8.3), $b'_{\mathbb{Q}} = 0$ (Proposition 8.4) and $b'_{\mathbb{Q}} < 0$ (Proposition 8.5).

Proposition 8.3. *If* $b'_{\mathbb{Q}} > 0$ *in* (8.14), *then* (8.15) *holds if and only if* $\mathbf{A}'_{\mathbb{Q}}b \geq \mathbf{A}b'_{\mathbb{Q}}$.

Proof. We only prove the 'only if'-part as the 'if'-part is obtained immediately. By Theorem 8.2 and $b'_{\mathbb{Q}} > 0$, (8.15) is equivalent to

$$\exists l, 0 \leq l \leq \frac{b}{b'_{\mathbb{Q}}} : \mathbf{A}'_{\mathbb{Q}}l \geq \mathbf{A}. \tag{8.23}$$

Define for $\mathbf{A}$ the subsets J and J' of $\{1, \ldots, n\}$ as:

$$J := \{j \in \{1, \ldots, n\} \mid a_j > 0\},$$
$$J' := \{j \in \{1, \ldots, n\} \mid a_j = 0\}.$$

From $\mathbf{A} \geq \mathbf{0}$ and $\mathbf{A} \neq \mathbf{0}$, it follows that $J \cup J' = \{1, \ldots, n\}$ and $J \neq \emptyset$. From (8.23) it follows that (recall the notation $\mathbf{A}'_{\mathbb{Q}} = (q_1, \ldots, q_n)$)

$$q_j l \geq a_j > 0 \quad \text{for each } j \in J. \tag{8.24}$$

As $l \geq 0$, this implies that

$$q_j > 0 \quad \text{for each } j \in J, \tag{8.25}$$

and that $l > 0$ as $J \neq \emptyset$. From

$$l \leq \frac{b}{b'_{\mathbb{Q}}}$$

and (8.25) we have that

$$q_j l \leq q_j \frac{b}{b'_{\mathbb{Q}}} \quad \text{for each } j \in J,$$

or, with (8.24), that

$$a_j \leq q_j \frac{b}{b'_{\mathbb{Q}}} \quad \text{for each } j \in J. \tag{8.26}$$

On the other hand, (8.23) also implies that $q_j l \geq a_j = 0$ for each $j \in J'$, and with $l > 0$ that $q_j \geq 0 = a_j$ for each $j \in J'$. Finally, as $b > 0$ we obtain

$$q_j b \geq 0 = a_j b'_{\mathbb{Q}} \quad \text{for each } j \in J'. \tag{8.27}$$

From (8.26), (8.27) and $J \cup J' = \{1, \ldots, n\}$ it follows that $\mathbf{A}'_{\mathbb{Q}} b \geq \mathbf{A} b'_{\mathbb{Q}}$. □

Proposition 8.4. *If* (8.15) *holds with* $b'_{\mathbb{Q}} = 0$ *in* (8.14), *then there exists an* $\epsilon > 0$ *such that*

$$\mathcal{A}'_{\mathbb{Q}} \subseteq \{\mathbf{x} \in \mathbb{Q}^n_+ \mid (\mathbf{A}'_{\mathbb{Q}})^T \mathbf{x} \leq \epsilon\} \subseteq \mathcal{A}_{\mathbb{Q}}. \tag{8.28}$$

Proof. As $\epsilon > 0$, the first subset relation in (8.28) is immediate. By Theorem 8.2, $b'_{\mathbb{Q}} = 0$ and $b > 0$, (8.15) is equivalent to

$$\exists l \geq 0 : \mathbf{A}'_{\mathbb{Q}} l \geq \mathbf{A}.$$

As $\mathbf{A} \neq \mathbf{0}$ (see (8.21)), this is equivalent to

$$\exists l > 0 : \mathbf{A}'_{\mathbb{Q}} l \geq \mathbf{A}.$$

This implies that

$$\forall j \in \{1, \ldots, n\} : (a_j > 0 \Rightarrow q_j > 0) \wedge (a_j = 0 \Rightarrow q_j \geq 0), \tag{8.29}$$

and, with $b > 0$, that

$$\forall j \in \{1, \ldots, n\} : (a_j > 0 \Rightarrow \frac{q_j b}{a_j} > 0) \wedge (a_j = 0 \Rightarrow q_j b \geq 0). \tag{8.30}$$

On the other hand, there exists an $\epsilon > 0$ such that $\{\mathbf{x} \in \mathbb{Q}^n_+ \mid (\mathbf{A}'_{\mathbb{Q}})^T \mathbf{x} \leq \epsilon\} \subseteq \mathcal{A}_{\mathbb{Q}}$ if and only if (Proposition 8.3)

$$\begin{aligned} & \exists \epsilon > 0 : \mathbf{A}'_{\mathbb{Q}} b \geq \mathbf{A} \epsilon, \\ \Leftrightarrow \quad & \exists \epsilon > 0 : \forall j \in \{1, \ldots, n\} : q_j b \geq a_j \epsilon. \end{aligned}$$

As $\mathbf{A} \geq \mathbf{0}$, this is equivalent to

$$\begin{aligned}
& \exists \epsilon > 0 : \forall j \in \{1, \ldots, n\} : (a_j > 0 \Rightarrow \frac{q_j b}{a_j} \geq \epsilon) \\
& \qquad\qquad\qquad\qquad\qquad \wedge (a_j = 0 \Rightarrow q_j b \geq 0), \\
\Leftrightarrow \quad & (\forall j \in \{1, \ldots, n\} : a_j = 0 \Rightarrow q_j b \geq 0) \\
& \qquad \wedge (\exists \epsilon > 0 : \forall j \in \{1, \ldots, n\} : a_j > 0 \Rightarrow \frac{q_j b}{a_j} \geq \epsilon), \\
\Leftrightarrow \quad & \forall j \in \{1, \ldots, n\} : (a_j = 0 \Rightarrow q_j b \geq 0) \wedge (a_j > 0 \Rightarrow \frac{q_j b}{a_j} > 0).
\end{aligned} \tag{8.31}$$

Conditions (8.30) and (8.31) are identical. □

Proposition 8.5. *If $b'_{\mathbb{Q}} < 0$ in* (8.14), *then* (8.15) *implies $\mathcal{A}'_{\mathbb{Q}} = \emptyset$.*

Proof. By Theorem 8.2, $b'_{\mathbb{Q}} < 0$ and $b > 0$, (8.15) is equivalent to

$$\exists l \geq 0 : \mathbf{A}'_{\mathbb{Q}} l \geq \mathbf{A}. \tag{8.32}$$

As in Proposition 8.4, this implies that

$$\forall j \in \{1, \ldots, n\} : (a_j > 0 \Rightarrow q_j > 0) \wedge (a_j = 0 \Rightarrow q_j \geq 0), \tag{8.33}$$

and thus that $\mathbf{A}'_{\mathbb{Q}} \geq \mathbf{0}$. Together with $b'_{\mathbb{Q}} < 0$ and Proposition 8.1 this implies that $\mathcal{A}'_{\mathbb{Q}} = \emptyset$. □

As we are looking for the maximal subsets $\mathcal{A}'_{\mathbb{Q}}$ of $\mathcal{A}_{\mathbb{Q}}$ which satisfy $(\mathbf{A}'_{\mathbb{Q}})^T \mathbf{D} \leq \mathbf{0}$, it follows from Proposition 8.5 that the case $b'_{\mathbb{Q}} < 0$ does not need to be considered since it only provides empty sets $\mathcal{A}'_{\mathbb{Q}}$. Moreover, it follows from Proposition 8.4 that for each set $\bar{\mathcal{A}}_{\mathbb{Q}} = \{\mathbf{x} \in \mathbb{Q}^n \mid (\mathbf{A}'_{\mathbb{Q}})^T \mathbf{x} \leq \mathbf{0}\}$ which satisfies (8.15)–(8.16) there exists a superset $\{\mathbf{x} \in \mathbb{Q}^n \mid (\mathbf{A}'_{\mathbb{Q}})^T \mathbf{x} \leq \epsilon\}$ of $\bar{\mathcal{A}}_{\mathbb{Q}}$ which satisfies (8.15)–(8.16). Therefore, the case $b'_{\mathbb{Q}} = 0$ does not need to be considered either. This leaves us with the case $b'_{\mathbb{Q}} > 0$ (Proposition 8.3). As then both b and $b'_{\mathbb{Q}}$ are positive (see (8.22)), they can be chosen equal. Thus in (8.14), take $\mathcal{A}'_{\mathbb{Q}} = \{\mathbf{x} \in \mathbb{Q}^n_+ \mid (\mathbf{A}'_{\mathbb{Q}})^T \mathbf{x} \leq b\}$. Then by Proposition 8.3, $\mathcal{A}'_{\mathbb{Q}} \subseteq \mathcal{A}_{\mathbb{Q}}$ if and only if

$$\mathbf{A}'_{\mathbb{Q}} \geq \mathbf{A}. \tag{8.34}$$

4.2. The case $\mathbf{A} \leq 0$

Assuming $\mathbf{A} \leq \mathbf{0}$ is equivalent to $-\mathbf{A} \in \mathbb{N}^n$. This implies that if $\mathbf{m} \in \mathcal{A}$ then each $\mathbf{m}' \geq \mathbf{m}$ ($\mathbf{m}' \in \mathbb{N}^n$) is also in $\mathcal{A}$. By Proposition 4.21, $\mathcal{A} \neq \mathbb{N}^n$ (see (8.3)) is equivalent to $\mathbf{A} \not\leq \mathbf{0} \vee b < 0$, which is, by (8.35), equivalent to $b < 0$. By Proposition 4.20 and $b < 0$, $\mathcal{A} \neq \emptyset$ (see (8.2)) is equivalent to $\mathbf{A} \neq \mathbf{0}$. To summarise, we assume in this subsection that:

$$-\mathbf{A} \in \mathbb{N}^{1\times n}, \tag{8.35}$$

$$\mathbf{A} \neq \mathbf{0}, \tag{8.36}$$

$$b < 0. \tag{8.37}$$

Proposition 8.6. *Consider $\mathcal{A}_{\mathbb{Q}}$ and $\mathcal{A}'_{\mathbb{Q}}$ as in* (8.13)–(8.14) *and* (8.35)–(8.37). *Then* (8.15) *holds if and only if $b'_{\mathbb{Q}} < 0$ and $\mathbf{A}'_{\mathbb{Q}} b \leq \mathbf{A} b'_{\mathbb{Q}}$.*

Proof. ($\Rightarrow$) From (8.15) it follows that $b'_{\mathbb{Q}} < 0$, because otherwise, (8.15) would lead to a contradiction by Theorem 8.2, $b < 0$ and $l \geq 0$. By Theorem 8.2, $\mathcal{A}'_{\mathbb{Q}} \subseteq \mathcal{A}_{\mathbb{Q}}$ if and only if

$$\exists l \geq \frac{b}{b'_{\mathbb{Q}}} : \mathbf{A}'_{\mathbb{Q}} l \geq \mathbf{A}, \tag{8.38}$$

$$\Leftrightarrow \quad \exists l \geq \frac{b}{b'_{\mathbb{Q}}} : \forall j \in \{1, \ldots, n\} : q_j l \geq a_j. \tag{8.39}$$

If $q_j < 0$, then it follows from (8.39) that $a_j \leq q_j l \leq q_j \frac{b}{b'_{\mathbb{Q}}}$. If $q_j \geq 0$, then it follows from (8.39), $\frac{b}{b'_{\mathbb{Q}}} > 0$ and $a_j \leq 0$ (as $\mathbf{A} \leq \mathbf{0}$) that $a_j \leq q_j \frac{b}{b'_{\mathbb{Q}}}$. Consequently, $\mathbf{A}'_{\mathbb{Q}} b \leq \mathbf{A} b'_{\mathbb{Q}}$ since $b'_{\mathbb{Q}} < 0$.
($\Leftarrow$) From $b'_{\mathbb{Q}} < 0$, $(\mathbf{A}'_{\mathbb{Q}})^T \mathbf{x} \leq b'_{\mathbb{Q}}$ and $b < 0$ (see (8.37)) it follows that

$$(\mathbf{A}'_{\mathbb{Q}})^T \frac{b}{b'_{\mathbb{Q}}} \mathbf{x} \leq b.$$

As $\mathbf{0} \leq \mathbf{x}$, this implies together with $\mathbf{A}'_{\mathbb{Q}} b \leq \mathbf{A} b'_{\mathbb{Q}}$ that $\mathbf{A}^T \mathbf{x} \leq b$. □

As both b and $b'_{\mathbb{Q}}$ are negative, they can be chosen equal. Thus, take $\mathcal{A}'_{\mathbb{Q}} = \{\mathbf{x} \in \mathbb{Q}^n \mid (\mathbf{A}'_{\mathbb{Q}})^T \mathbf{x} \leq b\}$ in (8.14). Consequently, by Proposition 8.6, $\mathcal{A}'_{\mathbb{Q}} \subseteq \mathcal{A}_{\mathbb{Q}}$ if and only if

$$\mathbf{A}'_{\mathbb{Q}} \geq \mathbf{A}. \tag{8.40}$$

Note that this condition is the same as (8.34).

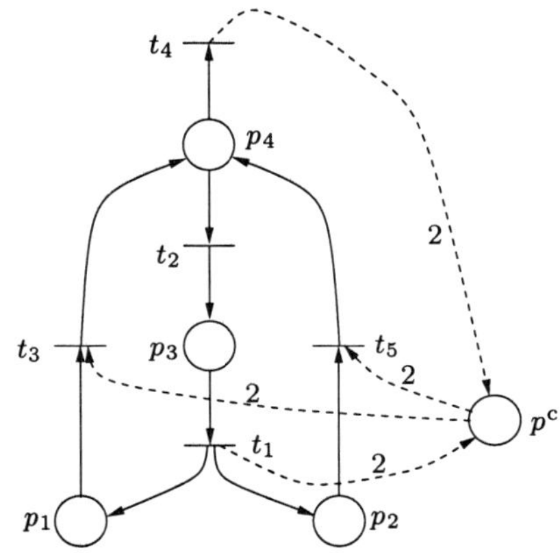

Figure 8.1. The Petri net of Example 8.2.

5. Maximal sets $\mathcal{A}'_{\mathbb{Q}}$

If $\mathbf{A} \geq \mathbf{0}$ or $\mathbf{A} \leq \mathbf{0}$, then it follows from the previous section that we can take $\mathcal{A}'_{\mathbb{Q}} = \{\mathbf{x} \in \mathbb{Q}^n \mid (\mathbf{A}'_{\mathbb{Q}})^T\mathbf{x} \leq b\}$ in (8.14). Furthermore, $\mathcal{A}'_{\mathbb{Q}} \subseteq \mathcal{A}_{\mathbb{Q}}$ if and only if $\mathbf{A}'_{\mathbb{Q}} \geq \mathbf{A}$. This provides an easy way to find maximal acceptable subsets $\mathcal{A}'_{\mathbb{Q}}$. Recall that for a partially ordered set $(Z, \leq)$, $\min(Z, \leq)$ denotes the set of minimal elements in $(Z, \leq)$. Define

$$\mathcal{Y} := \{\mathbf{y} \in \mathbb{Q}^n \mid \mathbf{y} \geq \mathbf{A}, \mathbf{y}^T\mathbf{D} \leq \mathbf{0}\}.$$

Define $(\mathcal{Y}, \leq)$ as the set $\mathcal{Y}$ with the componentwise partial order relation. Then it follows from the above that $\min(\mathcal{Y}, \leq)$ equals the set of vectors $\mathbf{A}'_{\mathbb{Q}}$ of maximal subsets $\mathcal{A}'_{\mathbb{Q}}$ of $\mathcal{A}_{\mathbb{Q}}$ which satisfy the conditions for uncontrollable and unobservable transitions. To each $\mathbf{y} \in \min(\mathcal{Y}, \leq)$ corresponds a maximal —within the limitations of the approach— set

$$\mathcal{A}'_{\mathbf{y}} := \{\mathbf{x} \in \mathbb{N}^n \mid \mathbf{y}^T\mathbf{x} \leq b\} \subseteq \mathcal{A}$$

for which the supervisor —after properly multiplying the defining inequality $\mathbf{y}^T\mathbf{x} \leq b$ with a positive integer— does not control (observe) any uncontrollable (unobservable) transitions.

Example 8.2. Consider the Petri net in Figure 8.1 with

$$\mathbf{F} = \begin{bmatrix} 1 & 0 & -1 & 0 & 0 \\ 1 & 0 & 0 & 0 & -1 \\ -1 & 1 & 0 & 0 & 0 \\ 0 & -1 & 1 & -1 & 1 \end{bmatrix}.$$

The legal set is $\mathcal{A} = \left\{\mathbf{m} \in \mathbb{N}^4 \mid \begin{bmatrix} 0 & 0 & 1 & 2 \end{bmatrix} \mathbf{m} \leq 4\right\}$. Transition t_2 is unobservable, all others are controllable, i.e. $T^c = T^o = \{t_1, t_3, t_4, t_5\}$

and $T^{\mathrm{uc}} = T^{\mathrm{uo}} = \{t_2\}$. Thus for the matrix $\mathbf{D}$ in (8.10) one has:

$$\mathbf{D}^T = \begin{bmatrix} 0 & 0 & 1 & -1 \\ 0 & 0 & -1 & 1 \end{bmatrix},$$

and $\mathcal{Y} = \{(q_1, q_2, q_3, q_4) \in \mathbb{Q}^4 \mid q_1 \geq 0, q_2 \geq 0, q_3 \geq 1, q_4 \geq 2, q_3 = q_4\}$. One easily obtains that $\min(\mathcal{Y}, \leq) = \{\begin{bmatrix} 0 & 0 & 2 & 2 \end{bmatrix}\}$. The corresponding set of states is

$$\mathcal{A}'_{\mathbf{y}} = \{\mathbf{m} \in \mathbb{N}^4 \mid \begin{bmatrix} 0 & 0 & 2 & 2 \end{bmatrix} \mathbf{m} \leq 4\}.$$

The supervisor for $\mathcal{A}'_{\mathbf{y}}$ can be realised by a Petri net and is given by $-\begin{bmatrix} 0 & 0 & 2 & 2 \end{bmatrix} \mathbf{F} = \begin{bmatrix} 2 & 0 & -2 & 2 & -2 \end{bmatrix}$ which is indicated in dashed lines in Figure 8.1. The control place is p^{c}. Note that there are no arcs between t_2 and p^{c}. △

6. Reduction of controllers with disjunctions

In contrast to Example 8.2, $\min(\mathcal{Y}, \leq)$ is not a singleton in general. If $\min(\mathcal{Y}, \leq)$ contains two elements $\mathbf{y}_1$ and $\mathbf{y}_2$ for which the corresponding sets $\mathcal{A}'_{\mathbf{y}_1}$ and $\mathcal{A}'_{\mathbf{y}_2}$ are not equal then the supervisor for $\mathcal{A}'_{\mathbf{y}_1}$ considers states in $\mathcal{A}'_{\mathbf{y}_2} \setminus \mathcal{A}'_{\mathbf{y}_1}$ as illegal, but they are not. So, if $\mathcal{A}'_{\mathbf{y}_1} \neq \mathcal{A}'_{\mathbf{y}_2}$ then these supervisors can be made more permissive by extending the Petri net model with disjunctions in the enabling condition. The supervisor enabling condition can be chosen among those which correspond to an element $\mathbf{y}$ of $\min(\mathcal{Y}, \leq)$, i.e. the enabling condition is:

$$\bigvee_{\mathbf{y} \in \min(\mathcal{Y}, \leq)} b - \mathbf{y}^T \mathbf{m} - \mathbf{y}^T \mathbf{F} \boldsymbol{\delta} \geq 0. \tag{8.41}$$

In general, $\min(\mathcal{Y}, \leq)$ is not a finite set. Next, we show that (8.41) is equivalent to a finite disjunction of linear inequalities $b - \mathbf{y}^T \mathbf{m} - \mathbf{y}^T \mathbf{F} \boldsymbol{\delta} \geq 0$.

By the decomposition theorem for polyhedra (see e.g. [4] or [34]), each element in a polyhedron that contains at least one vertex, can be written as a nonnegative sum of the *rays* and a convex sum of the *vertices* of the polyhedron. Theorem 8.3 can be found in [4].

Theorem 8.3. *A nonempty polyhedron $\mathcal{P}$ has at least one vertex if and only if $\mathcal{P}$ does not contain a line.*

As $\mathcal{Y}$ is a subset of the orthant $\mathbf{y} \geq \mathbf{A}$, one easily verifies that $\mathcal{Y}$ does not contain a line. Therefore, each $\mathbf{y} \in \mathcal{Y}$ can be written as a convex sum of the *vertices* $\mathbf{v}_1, \ldots, \mathbf{v}_{k_{\mathrm{v}}}$ and a nonnegative sum of the *rays* $\mathbf{r}_1, \ldots, \mathbf{r}_{k_{\mathrm{r}}}$

of $\mathcal{Y}$:

$$\mathbf{y} = \sum_{i=1}^{k_v} \lambda_i \mathbf{v}_i + \sum_{i=1}^{k_r} \mu_i \mathbf{r}_i,$$

where $\lambda_i \geq 0$ $(i = 1, \ldots, k_v)$, $\sum_{i=1}^{k_v} \lambda_i = 1$ and $\mu_i \geq 0$ $(i = 1, \ldots, k_r)$, and vice versa. As $\mathbf{y} \geq \mathbf{A}$ for all $\mathbf{y} \in \mathcal{Y}$, it follows that $\mathbf{r}_i \geq \mathbf{0}$ for $i = 1, \ldots, k_r$. Therefore, if

$$\bar{\mathcal{Y}} := \left\{ \sum_{i=1}^{k_v} \lambda_i \mathbf{v}_i \middle| \sum_{i=1}^{k_v} \lambda_i = 1 \text{ and } \lambda_i \geq 0, \forall i \in \{1, \ldots, k_v\} \right\},$$

then $\min(\mathcal{Y}, \leq) = \min(\bar{\mathcal{Y}}, \leq)$. Consequently, (8.41) is equivalent to

$$\bigvee_{\mathbf{y} \in \min(\bar{\mathcal{Y}}, \leq)} b - \mathbf{y}^T \mathbf{m} - \mathbf{y}^T \mathbf{F} \boldsymbol{\delta} \geq 0. \tag{8.42}$$

Lemma 8.1. *Let* $\mathbf{y} \in \min(\bar{\mathcal{Y}}, \leq)$ *and* $\mathbf{y} = \sum_{k=1}^{k_v} \lambda_k \mathbf{v}_k$ *with* $\lambda_1, \ldots, \lambda_{k_v} \geq 0$ *and* $\sum_{k=1}^{k_v} \lambda_k = 1$. *Then* $b - \mathbf{y}^T \mathbf{m} - \mathbf{y}^T \mathbf{F} \boldsymbol{\delta} \geq 0$ *implies*

$$\bigvee_{\substack{k=1,\ldots,k_v, \\ \lambda_k > 0}} b - \mathbf{v}_k^T \mathbf{m} - \mathbf{v}_k^T \mathbf{F} \boldsymbol{\delta} \geq 0. \tag{8.43}$$

Proof. It suffices to prove that the set of inequalities

$$b - \mathbf{y}^T \mathbf{m} - \mathbf{y}^T \mathbf{F} \boldsymbol{\delta} \geq 0, \tag{8.44}$$

$$-b + \mathbf{v}_k^T \mathbf{m} + \mathbf{v}_k^T \mathbf{F} \boldsymbol{\delta} > 0 \qquad (k = 1, \ldots, k_v; \lambda_k > 0), \tag{8.45}$$

never has a solution $\boldsymbol{\delta} \in \mathbb{N}^m$ for any $b \in \mathbb{Z}, \mathbf{m} \in \mathbb{N}^n$ and $\mathbf{F} \in \mathbb{Z}^{n \times m}$. Add all inequalities in (8.45) —after multiplying each of them with $\lambda_k > 0$— and add the result to (8.44). One obtains:

$$b \left(1 - \sum_{\substack{k=1,\ldots,k_v, \\ \lambda_k > 0}} \lambda_k \right) - \left(\mathbf{y} - \sum_{\substack{k=1,\ldots,k_v, \\ \lambda_k > 0}} \lambda_k \mathbf{v}_k \right) \mathbf{m}$$
$$- \left(\mathbf{y} - \sum_{\substack{k=1,\ldots,k_v, \\ \lambda_k > 0}} \lambda_k \mathbf{v}_k \right) \mathbf{F} \boldsymbol{\delta} > 0 \tag{8.46}$$

As the left hand side of (8.46) is zero, the set of inequalities (8.44)–(8.45) never has a solution. □

Lemma 8.2. *Consider a polytope $\mathcal{P}$ with vertices $\mathbf{w}_1, \ldots, \mathbf{w}_s$. Let $\mathbf{x} \in \min(\mathcal{P}, \leq)$ with $\mathbf{x} = \sum_{k=1}^{s} \omega_k \mathbf{w}_k$, $\omega_1, \ldots, \omega_s \geq 0$ and $\sum_{k=1}^{s} \omega_k = 1$. If $\omega_j > 0$, then*

$$\mathbf{w}_j \in \min(\mathcal{P}, \leq) \cap \min(\{\mathbf{w}_1, \ldots, \mathbf{w}_s\}, \leq). \tag{8.47}$$

Proof. 1. First we show that $\mathbf{w}_j \in \min(\mathcal{P}, \leq)$ by contraposition. Assume $\mathbf{w}_j \notin \min(\mathcal{P}, \leq)$. Then there exists an $\mathbf{x}' \in \mathcal{P}$ such that $\mathbf{x}' \leq \mathbf{w}_j$ and $\mathbf{x}' \neq \mathbf{w}_j$. Define

$$\bar{\mathbf{x}} = \sum_{\substack{k=1,\ldots,s \\ k \neq j}} \omega_k \mathbf{w}_k + \omega_j \mathbf{x}'.$$

As $\omega_j > 0$, it is then true that $\bar{\mathbf{x}} \leq \mathbf{x}$ and $\bar{\mathbf{x}} \neq \mathbf{x}$. Thus, to contradict $\mathbf{x} \in \min(\mathcal{P}, \leq)$, it suffices to show that $\bar{\mathbf{x}} \in \mathcal{P}$.

As $\mathbf{x}' \in \mathcal{P}$, there exist $\nu_1, \ldots, \nu_s \geq 0$ with $\sum_{k=1}^{s} \nu_k = 1$ such that $\mathbf{x}' = \sum_{k=1}^{s} \nu_k \mathbf{w}_k$. Thus,

$$\bar{\mathbf{x}} = \sum_{\substack{k=1,\ldots,s \\ k \neq j}} \omega_k \mathbf{w}_k + \omega_j \sum_{k=1}^{s} \nu_k \mathbf{w}_k.$$

It follows from $\sum_{k=1}^{s} \nu_k = 1$ and $\sum_{k=1}^{s} \omega_k = 1$ that

$$\sum_{\substack{k=1,\ldots,s \\ k \neq j}} \omega_k + \omega_j \sum_{k=1}^{s} \nu_k = 1.$$

This finishes the proof.
2. The fact that $\mathbf{w}_j \in \min(\{\mathbf{w}_1, \ldots, \mathbf{w}_s\}, \leq)$ follows from part 1 of this proof and $\{\mathbf{w}_1, \ldots, \mathbf{w}_s\} \subseteq \mathcal{P}$. □

Define

$$\mathcal{Y}_{\mathrm{v}} := \{\mathbf{v}_1, \ldots, \mathbf{v}_{k_{\mathrm{v}}}\}$$

as the set of vertices of $\mathcal{Y}$. Consequently, from Lemmas 8.1 and 8.2 and $\min(\mathcal{Y}, \leq) = \min(\bar{\mathcal{Y}}, \leq)$ it follows that (8.42) is a sufficient condition for

$$\bigvee_{\mathbf{y} \in \min(\mathcal{Y}_{\mathrm{v}}, \leq) \cap \min(\mathcal{Y}, \leq)} b - \mathbf{y}^T \mathbf{m} - \mathbf{y}^T \mathbf{F} \delta \geq 0. \tag{8.48}$$

Because of the disjunctions and because of

$$\min(\mathcal{Y}_{\mathrm{v}}, \leq) \cap \min(\mathcal{Y}, \leq) \subseteq \min(\mathcal{Y}, \leq) = \min(\bar{\mathcal{Y}}, \leq),$$

it is also a necessary condition. Consequently, conditions (8.41) and (8.48) are equivalent. The supervisor enabling condition in (8.48) needs to be added to the plant enabling condition $\mathbf{m} + \mathbf{F}\boldsymbol{\delta} \geq \mathbf{0}$. Note that $\min(\mathcal{Y}_{\mathrm{v}}, \leq)$ is not necessarily a subset of $\min(\mathcal{Y}, \leq)$. The control law in (8.48) is the main result of this chapter.

To construct the supervisor, only the set $\min(\mathcal{Y}_{\mathrm{v}}, \leq) \cap \min(\mathcal{Y}, \leq)$ is required. Constructing the set of vertices of a polyhedron is a standard problem in linear optimisation. This problem is NP-complete (see [13]). Furthermore, $\mathbf{y} \in \min(\mathcal{Y}_{\mathrm{v}}, \leq)$ belongs to $\min(\mathcal{Y}, \leq)$ if and only if $\{\mathbf{x} \in \mathcal{Y} \mid \mathbf{x} \leq \mathbf{y}\} = \{\mathbf{y}\}$.

Because of the linear structure of the supervisor in (8.48), existing algorithms for constructing the reachability graph/tree of a Petri net can easily be extended to cope with the disjunction operator in (8.48). This allows to study the behaviour of the closed-loop system.

So far, we used the no concurrency assumption. Under the transition bag assumption the reduction of the supervisor after (8.41) does not hold, because there does not yet exist a method that constructs an enabling condition under the transition bag assumption for a legal set which is described by the union of linear inequalities. However, if $\min(\mathcal{Y}, \leq)$ is a singleton $\{\mathbf{y}\}$, there is only one supervisor condition and no reduction is needed. The supervisor condition then is $b - \mathbf{y}^T\mathbf{m} \geq \mathbf{B}_{\mathbf{y}}^T\boldsymbol{\delta}$ (with $(\mathbf{B}_{\mathbf{y}})(t) = \max((\mathbf{y}^T\mathbf{F})(t), 0), \forall t \in T$) and can be realised by a Petri net (cfr. the result of Example 8.2).

7. A subset of the supremal controllable subset

Because of the no concurrency assumption and the self-loop freeness, (8.48) is equivalent to guaranteeing that the state of the Petri net plant never leaves the set (cfr. Section 7 in Chapter 2)

$$\hat{\mathcal{A}} := \bigcup_{\mathbf{y} \in \min(\mathcal{Y}_{\mathrm{v}}, \leq) \cap \min(\mathcal{Y}, \leq)} \{\mathbf{m} \in \mathbb{N}^n \mid \mathbf{y}^T\mathbf{m} \leq b\}. \tag{8.49}$$

Consequently, if $\mathcal{A}$ is as in (8.1) with $\mathbf{A} \geq \mathbf{0}$ or $\mathbf{A} \leq \mathbf{0}$, we have constructed a nontrivial subset $\hat{\mathcal{A}}$ of the supremal controllable subset $\mathcal{A}^*$. In a lot of cases this approximation is exact, as illustrated in the following example, and the resulting supervisor is thus maximally permissive.

Example 8.3. Consider the Petri net in Figure 8.2 with

$$\mathbf{F} = \begin{bmatrix} 1 & 0 & 0 & -1 & 0 & 0 \\ -1 & 1 & 0 & 0 & -1 & 1 \\ -1 & 0 & 1 & 0 & 1 & 0 \\ 0 & -1 & -1 & 2 & 0 & 0 \end{bmatrix}.$$

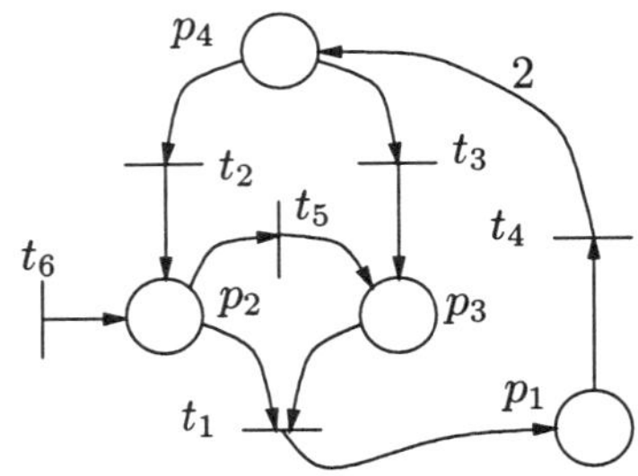

Figure 8.2. The Petri net of Example 8.3.

The legal set is $\mathcal{A} = \{\mathbf{m} \in \mathbb{N}^4 \mid [1 \quad 0 \quad 0 \quad 0]\,\mathbf{m} \leq 2\}$. Transition t_1 is uncontrollable and observable, all others are controllable, i.e. $T^{\mathrm{c}} = \{t_2, t_3, t_4\}$, $T^{\mathrm{uc}} = \{t_1\}$ and $T^{\mathrm{uo}} = \emptyset$. Thus for $\mathbf{D}$ in (8.10):

$$\mathbf{D}^T = [1 \quad -1 \quad -1 \quad 0].$$

One obtains that

$$\mathcal{Y} = \{(q_1, q_2, q_3, q_4) \in \mathbb{Q}^4 \mid q_1 \geq 1, q_2 \geq 0, q_3 \geq 0, q_4 \geq 0, q_2 + q_3 \geq q_1\}.$$

It follows that $\min(\mathcal{Y}_{\mathrm{v}}, \leq) \cap \min(\mathcal{Y}, \leq) = \{\mathbf{v}_1, \mathbf{v}_2\}$ with $\mathbf{v}_1 = (1, 1, 0, 0)$ and $\mathbf{v}_2 = (1, 0, 1, 0)$. Thus,

$$\hat{\mathcal{A}} = \{\mathbf{m} \in \mathbb{N}^4 \mid [1 \quad 1 \quad 0 \quad 0]\,\mathbf{m} \leq 2\} \cup \{\mathbf{m} \in \mathbb{N}^4 \mid [1 \quad 0 \quad 1 \quad 0]\,\mathbf{m} \leq 2\}$$

in (8.49). This set is equal to $\mathcal{A}^*$; the approximation is exact. The set $\mathcal{A}^*$ is retrieved without doing any reachability analysis. △

Notice that both approximations for $\mathcal{A}^*$ in Chapter 7 and this chapter are of the same form. However, the conditions under which they are valid, are different. One can verify that the approximation of $\mathcal{A}^*$ in Example 7.3 (p. 157) —which showed to be exact in Example 7.6 (p. 167)— can be recovered with the here proposed technique.

As the definition of $\mathcal{A}^*$ is independent of the concurrency assumption being used, this result applies to all concurrency assumptions.

8. Notes and references

Other references on supervisory control of Petri nets with unobservable transitions are the following. In [16] and [43] the authors construct, by observing the observable events, first the set of all possible present states and then design control laws for this set. In [26] another approach is used. The authors try to extend the result of [42] to Petri nets with uncontrollable and unobservable transitions. The legal set, which is described by a finite conjunction of linear inequalities, is replaced by another set which satisfies a number of structural conditions. These conditions take into account the presence of uncontrollable and unobservable transitions. Our approach is related to this one. However, as we do not fix the form of the linear inequality which describes $\mathcal{A}'$, the here obtained control laws are more permissive. In a similar manner, in [12] the original Petri net is extended with an *observer net.*

A linear algebraic approach to the analysis of Petri nets can be found in [11] and [35].

References

[1] H. Alla and R. David. A modelling and analysis tool for discrete event systems: continuous Petri nets. *Performance Evaluation*, 33(3):175–199, 1998.

[2] K. Barkaoui, A. Chaoui, and B Zouari. Supervisory control of discrete event systems based on structure theory of Petri nets. In *Proceedings of the IEEE Systems Man and Cybernetics*, Orlando, Florida, 1997.

[3] K. Barkaoui and J.-P. Pradat-Peyre. On liveness and controlled siphons in Petri nets. In *17th International Conference on Application and Theory of Petri Nets*, Lecture Notes in Computer Science. Springer, 1996.

[4] D. Bertsimas and J. Tsitsiklis. *Introduction to linear optimization.* Athena Scientific, 1997.

[5] R.K. Boel, L. Ben-Naoum, and V. Van Breusegem. On forbidden state problems for a class of controlled Petri nets. *IEEE Transactions on Automatic Control*, 40(10):1717–1731, 1995.

[6] C.G. Cassandras and S. Lafortune. *Introduction to Discrete Event Systems.* Kluwer Academic Publishers, 1999.

[7] H. Chen. Net structure and control logic synthesis for controlled Petri nets. *IEEE Transactions on Automatic Control*, 43(10):1446–1450, 1998.

[8] H. Chen. Control synthesis of Petri nets based on S-decreases. *Discrete Event Dynamic Systems: Theory and Applications*, 10(3), 2000.

[9] B.A. Davey and H.A. Priestley. *Introduction to lattices and order.* Cambridge University Press, 1990.

[10] R. David and H. Alla. Petri nets for modeling of dynamic systems — a survey. *Automatica*, 30(2):175–202, 1994.

[11] J. Desel. *Petrinetze, lineare Algebra und lineare Programmierung*, volume 26 of *Teubner-Texte zur Informatik.* B.G. Teubner Stuttgart - Leipzig, 1998.

[12] A. Fanni, A. Giua, and N. Sanna. Control and error recovery of Petri net models with event observers. In *Proceedings of the 2nd International Workshop on Manufacturing and Petri Nets*, pages 53–68, 1997.

[13] K. Fukuda, T.M. Liebling, and F. Margot. Analysis of backtrack algorithms for listing all vertices and all faces of a convex polyhedron. *Computational Geometry*, 8:1–12, 1997.

[14] A. Giua and F. DiCesare. Supervisory design using Petri nets. In *Proceedings of the 30th Conference on Decision and Control*, pages 92–97, 1991.

[15] A. Giua, F. DiCesare, and M. Silva. Generalized mutual exclusion constraints on petri nets with uncontrollable transitions. In *Proceedings of the 1992 IEEE International Conference on Systems, Man and Cybernetics*, pages 974–979, 1992.

[16] A. Giua and C. Seatzu. Manufacturing control with Petri net observers. In *Proceedings of the 6th IEEE Mediterranean Conference on Control and Automation*, pages 759–764, 1998.

[17] L.E. Holloway, X. Guan, and L. Zhang. A generalization of state avoidance policies for controlled Petri nets. *IEEE Transactions on Automatic Control*, 41(6):804–815, 1996.

[18] L.E. Holloway and B.H. Krogh. Synthesis of feedback control logic for a class of controlled Petri nets. *IEEE Transactions on Automatic Control*, 35(5):514–523, 1990.

[19] L.E. Holloway, B.H. Krogh, and A. Giua. A survey of Petri net methods for controlled discrete event systems. *Discrete Event Dynamic Systems: Theory and Applications*, 7:151–190, 1997.

[20] A. Ichikawa and K. Hiraishi. Analysis and control of discrete event systems represented by Petri nets. In *Discrete Event Systems: Models and Applications*, volume 103 of *Lecture Notes in Control and Information Sciences*, pages 115–134. Springer-Verlag, 1987.

[21] B.H. Krogh and L.E. Holloway. Synthesis of feedback control logic for discrete manufacturing systems. *Automatica*, 27(4):641–651, 1991.

[22] R. Kumar and L.E. Holloway. Supervisory control of Petri nets with regular specification languages. *IEEE Transactions on Automatic Control*, 41(2):245–249, 1996.

[23] K. Lautenbach and H. Ridder. The linear algebra of deadlock avoidance — a Petri net approach. Technical Report 25-96, Institut für Informatik, Universität Koblenz-Landau, 1996.

[24] Y. Li and W.M. Wonham. Control of vector discrete-event systems—Part 2: controller synthesis. *IEEE Transactions on Automatic Control*, 39(3):512–531, 1994.

[25] Y. Li and W.M. Wonham. Concurrent vector discrete-event systems. *IEEE Transactions on Automatic Control*, 40(4):628–638, 1995.

[26] J.O. Moody and P.J. Antsaklis. Petri net superviosrs for DES with uncontrollable and unobservable transitions. *IEEE Transactions on Automatic Control*, 45, 2000.

[27] T.S. Motzkin. Two consequences of the transition theorem on linear inequalities. *Econometria*, 19:184–185, 1951.

[28] T. Murata. Petri nets: properties, analysis and applications. *Proceedings IEEE*, 77(4):541–580, 1989.

[29] C.H. Papadimitriou and K. Steiglitz. *Combinatorial Optimization: Algorithms and Complexity*. Prentice-Hall, 1982.

[30] J.L. Peterson. *Petri Net Theory and the Modelling of Systems*. Prentice-Hall, 1981.

[31] P. Ramadge and W.M. Wonham. The control of discrete event systems. *Proceedings IEEE*, 77(1):81–98, 1988.

[32] W. Reisig. *Petri nets*. Springer-Verlag, 1985.

[33] E. Schechter. *Handbook of analysis and its foundations*. Academic Press, 1997.

[34] A. Schrijver. *Theory of linear and integer programming*. Wiley, 1998.

[35] M. Silva, E. Teruel, and J.M. Colom. Linear algebraic and linear programming techniques for the analysis of place/transition net systems. In W. Reisig and G. Rozenberg, editors, *Lectures on Petri nets: Basic Models*, volume 1491, pages 309–373. Springer, 1998.

[36] I.A. Stewart. Reachability in some classes of acyclic Petri nets. *Fundamenta Informaticae*, 23:91–100, 1995.

[37] G. Stremersch and R.K. Boel. Reduction of the supervisory control problem for Petri nets. *IEEE Transactions on Automatic Control*, 45(12):2358–2363, 2000.

[38] G. Stremersch and R.K. Boel. Decomposition of the supervisory control problem for Petri nets under preservation of maximal permissiveness. *IEEE Transactions on Automatic Control*, 2001.

[39] G. Stremersch and R.K. Boel. Structuring acyclic Petri nets for reachability analysis and control. *Discrete Event Dynamic Systems: Theory and Applications*, 2001.

[40] S. Vajda. *Theory of linear and non-linear programming*. Longman, 1974.

[41] K. Xing, B. Hu, and H. Chen. Deadlock avoidance policy for Petri-net modeling of flexible manufacturing systems with shared resources. *IEEE Transactions on Automatic Control*, 41(2):289–295, 1996.

[42] K. Yamalidou, J. Moody, M. Lemmon, and P. Antsaklis. Feedback control of Petri nets based on place invariants. *Automatica*, 32(1):15–28, 1996.

[43] L. Zhang and L.E. Holloway. Forbidden state avoidance in controlled Petri nets under partial observation. In *Proceedings of the 33rd Annual Allerton Conference on Communications, Control, and Computing*, pages 146–155, 1995.

Index